Shuichi Iwata, Yukio Ohsawa, Shusaku Tsumoto, Ning Zhong, Yong Shi
and Lorenzo Magnani (Eds.)

Communications and Discoveries from Multidisciplinary Data

Studies in Computational Intelligence, Volume 123

Editor-in-chief
Prof. Janusz Kacprzyk
Systems Research Institute
Polish Academy of Sciences
ul. Newelska 6
01-447 Warsaw
Poland
E-mail: kacprzyk@ibspan.waw.pl

Further volumes of this series can be found on our
homepage: springer.com

Vol. 102. Carlos Cotta, Simeon Reich, Robert Schaefer and
Antoni Ligeza (Eds.)
Knowledge-Driven Computing, 2008
ISBN 978-3-540-77474-7

Vol. 103. Devendra K. Chaturvedi
*Soft Computing Techniques and its Applications in Electrical
Engineering,* 2008
ISBN 978-3-540-77480-8

Vol. 104. Maria Virvou and Lakhmi C. Jain (Eds.)
*Intelligent Interactive Systems in Knowledge-Based
Environment,* 2008
ISBN 978-3-540-77470-9

Vol. 105. Wolfgang Guenthner
*Enhancing Cognitive Assistance Systems with Inertial
Measurement Units,* 2008
ISBN 978-3-540-76996-5

Vol. 106. Jacqueline Jarvis, Dennis Jarvis, Ralph Rönnquist
and Lakhmi C. Jain (Eds.)
Holonic Execution: A BDI Approach, 2008
ISBN 978-3-540-77478-5

Vol. 107. Margarita Sordo, Sachin Vaidya and Lakhmi C. Jain
(Eds.)
*Advanced Computational Intelligence Paradigms
in Healthcare - 3,* 2008
ISBN 978-3-540-77661-1

Vol. 108. Vito Trianni
Evolutionary Swarm Robotics, 2008
ISBN 978-3-540-77611-6

Vol. 109. Panagiotis Chountas, Ilias Petrounias and Janusz
Kacprzyk (Eds.)
*Intelligent Techniques and Tools for Novel System
Architectures,* 2008
ISBN 978-3-540-77621-5

Vol. 110. Makoto Yokoo, Takayuki Ito, Minjie Zhang,
Juhnyoung Lee and Tokuro Matsuo (Eds.)
Electronic Commerce, 2008
ISBN 978-3-540-77808-0

Vol. 111. David Elmakias (Ed.)
New Computational Methods in Power System Reliability,
2008
ISBN 978-3-540-77810-3

Vol. 112. Edgar N. Sanchez, Alma Y. Alanís and Alexander
G. Loukianov
*Discrete-Time High Order Neural Control: Trained with
Kalman Filtering,* 2008
ISBN 978-3-540-78288-9

Vol. 113. Gemma Bel-Enguix, M. Dolores Jiménez-López
and Carlos Martín-Vide (Eds.)
New Developments in Formal Languages and Applications,
2008
ISBN 978-3-540-78290-2

Vol. 114. Christian Blum, Maria José Blesa Aguilera, Andrea
Roli and Michael Sampels (Eds.)
Hybrid Metaheuristics, 2008
ISBN 978-3-540-78294-0

Vol. 115. John Fulcher and Lakhmi C. Jain (Eds.)
Computational Intelligence: A Compendium, 2008
ISBN 978-3-540-78292-6

Vol. 116. Ying Liu, Aixin Sun, Han Tong Loh, Wen Feng Lu
and Ee-Peng Lim (Eds.)
*Advances of Computational Intelligence in Industrial
Systems,* 2008
ISBN 978-3-540-78296-4

Vol. 117. Da Ruan, Frank Hardeman and Klaas van der Meer
(Eds.)
Intelligent Decision and Policy Making Support Systems,
2008
ISBN 978-3-540-78306-0

Vol. 118. Tsau Young Lin, Ying Xie, Anita Wasilewska
and Churn-Jung Liau (Eds.)
Data Mining: Foundations and Practice, 2008
ISBN 978-3-540-78487-6

Vol. 119. Slawomir Wiak, Andrzej Krawczyk and Ivo Dolezel
(Eds.)
*Intelligent Computer Techniques in Applied
Electromagnetics,* 2008
ISBN 978-3-540-78489-0

Vol. 120. George A. Tsihrintzis and Lakhmi C. Jain (Eds.)
Multimedia Interactive Services in Intelligent Environments,
2008
ISBN 978-3-540-78491-3

Vol. 121. Nadia Nedjah, Leandro dos Santos Coelho
and Luiza de Macedo Mourelle (Eds.)
Quantum Inspired Intelligent Systems, 2008
ISBN 978-3-540-78531-6

Vol. 122. Tomasz G. Smolinski, Mariofanna G. Milanova
and Aboul-Ella Hassanien (Eds.)
Applications of Computational Intelligence in Biology, 2008
ISBN 978-3-540-78533-0

Vol. 123. Shuichi Iwata, Yukio Ohsawa, Shusaku Tsumoto,
Ning Zhong, Yong Shi and Lorenzo Magnani (Eds.)
*Communications and Discoveries from Multidisciplinary
Data,* 2008
ISBN 978-3-540-78732-7

Shuichi Iwata
Yukio Ohsawa
Shusaku Tsumoto
Ning Zhong
Yong Shi
Lorenzo Magnani
(Eds.)

Communications and Discoveries from Multidisciplinary Data

With 121 Figures and 57 Tables

 Springer

Shuichi Iwata
Graduate School of Engineering
The University of Tokyo
7-3-1 Hongo, Bunkyo-ku
Tokyo 113-8656
Japan
iwata@k.u-tokyo.ac.jp

Yukio Ohsawa
Department of Systems Innovation
School of Engineering
The University of Tokyo
7-3-1 Hongo, Bunkyo-ku
Tokyo 113-8656
Japan
ohsawa@sys.t.u-tokyo.ac.jp

Shusaku Tsumoto
Department of Medical Informatics
School of Medicine
Shimane University
Enya-cho Izumo City
Shimane 693-8501
Japan
tsumoto@computer.org

Ning Zhong
Department of Information Engineering
Maebashi Institute of Technology
460-1, Kamisadori-Cho
Maebashi-City, 371-0816
Japan
zhong@maebashi-it.ac.jp
and Director of the WICI/BJUT

Yong Shi
Director of Research Center on Data
Technology and Knowledge Economy
Chinese Academy of Sciences
Beijing 100080, PR China
and
College of Information
Science and Technology
University of Nebraska at Omaha
NE 68182, USA
yshi@mail.unomaha.edu

Lorenzo Magnani
Department of Philosophy
University of Pavia
Piazza Botta 6, 27100 Pavia, Italy
lmagnani@unipv.it

ISBN 978-3-540-78732-7   e-ISBN 978-3-540-78733-4
DOI: 10.1007/978-3-540-78733-4

Studies in Computational Intelligence ISSN 1860-949X

Library of Congress Control Number: 2008923733

Cover design: Deblik, Berlin, Germany

Printed on acid-free paper

9 8 7 6 5 4 3 2 1

springer.com

For make your community creative

Scientific "Agendum" of Data Science

Shuichi Iwata

The University of Tokyo, Kashiwa, Japan
iwata@k.u-tokyo.ac.jp

Data on nuclear cross section have integrated metals, ceramics, waters and other materials to form various nuclear reactors, which is a basic for nuclear engineering. Data on intrinsic properties of substances are used to link extrinsic properties of substances and structure-sensitive engineering properties of defects as fundamental constants have been defined in a network of elaborate experiments and models in science. These understandings have driven me to data activities since 1970 when I at first became to know CODATA by name, and it is a good time now for me review the committed works in these 37 years and write down my personal gagendumh reflecting the re-view so as to have more collaborators for the next milestone.

As scientific agendum of CODATA, we have added an aspect gdata and societyh explicitly through such opportunities of WSIS(World Summit on the Information Society)s and lessons on tragedies, namely, suffering from natural disasters and epidemics. Not only by global warming problems pointed out clearly by IPCC (Intergovernmental Panel on Climate Change) but also by other global issues as known well partly by UN MDGs(the eight Millennium Development Goals), we have become to know the necessity of linking scientific, technical, economic, social and political agenda with proper missions and guidelines for the society. Here human-centered reorganizations of domain-differentiated sciences from natural sciences to social sciences, are requested to be carried out, where and when we need common data for the proper and holistic reorganization to reach right decisions and consensus of society. If we do not share common data on global warming effects, we cannot write down our remedies against inconvenient truths and establish flexible and steady roadmaps for the sustainable society. It is necessary for us experts to write down proper remedies together by linking such associated scientific domains as politics, laws, ethics, economics, environmental sciences, ecologies, civil engineering, manufacturing, waste management and so on and complimenting missing links there for better solutions. And as a consequence of such efforts we may come to create a new scientific domain gdata-driven sustainable scienceh to design and manage the society properly. It is really requiring us continuous big efforts with challenging spirits to start everything from facts and data, which is our raison dfetre and concerns our identity as data scientists.

However, the above agendum essentially has already been associated to our core activities. Data and knowledge corresponding missing links of domain specific sciences have been daily works of CODATA. For example,

- fundamental constants have been compiled reflecting advances of precise measurements and basic sciences, and also higher coherences of scientific models, which has resulted in creating new sciences like nano-sciences, spintronics and other specialized scientific domains, and also several key standards for the information society, namely, radio wave standards, current, voltage and so on,
- spectra data and diffraction data have guided us to get microstructural information on substances and materials, and also on life. Together with models and interatomic potentials derived from first principles calculation, new scientific and technical fields have been exploited rapidly, and application areas are spreading widely over drug design, defect theory, fracture dynamics, materials design, process design, earth sciences, and even in bioscience to see the origin of life and medical diagnostics.
- many exemplars in biosciences to get insight through using common data: arabidopsis thaliana data for evolving botany; RNA data to link DNA and protein, and consequently associated with disease and health; data on Tradescantia ohiensis and/or nude mice for irradiation effects, and by taking advantage of recent advances in genomics, proteomics biophysics and biochemistry, an breakthrough is going on to overcome limitations of epidemiological survey based on statistics.

Data-oriented statistical approaches are combined with scientific models and practical monitoring, and traditional established safety/risk/reliability standards are changing into proactive and dynamic adaptive standards. Safety/risk issues in medical services, nuclear reactors, aircrafts, company managements, energy resource security and so on can be dealt with in a similar way. Openness and transparency of many disciplines and scientific domains promoted by e-science projects and so called global information commons are prerequisites for a revolution of sciences by the 7 billion people. Devices for the revolution might be something creative of emerging wisdoms and welling up emotions in the internet, which may be more than such knowledge management approaches as ontology, metadata, object-oriented approach, semantic-web, common sense reasoning and so on.

Through evaluation of fundamental constants we are integrating quantum worlds, atomistic worlds and macroscopic worlds quantitatively, where scientific disciplines and domains are networked with a certain consistency. Through preparation of geometric data of parts with properties we can design an artifacts and assemble the available parts into the artifacts of integrity and cost-effective performances, where domain differentiated engineering disciplines are integrated to establish manufacturing industries. Design and maintenance of landscapes, cities, countries, regional environments and global climates are used to be carried out in a similar way, namely, sharing data by stakeholders and coordinating different views and opinions by the shared common data. Quality of data matters every time and everywhere. Gold in, gold out. Reasonable estimations of uncertainties on data may produce better results and outcomes. The more the problem to be solved is uncertain, the more we should become flexible. Evidence-based deterministic approaches do not work effectively, and adaptive and heuristic approaches work better coupled with in situ data capture, evaluation, and quick decision and timely actions. Holistic creativity as a group is a key for a success of the group, where practical maintenance of data quality for proper decision is important. Time constants of data life cycle are becoming shorter, and diversities of stakeholders and complexities of data are increasing. New disciplines are to be continuously created by taking advantage of available data and devices so as to prepare solutions on time. Without proper managements of continuously produced big data, and without productivity of new disciplines based on data, we cannot solve important problems of the world. Data science may play an important role there.

Shuichi Iwata

Contents

Part I

Thought, Communication, and Actions

1

Sharing Representations and Creating Chances through Cognitive Niche Construction.
The Role of Affordances and Abduction

Lorenzo Magnani[1] and Emanuele Bardone[2]

[1] Department of Philosophy, University of Pavia, Pavia, Italy and Sun Yat-sen University, Guangzhou, P.R. China
`lmagnani@unipv.it`
[2] Department of Philosophy, University of Pavia, Pavia, Italy
`bardone@unipv.it`

Summary. As a matter of fact, humans continuously delegate and distribute cognitive functions to the environment to lessen their limits. They build models, representations, and other various mediating structures, that are considered to aid thought. In doing these, humans are engaged in a process of cognitive niche construction. In this sense, we argue that a *cognitive niche* emerges from a network of continuous interplays between individuals and the environment, in which people alter and modify the environment by mimetically externalizing fleeting thoughts, private ideas, etc., into external supports. For cognitive niche construction may also contribute to make available a great portion of knowledge that otherwise would remain simply unexpressed or unreachable. This can turn to be useful especially for all those situations that require to transmit and share knowledge, information, and, more generally, cognitive resources. In dealing with the exploitation of cognitive resources embedded in the environment, the notion of *affordance*, originally proposed by Gibson [1] to illustrate the hybrid character of visual perception, together with the proximal/distal distinction described by Brunswik [2], are extremely relevant. In order to solve various controversies on the concept of affordance and on the status of the proximal/distal dichotomy, we will take advantage of some useful insights that come from the study on *abduction*. Abduction may also fruitfully describe all those human and animal hypothetical inferences that are operated through actions which consist in smart manipulations to both detect new affordances and to create manufactured external objects that offer new affordances/cues.

1.1 Introduction

As a matter of fact, humans continuously delegate and distribute cognitive functions to the environment to lessen their limits. They build models, representations, and other various mediating structures, that are thought to be good to think. The aim of this paper is to shed light on these *designing* activities. In the first part of the paper we will argue that these designing activities are closely related to the process of niche

L. Magnani and E. Bardone: *Sharing Representations and Creating Chances through Cognitive Niche Construction. The Role of Affordances and Abduction*, Studies in Computational Intelligence (SCI) **123**, 3–40 (2008)
`www.springerlink.com`

construction. We will point out that in building various mediating structures, such as models or representations, humans alter the environment and thus create *cognitive niches*.

In dealing with the exploitation of cognitive resources embedded in the environment, the notion of *affordance*, originally proposed by Gibson [1] to illustrate the hybrid character of visual perception, together with the proximal/distal distinction described by Brunswik [2], can be extremely relevant. The analysis of the concept of affordance also provides an alternative account about the role of external – also artifactual – objects and devices. Artifactual cognitive objects and devices extend, modify, or substitute "natural" affordances actively providing humans and many animals with new opportunities for action [3].[1] In order to solve various controversies on the concept of affordance and on the status of the proximal/distal dichotomy, we will take advantage of some useful insights that come from the study on *abduction*. Abduction may also fruitfully describe all those human and animal hypothetical inferences that are operated through actions which consist in smart manipulations to both detect new affordances and to create manufactured external objects that offer new affordances/cues. After presenting some theoretical muddles concerning affordance and proximal/distal distinction, in the last part of the paper we will refer to abduction as the process which humans lean on in order to detect and design affordances and thus modify or even create cognitive niches.

1.2 Humans as Chance Seekers

1.2.1 Incomplete Information and Human Cognition

Humans usually make decisions and solve problems relying on incomplete information [7]. Having incomplete information means that 1) our deliberations and decisions are never *the best* possible answer, but they are at least *satisficing*; 2) our conclusions are always *withdrawable* (i.e. questionable, or never final). That is, once we get more information about a certain situation we can always revise our previous decisions and think of alternative pathways that we could not "see" before; 3) a great part of our job is devoted to elaborating conjectures or hypotheses in order to obtain more adequate information. Making conjectures is essentially an act that in most cases consist in manipulating our problem, and the representation we have of it, so that we may eventually acquire/create more "valuable" knowledge resources. Conjectures can be either the fruit of an abductive selection in a set of pre-stored hypotheses or the creation of new ones, like in scientific discovery (see section 1.3). To make conjectures humans often need more evidence/data: in many cases this further cognitive action is the only way to simply make possible (or at least enhance) a thought to "hypotheses" which are hard to successfully produce.

---

[1] Neuropsychological and physiological evidence on how visual information affords and potentiates action – and the highly integrated nature of visual and motor representations – is described in Tucker and Ellis [4], Knoblich and Flach [5], and Derbyshire, Ellis, and Tucker [6].

Consider, for instance, diagnostic settings: often the information available does not allow a physician to make a precise diagnosis. Therefore, he/she has to perform additional tests, or even try some different treatments to uncover symptoms otherwise hidden. In doing so he/she is simply aiming at increasing the *chances* of making the appropriate decision. There are plenty of situations of that kind. For example, scientists are continuously engaged in a process of manipulating their research settings in order to get more valuable information, as illustrated by Magnani [8]. Most of this work is completely tacit and embodied in practice. The role of various laboratory artifacts is a clear example, but also in everyday life people daily face complex situations which require knowledge and manipulative expertise of various kinds no matter who they are, whether teachers, policy makers, politicians, judges, workers, students, or simply wives, husbands, friends, sons, daughters, and so on. In this sense, humans can be considered *chance seekers*, because they are continuously engaged in a process of building up and then extracting latent possibilities to uncover new valuable information and knowledge.

The idea we will try to deepen in the course of this paper is the following: as chance seekers, humans are *ecological engineers*. That is: humans like other creatures do not simply live their environment, but they actively shape and change it looking for suitable chances. In doing so, they construct *cognitive niches* (cf. Tooby and DeVore [9] and Pinker [10, 11]) through which the offerings provided by the environment in terms of cognitive possibilities are appropriately selected and/or manufactured to enhance their fitness as chance seekers. Hence, this ecological approach aims at understanding cognitive systems in terms of their *environmental situatedness* (cf. Clancey [12] and Magnani [13]). Within this framework, chances are that "information" which is not stored internally in memory or already available in an external reserve but that has to be "extracted" and then *picked up* upon occasion.

Related to this perspective is also the so-called *Perceptual Activity Theory* (PA) (cf. Ellis [14] and Ramachandran [15]). What these studies suggest is that an observer actively selects the perceptual information it needs to control its behavior in the world [16]. In this sense, we do not store descriptions of pictures, objects or scenes we perceive in a static way: we continuously adjust and refine our perspective through further *perceptual exploration* that allows us to get a more detailed understanding. As Thomas [16] put it, "PA theory, like *active vision* robotics, views it [perception] as a continual process of active interrogation of the environment". As we will show in the following sections "the active interrogation of the environment" is also at the root of the evolution of our organism and its cognitive system. We will also describe this ecological activity by introducing the notion of abduction (cf. section 1.3) and its semiotic dimension.

1.2.2 Cognitive Niche Construction and Human Cognition as a Chance-Seeker System

It is well-known that one of the main forces that shape the process of adaptation is natural selection. That is, the evolution of organisms can be viewed as the result of a selective pressure that renders them well-suited to their environments. Adaptation is therefore considered as a sort of *top-down process* that goes from the environment to the living creature (cf. [17]). In contrast to that, a small fraction of evolutionary biologists have recently tried to provide an alternative theoretical framework by emphasizing the role of niche construction (cf. [18–20]).

According to this view, the environment is a sort of "global market" that provides living creatures with unlimited possibilities. Indeed, not all the possibilities that the environment offers can be exploited by the human and non-human animals that act on it. For instance, the environment provides organisms with water to swim in, air to fly in, flat surfaces to walk on, and so on. However, no creatures are fully able to take advantage of all of them. Moreover, all organisms try to modify their surroundings in order to better exploit those elements that suit them and eliminate or mitigate the effect of the negative ones.

This process of *environmental selection* (cf. [21]) allows living creatures to build and shape the "ecological niches". An ecological niche can be defined, following Gibson, as a "setting of environmental features that are suitable for an animal" [1]. It differs from the notion of habitat in the sense that the niche describes *how* an organism lives its environment, whereas habitat simply describes *where* an organism lives.

In any ecological niche, the selective pressure of the *local* environment is drastically modified by organisms in order to lessen the negative impacts of all those elements which they are not suited to. This new perspective constitutes a radical departure from traditional theory of evolution introducing a second inheritance system that Odling-Smee et al. [20] call *ecological inheritance system*. According to this view, acquired characters – discarded for such a long time – can enter evolutionary theories as far as they cause a modification to the environment that can persist and thus can modify the local selective pressure.[2] Ecological inheritance system is different from the genetic one in the following way [20]: 1) genetic materials can be inherited only from parents or relatives. Conversely, modifications on the environment can affect everyone, no matter who he/she is. It may regard unrelated organisms also belonging to other species. There are several global phenomena such as climate change that regard human beings, but also the entire ecosystem; 2) genes transmission is a one way transmission flow, from parents to offspring, whereas environmental information can travel backward affecting several generations. Pollution, for instance, affects young as well as old people; 3) genetic inheritance can happen once during one's life, at the time of reproductive phase. In contrast, ecological information can be transferred during the entire duration of life. Indeed, it depends on the eco-engineering capacities at play; 4) genetic inheritance system leans on the presence of *replicators*, whereas the ecological inheritance system leans on the *persistence* of whatsoever changes made upon the environment.

---

[2] This perspective has generated some controversies, since it is not clear the extent to which modifications count as niche-construction, and so enter the evolutionary scene. The main objection regards how far individual or even collective actions can really have ecological effects, whether they are integrated or merely aggregated changes. On this point, see [22] and the more critical view held by Dawkins [23]. For a reply to these objections, see [24].

Indeed, natural selection is somehow not halted by niche construction. Rather, this means adaptation cannot be only considered by referring to the agency of the environment, but also to that of the organism acting on it. In this sense, organisms are ecological engineers, because they do not simply live their environment, but they actively shape and change it (cf. [25]). Gibson clearly acknowledges the role of artifacts: "Perceiving is the simplest and best kind of knowing. But there are other kinds [...]. Knowing by means of instruments extends perceiving into the realm of the very distant and the very small; [...] Knowing by means of language makes knowing explicit instead of tacit" [1, p. 263].

Our contention is that the notion of niche construction can be also usefully applied to human cognition. More precisely, we claim that cognitive niche construction can be considered as one of the most distinctive traits of human cognition.[3] Organisms are equipped with various ontogenetic mechanisms that permit them to acquire information and thus better adapt to the environment: for instance, immune system in vertebrates and brain-based learning in animals and humans. Their role is to provide organisms with a supplementary mechanism to acquire information and thus face various environmental contingencies that are not – and cannot be – specified at the genetic level [20, p. 255]. A genetically specified initial set of behaviors is elaborated through experience of a relevant environment. These ontogenetic mechanisms are therefore a sort of *on-board* system allowing flexibility and plasticity of response to an ever-changing environment, which are at the core of the notion of cognition that is at the basis of our treatment [27][4].

In the case of human beings and other mammals, bigger brains allow to store information which could not be pre-defined by the genes [29, pp. 182–193]. Flexibility and plasticity of response to an ever-changing environment are connected to the necessity of having other means for acquiring information, more readily and quickly of the genetic one. We posit that niche construction plays a fundamental role to meet this requirement. Plasticity and flexibility depend on niche construction as far as various organisms may alter local selective pressure via niche construction, and thus increase their chances for surviving. More specifically, cognitive niches are crucial

---

[3] If we also recognize in animals, like many ethologists do, a kind of nonlinguistic thinking activity basically model-based (i.e. devoid of the cognitive functions provided by human language), their ecological niches can be called "cognitive", when for example complicate animal artifacts like landmarks of caches for food are fruit of "flexible" and learned thinking activities which indeed cannot be entirely connected with innate endowments (cf. [26,27]). The psychoanalyst Carl Gust Jung, who is aware that also animals make artifacts, nicely acknowledges their cognitive role proposing the expression "natural culture": "When the beaver fells trees and dams up a river, this is a performance conditioned by its differentiation. Its differentiation is a product of what one might call 'natural culture', which functions as a transformer of energy" [28, p. 42].

[4] Godfrey-Smith defined cognition as the capacity of coping with a range of possible behavioral options with different consequences for the organism's chance to survive. This definition allows him to embrace a broader notion of cognition which extends it to animal and plant behaviors. We will embrace this thesis in the sections devoted to abduction and affordance.

in developing more and more sophisticated forms of flexibility, because they constitute an additional source of information favoring behavior and development control. In this case, epigenesis is therefore augmented, and, at genetic level, it is favored by genes regulating *epigenetic openness* [30]. Epigenetic openness is closely related to what Godfrey-Smith [27] called *phenotypic plasticity*; the flexible response of living organisms (humans in particular) leans on sensitivity to environment clues, and this process of attunement to relevant aspects of the environment cannot be separated from niche construction.

In the case of human beings, the process of attunement leans on the continuous interplay between individuals and the environment, in which they more or less tacitly manipulate what is occurring outside at the level of the various structures of the environment in a way that is suited to them. It emerges from a network of continuous interplay between individuals and the environment, in which they more or less tacitly manipulate what is occurring outside at the level of the various structures of the environment in a way that is suited to them. Accordingly, we may argue that the creation of cognitive niches is *the* way cognition evolves, and humans can be considered as ecological cognitive engineers.

Recent studies on distributed cognition seem to support our claim.[5] According to this approach, cognitive activities like, for instance, problem solving or decision-making, cannot only be regarded as internal processes that occur within the isolated brain. Through the process of niche creation humans extend their minds into the material world, exploiting various external resources. Therefore, they exhibit a range of cognitive behaviors insofar as they are merged into a network of ecological interactions. For "external resources" we mean everything that is not inside the human brain, and that could be of some help in the process of deciding, thinking about, or using something. Therefore, external resources can be artifacts, tools, objects, and so on. Problem solving, such as general decision-making activity (cf. [37]), for example, are unthinkable without the process of connection between internal and external resources.

In other words, the exploitation of external resources is the process which allows the human cognitive system to be shaped by environmental (or contingency) elements. According to this statement, we may argue that external resources play a pivotal role in almost any cognitive process. Something important must still be added, and it deals with the notion of representation: in this perspective the traditional notion of representation as a kind of abstract mental structure becomes old-fashioned and misleading.[6] If some cognitive performances can be viewed as the result of a smart interplay between humans and the environment, the representation of a problem is partly internal but it also depends on the smart interplay between the individual and the environment.

As we have already said, an alternative definition of the ecological niche that we find appealing in treating our problem has been provided by Gibson [1]: he pointed out that a niche can be seen as a set of *affordances*. Our contention is that the notion

---

[5] Cf. [31–36].
[6] Cf. [31, 38, 39].

of affordance may help provide sound answers to the various questions that come up with the problem of ecological niches. The notion of affordance is fundamental for two reasons. First of all, it defines the nature of the relationship between an agent and its environment, and the mutuality between them. Second, this notion may provide a general framework to illustrate humans as chance seekers.

In order to illustrate the notion of affordance and its implications for our approach, let us introduce the concept of abduction. This discussion will provide the proper cognitive framework to tackle some theoretical muddles.

1.3 Abduction and its Cognitive Relevance

We have posited that humans continuously exploit latent possibilities and chances offered by the environment. This is carried out by an ecological activity called cognitive niche construction. A cognitive niche describes *how* humans exploit external resources and incorporate them into their cognitive systems. As mentioned above, in constructing cognitive niches, humans do not hold in their memory a rich model of the environment, but they pick up what they find outside upon occasion. In dealing with these activities that can be considered "semiotic", the role of abduction is a key point in our discussion (cf. Magnani [35]). The semiotic dimension of abduction has been clearly depicted by Peirce. As he put it, all thinking is in signs, and signs can be icons, indices, or symbols. In this sense, all *inference* is a form of sign activity, where the word sign includes "feeling, image, conception, and other representation" [40, 5.283]. We will come back to this point in the following section, devoted to the discussion of the concept of affordance.[7]

As maintained above, people usually think and act in presence of poor and incomplete information and knowledge. Consequently they have to find ways of getting new data and new knowledge, and of choosing alternative heuristic pathways to help in their activities of making hypotheses. They can be new or smartly selected among a set of pre-stored options in an internal or external reserve. Of course finding "new" solutions to problems that arise in theoretical and practical situations is fundamental. It is in this sense that creativity is a fundamental component of human cognition. The philosophical concept of abduction may be a candidate for illustrating these processes, because it offers a unified approach to model creative process, which can fruitfully integrate the real narrowness of a merely psychological approach, too experimentally human-oriented.

A hundred years ago, C.S. Peirce [40] coined the concept of abduction in order to illustrate that the process of scientific discovery is not irrational and that a methodology of discovery is possible. Peirce interpreted abduction essentially as an "inferential" *creative process* of generating a new hypothesis. Abduction has a logical

---

[7] The semiotic dimension of abduction is also connected to its *model-based* aspects (for examples visualizations, diagram exploitation, thought experiment), that certainly constitute a considerable part of the creative meaning processes (cf. Magnani [41]). Cf. also the following subsections.

form (fallacious, if we model abduction by using classical syllogistic logic) distinct from deduction and induction. Reasoning which starts from reasons and looks for consequences is called *deduction*; that which starts from consequences and looks for reasons is called *abduction*.

Abduction – a distinct form of reasoning – is the process of *inferring* certain facts and/or laws and hypotheses that render some sentences plausible, that *explain* or *discover* some (eventually new) phenomenon or observation; it is the process of reasoning in which explanatory hypotheses are formed and evaluated. There are two main epistemological meanings of the word abduction (cf. Magnani [8]): 1) abduction that only generates "plausible" hypotheses ("selective" or "creative") and 2) abduction considered as inference "to the best explanation", which also evaluates hypotheses. An illustration from the field of medical knowledge is represented by the discovery of a new disease and the manifestations it causes which can be considered as the result of a creative abductive inference. Therefore, "creative" abduction deals with the whole field of the growth of scientific knowledge. This is irrelevant in medical *diagnosis* where instead the task is to "select" from an encyclopedia of pre-stored diagnostic entities. We can call both inferences ampliative, selective and creative, because in both cases the reasoning involved amplifies, or goes beyond, the information incorporated in the premises [42].

1.3.1 Theoretical Abduction

Theoretical abduction certainly illustrates much of what is important in creative abductive reasoning, in humans and in computational programs, but fails to account for many cases of explanations occurring in science when the exploitation of environment is crucial. It fails to account for those cases in which there is a kind of "discovering through doing", cases in which new and still unexpressed information is codified by means of manipulations of some external objects (in the case of science, they are called *epistemic mediators*). Magnani introduced this concept of theoretical abduction in [8]: he maintains that there are two kinds of theoretical abduction, "sentential", related to logic and to verbal/symbolic inferences, and "model-based", related to the exploitation of internalized models of diagrams, pictures, etc.,[8] cf. below in this paper. The concept of *manipulative abduction*[9] captures a large part of thinking activities where the role of action is central, and where the features of this action are implicit and hard to be elicited: action can provide otherwise unavailable information that enables the agent to solve problems by starting and by performing a suitable abductive process of generation or selection of hypotheses.

---

[8] Following Nersessian [43], the term "model-based reasoning" is used to indicate the construction and manipulation of various kinds of representations, not mainly sentential and/or formal, but mental and/or related to external mediators.

[9] Manipulative abduction and epistemic mediators are introduced and illustrated in Magnani [8].

1.3.2 Manipulative Abduction: Thinking through Doing

Manipulative abduction occurs when many external things, usually inert from the semiotic point of view, can be transformed into "cognitive mediators" that give rise - for instance in the case of scientific reasoning - to new signs, new chances for interpretants, and new interpretations.

Gooding [44] refers to this kind of concrete manipulative reasoning when he illustrates the role in science of the so-called "construals" that embody tacit inferences in procedures that are often apparatus and machine based. The embodiment is of course an expert manipulation of meaningful semiotic objects in a highly constrained experimental environment, and is directed by abductive movements that imply the strategic application of old and new *templates* of behavior mainly connected with extra-rational components, for instance emotional, esthetical, ethical, and economic.

The hypothetical character of construals is clear: they can be developed to examine or discard further chances, they are provisional creative organization of experience and some of them become in their turn hypothetical *interpretations* of experience, that is more theory-oriented, their reference/meaning is gradually stabilized in terms of established observational practices.

It is difficult to establish a list of invariant behaviors that are able to describe manipulative abduction in science. As illustrated above, certainly the expert manipulation of objects in a highly semiotically constrained experimental environment implies the application of old and new *templates* of behavior that exhibit some regularities. The activity of building construals is highly conjectural and not immediately explanatory: these templates are hypotheses of behavior (creative or already cognitively present in the scientist's mind-body system, and sometimes already applied) that abductively enable a kind of epistemic "doing". Hence, some templates of action and manipulation can be *selected* in the set of the ones available and pre-stored, others have to be *created* for the first time to perform the most interesting creative cognitive accomplishments of manipulative abduction.

From the general point of view of epistemic situations manipulative abductive reasoning exhibits other very interesting templates. For example: 1. action elaborates a *simplification* of the reasoning task and a redistribution of effort across time when we "need to manipulate concrete things in order to understand structures which are otherwise too abstract" [45], or when we are in presence of *redundant* and unmanageable information; 2. action can be useful in presence of *incomplete* or *inconsistent* information - not only from the "perceptual" point of view - or of a diminished capacity to act upon the world: it is used to get more data to restore coherence and to improve deficient knowledge; 3. action as a *control of sense data* illustrates how we can change the position of our body (and/or of the external objects) and how to exploit various kinds of prostheses (Galileo's telescope, technological instruments and interfaces) to get various new kinds of stimulation: action provides some tactile and visual information (e.g., in surgery), otherwise unavailable; 4. action enables us to build *external artifactual models* of task mechanisms instead of the corresponding internal ones, that are adequate to adapt the environment to the agent's needs:

experimental manipulations exploit *artificial apparatus* to free new possible stable and repeatable sources of information about hidden knowledge and constraints.[10]

The whole activity of manipulation in science is devoted to build various external *epistemic mediators*[11] that function as versatile semiotic tools able to provide an enormous new source of information and knowledge. Therefore, manipulative abduction represents a kind of redistribution of the epistemic and cognitive effort to manage objects and information that cannot be immediately represented or found internally (for example exploiting the resources of visual imagery).[12] If the structures of the environment play such an important role in shaping our semiotic representations and, hence, our cognitive processes, we can expect that physical manipulations of the environment receive a cognitive relevance.

Several authors have pointed out the role that physical actions can have at a cognitive level. In this sense Kirsh and Maglio [51] distinguish actions into two categories, namely *pragmatic actions* and *epistemic actions*. Pragmatic actions are the actions that an agent performs in the environment in order to bring itself physically closer to a goal. In this case the action modifies the environment so that the latter acquires a configuration that helps the agent to reach a goal which is understood as physical, that is, as a desired state of affairs. Epistemic actions are the actions that an agent performs in a semiotic environment in order to discharge the mind of a cognitive load or to extract information that is hidden or that would be very hard to obtain only by internal computation.

If a manipulative action performed upon the environment is devoted to create a configuration of signs that carries relevant information, that action will well be able to be considered as a cognitive semiotic process and the configuration of elements it creates will well be able to be considered an external representation. In this case, we can really speak of an embodied cognitive process in which an action constructs an external representation by means of manipulations.

The entire process through which an agent arrives at a physical action that can count as cognitive manipulating can be understood by means of the concept of manipulative abduction we have just introduced. Manipulative abduction is a specific case of cognitive manipulating in which an agent, when faced with an external situation from which it is hard or impossible to extract new meaningful features of an object, selects or creates an action that structures the environment in such a way that it gives information chances which would be otherwise unavailable and which is used specifically to infer explanatory hypotheses.

---

[10] The problem of manipulative abduction and of its tacit features is strongly related to the whole area of recent research on embodied reasoning (cf. Anderson [46]), but also relates to studies on external representations and situated robotics (cf. Clancey [47], Agree and Chapman [48], Brooks and Stein [49]). The role of manipulative abduction in ethical reasoning is illustrated in Magnani [36].

[11] The expression if derived from the cognitive anthropologist Hutchins, that coins the expression "mediating structure" to refer to various external tools that can be built to cognitively help the activity of navigating in modern but also in "primitive" settings [32, 50].

[12] It is difficult to preserve precise spatial relationships using mental imagery, especially when one set of them has to be moved relative to another.

1.4 Affordances and Cognition: The Received View

As we have illustrated in the first part of this paper, humans and some animals manipulate and distribute cognitive meanings after having delegated them to suitable environmental supports. The activity of cognitive niche construction reveals something important about the human and animal cognitive system.

As already mentioned, human cognition and its evolutionary dimension can be better understood in terms of its environmental situatedness. This means humans do not retain in their memory rich representations of the environment and its variables, but they actively manipulate it by picking up information and resources upon occasion, already available, or extracted/created and made available: information and resources are not only given, but they are actively sought and even manufactured. In this sense, we consider human cognition as a chance-seeker system. Consequently, in our terminology, chances are not simply information, but they are "affordances", namely, environmental anchors that allow us to better exploit external resources.

1.4.1 The Notion of Affordance and its Inferential Nature

One of the most disturbing problems with the notion of affordance is that any examples provide different, and sometimes ambiguous insights on it. This fact makes very hard to give a conceptual account of it. That is to say, when making examples everybody grasps the meaning, but as soon as one tries to conceptualize it the clear idea one got from it immediately disappears. Therefore, we hope to go back to examples from abstraction without loosing the intuitive simplicity that such examples provide to the intuitive notion.

The entire debate during the last fifteen years about the notion of affordance is very rich and complicated, but also full of conflicts and ambiguities. This subsection aims at giving just an overview of some issues we consider central to introduce to our treatment.

Gibson defines "affordance" as what the environment offers, provides, or furnishes. For instance, a chair affords an opportunity for sitting, air breathing, water swimming, stairs climbing, and so on. By cutting across the subjective/objective frontier, affordances refer to the idea of agent-environment mutuality. Gibson did not only provide clear examples, but also a list of definitions (cf. [52]) that may contribute to generating possible misunderstanding:

1. affordances are opportunities for action;
2. affordances are the values and meanings of things which can be directly perceived;
3. affordances are ecological facts;
4. affordances imply the mutuality of perceiver and environment.

We contend that the Gibsonian ecological perspective originally achieves two important results. First of all, human and animal agencies are somehow hybrid, in the sense that they strongly rely on the environment and on what it offers. Secondly,

Gibson provides a general framework about how organisms directly perceive objects and their affordances. His hypothesis is highly stimulating: "[...] the perceiving of an affordance is not a process of perceiving a value-free physical object [...] it is a process of perceiving a value-rich ecological object", and then, "physics may be value free, but ecology is not" [1, p. 140]. These two issues are related, although some authors seem to have disregarded their complementary nature. It is important here to clearly show how these two issues can be considered two faces of the same medal. Let us start our discussion.

1.4.2 Affordances are Opportunities for Action

Several authors have been extensively puzzled by the claim repeatedly made by Gibson that "an affordance of an object is directly perceived".[13] During the last few years an increasing number of contributions has extensively debated the nature of affordance as opportunity for action. Consider for instance the example "stairs afford climbing". In this example, stairs provide us with the opportunity of climbing; we climb stairs because we perceive the property of "climbability", and that affordance emerges in the interaction between the perceiver and stairs (cf. [54, 56]). In order to prevent from any possible misunderstanding, it is worth distinguishing between "affordance property" and "what" and object affords [57]. In the former sense, the system "stairs-plus-perceiver" exhibits the property of climbability, which is an *affordance property*. Whereas in the latter the possibility of climbing is clearly *what* an object affords.

1.4.3 Affordances are Ecological Facts

Concerning this point, Gibson argued that affordances are ecological facts. Consider, for instance, a block of ice. Indeed, from the perspective of physics a block of ice melting does not cease to exist. It simply changes its state from solid to liquid. Conversely, to humans a block of ice melting does go out of existence, since that drastically changes the way we can interact with it. A block of ice can chill a drink the way cold water cannot. Now, the point made by Gibson is that we may provide alternative descriptions of the world: the one specified by affordances represents the environment in terms of action possibilities. As Vicente put it, affordances "[...] are a way of measuring or representing the environment with respect to the action capabilities of an individual [...] one can also describe it [a chair] with respect to the possibilities for action that it offers to an organism with certain capabilities" [58]. Taking a step further, we may claim that affordances are chances that are *ecologically rooted*. They are ecological rooted because they rely on the mutuality between an agent (or a perceiver) and the environment. As ecological chances, affordances are the result of a hybridizing process in which the perceiver meets the environment. The emphasized stress on the mutuality between the perceiver and the environment provides a clear evidence of this point.

---

[13] Cf. [53–56].

1.4.4 Affordances Imply the Mutuality of Perceiver and Environment

Recently, Patel and Zhang [59], also going beyond the ecological concept of affordance in animals and wild settings by involving its role in human cognition and artifacts, in an unorthodox perspective, connect the notion of affordance to that of distributed representation. They maintain that affordances can be also related to the role of distributed representations extended across the environment and the organism. These kinds of representation come about as the result of a blending process between two different domains: on one hand the internal representation space, that is the physical structure of an organism (biological, perceptual, and cognitive faculties); on the other the external representation of space, namely, the structure of the environment and the information it provides. Both these two domains are described by constraints so that the blend consists of the allowable actions. Consider the example of an artifact like a chair. On one hand the human body constrains the actions one can make; on the other the chair has its constraints as well, for instance, its shape, weight, and so on. The blend consists of the allowable actions given both *internal* and *external* constraints.[14]

Patel and Zhang's idea tries to clarify that affordances result from a hybridizing process in which the environmental features and the agent's ones in terms of constraints are blended into a new domain which they call *affordance space*. Taking a step further, Patel and Zhang define affordances as allowable actions. If this approach certainly acknowledges the hybrid character of affordance we have described above and the mutuality between the perceiver and the environment, it seems however lacking with regard to its conceptual counterpart. As already argued, affordances are action-based opportunities.

1.5 Affordances as Eco-Cognitive Interactional Structures

Taking advantage of some recent results in the areas of distributed and animal cognition, we can find that a very important aspect that is not sufficiently stressed in literature is the dynamic one, related to designing affordances, with respect to their evolutionary framework: human and non-human animals can "modify" or "create" affordances by manipulating their cognitive niches. Moreover, it is obvious to note that human, biological bodies themselves evolve: and so we can guess that even the more basic and wired perceptive affordances available to our ancestors were very different from the present ones.[15] Of course different affordances can also be detected in children, and in the whole realm of animals.

---

[14] This idea can also be connected to the concept of cognitive fluidity argued by Mithen [60]. From the perspective of cognitive palaeoanthropology, Mithen claimed that the modern mind is characterized by the capacity of applying to heterogeneous domains forms of thinking originally designed for specific tasks. He also contends that in hominids this change originated through a blend of two different intelligence domains, namely, that of internal representations entities and external artifacts [35].

[15] The term "wired" can be easily misunderstood. Generally speaking, we accept the distinction between cognitive aspects that are "hardwired" and those which are simply

1.5.1 Pseudothoughts and Model-Based Thinking in Humans and Animals: Affordances as Chances

Gibson's "affordances" [1] and Brunswik's interplay between proximal and distal environment [2] (cf. the following section) also deal with the problem of the so-called model-based protothoughts (cf. [61]) (we prefer to call them pseudothoughts) which concern any kind of cognition far from the cognitive features granted by human language. These kinds of cognitive tools typical of infants and of many animals (and still operating in human adults in various forms of more or less unexpressed thinking) are hypothesized to express the organic beings' implicit skills to act in a perceived environment as a distal environment ordered in terms of the possibilities to *afford* the action in response to local changes.

Different actions will be suitable to different ways of apprehending aspects of the external world. The objectification of the world made possible by language and other highly abstract organizing cognitive techniques (like mathematics) is not needed. A Gibsonian affordance is a resource or chance that the environment presents to the "specific" organism, such as the availability of water or of finding recovery and concealment. Of course the same part of the environment offers different affordances to different organisms. The concept of affordance can be also extended to animals and artificial environments built by human and non-human animals. My cat affords her actions in the kitchen of my house differently than me, obviously I do not find affordable to easily jump through the window or on the table! I simply cannot imagine the number of things that my cat Sheena is possibly "aware" of (and her way of being aware) in a precise moment, such as the taste of the last mouse she caught and the type of memory she has of her last encounter with a lizard:[16] "Only a small part of the network within which mouseness is nested for us extends into the cat's world" (cf. [64, p. 203]). It is possible to imagine – but this is just science fiction - that we could "afford" the world like a dolphin only artificializing it by using prosthetic instruments and tools like sonar, fruit of modern scientific knowledge.

"pre-wired". By the former term we refer to those aspects of cognition which are fixed in advance and not modifiable. Conversely, the latter term refers to those abilities that are built-in prior the experience, but that are modifiable in later individual development and through the process of attunement to relevant environmental cues: the importance of development, and its relation with plasticity, is clearly captured thanks to the above distinction. Not all aspects of cognition are pre-determined by genes and hard-wired components.

[16] The point of view of Gibson has been taken into account by several people in the computational community, for example by Brooks in robotics [62]. "Vision is not delivering a high level representation of the world, instead it cooperates with motor controls enabling survival behavior in the environment. [...] While it is very sensible that the main goal of vision in humans is to contribute to moving and acting with objects in the word, it is highly improbable that a set of actions can be identified as the output of vision. Otherwise, vision must include all sort of computations contributing to the acting behavior in that set: it is like saying that vision should cover more or less the whole brain activity" (cf. Domenella and Plebe [63, pp. 369-370]).

1.5.2 Proximal and Distal Environment

In the extended perspective above we can consider an artifact with its affordances in the framework of a distributed cognitive system where people interact with external cognitive representations.[17] Internal representations are the knowledge and structure in individual's brains, external representations are the knowledge and structure in the external environment, for example in a specific artifact (cf. [66]). These external representations have many non-trivial properties (symbolic/semiotic patterns) depending on the kind of cognitive delegations operated when building them, the structure of the artifact itself (physical and chemical configurations), and the constraints intrinsic to its materiality.

Following Clark's perspective on language as an external tool [67] Wheeler speaks of *on-line* – like in the case of manipulative abduction, which involves both internal and external representations – and *off-line* (also called *inner rehearsal*, based on internal representations alone. A true situation of distributed cognition is occurring in the case of on-line thinking, like in our case of manipulative abduction and in other less expert and less creative cases, where the resources are not merely inner (neurally-specified) and embodied but hybridly intertwined with the environment: in this case we face with an abductive/adaptive process produced in the dynamical inner/outer coupling where internal elements are "directly *causally locked onto* the contributing external elements" [68, p. 705].

It is extremely important to note that our epistemological distinction between theoretical and manipulative abduction is certainly also based on the possibility of separating the two aspects in real cognitive processes, that resort to the distinction between off-line (theoretical, when only inner aspects are at stake) and on-line (manipulative, where the interplay between internal and external aspects is fundamental. As Wheeler has recently observed some thinkers like Esther Thelen and Andy Clark have raised doubts about the on-line/off-line distinction "[...] on the grounds that no intelligent agent is (they claim) ever wholly on-line or wholly off-line. On this view, intelligence is always a dynamic negotiation between on-line and off-line processes" [68, p. 707, footnote 14]. We contend that, even if manipulative/on-line cases exist in great numbers, there are also cognitive processes that seem to fall into the class of off-line thinking, as we can simply introspectively understand.

In this perspective affordances can be considered as related to distributed representations extended across the environment and the organism. In the case of the affordances possibly offered by an artifact a basic problem is that the "cognitive" properties of the components of it cannot be inferred from the properties of those components themselves [32]. Following Zhang and Patel [59, p. 336] we can say that

> The external and internal representation spaces can be described by either constraints or allowable actions. Constraints are the negations of allowable

---

17 Of course cognition can also be distributed across a collective of individuals. This is in line with the so-called "social manifestation thesis" put forward by Wilson. As he put it: "[...] individuals engage in some forms of cognition only insofar as they constitute part of a social group. [...] These psychological traits are not simply properties of the individual members of the group, but features of the group itself" [65, p. 228].

actions. That is, the allowable actions are those satisfying the constraints, and the constraints set the range of the allowable actions. If the external and internal representation spaces are described by constraints, then the affordances are the disjunction of the constraints of the two spaces. If the external and internal representation spaces are described by allowable actions, then the affordances are the conjunction of the allowable actions of the two spaces.

Brunswik's [2] ecological theory (where agents and their environments are still seen, like for Gibson, as strongly intertwined) in terms of proximal and distal environment (cf. his famous lens-model), suitably combined with the theory of affordances, can further clarify the role in cognition of artifacts and of high-level cognitive processes, especially in human beings. Following Brunswik's theory an organism is not able to perceive distal stimuli directly but instead must *infer* what is going on in its ecological niche from the cues available (cues provided by proximal stimuli). The success (Brunswik says the *ecological validity*) of this "vicarious" inference[18] is of course jeopardized by the constitutive incompleteness (and unreliability, ambiguity, and equivocality) of the cues available and by their highly variable diagnostic character: Brunswik, implicitly professing an abductive attitude worthy of Peirce, says: "[…] ordinarily organisms must behave as if in a semierratic ecology" [69, p. 209], given the intrinsic "ambiguity in the causal texture of the environment" [70, p. 255].[19] In this sense he adds that the cues and the mediating inference are both "probabilistic", like in the case of abduction where it is always the case that: "Both the object-cue and the means-end relationship are relations between probable partial causes and probable causal effects" [70, p. 255].

As we will better describe in subsection 1.6 and in section 1.7 the inference above is of course abductive, and *mediates* the relationship between an organism's central desires and needs and the distal state of affairs, enabling it to (provisionally) *stabilize* the relationship itself by reaching goals. Of course this mediating process involves different means (for example selecting, combining, and substituting cues to overcome their limitations) [72] in different environmental circumstances [73, p. 12], where time is often a key element which strongly affects information incompleteness and the degree of success of the various means and inferential procedures used.

The organism's "end stage" of being afforded, which in the case of humans, involves all aspects of what Magnani has called *multimodal abduction* [74], is of course reached, like Gibson contends, through *perception*, as a suitable collections of sensory information rather than through the organism's overt behavior (in this case perception resorts to a spontaneous abduction performed through various sensory mechanisms and their interplay, cf. section 1.3).

It has to be said that recent research in human and not human animals has jeopardized the assumption concerning the basic cognitive impenetrability of perception

---

[18] Here the word inference has to be intended in the expanded semiotic Peircean sense, i. e. the term must not collapse in the restricted logical meaning (cf. below section 1.6).

[19] A detailed illustration of these aspects of Brunswikian theory are given in Hammond and Steward [71, part I].

(an analysis of this problem if presented in [26]) in the framework of animal cognition). In sum, perception is informationally "semi-encapsulated", and also pre-wired, i.e., despite its bottom-down character, it is not insulated from plastic cognitive processes and contents acquired through learning and experience. Many results in the field of cognitive experimental psychology and of neuroscience corroborate the assumption above: 1) the role of emotions (anxiety) in perceiving affordances and the role of attentional narrowing mechanisms in the related changes [75]; 2) the nonconscious effects of perceptual accuracy and on consequent actions of tool use (reachability influences perceived distance [76]); 3) the role of moral and social weights in people that carry babies or groceries [77]; 4) the role of motivational states (desires, wishes, preferences) [78]; 5) the effect of the adoption of the actor's perspective while observing and understanding actions through mentalistic or motor schemes (for experimental results cf. [79] and [80]; for neurobiological correlates cf. [81] and [82]).

In this updated perspective the "end stage" of perception is also reached through a more complicated *inferential processing* which in turn occurs, either by

1. a mere *internal* cognitive processing (for example expressed by model-based abduction and, in the case of human beings, also by sentential abduction, cf. section 1.3)

   or

2. a (cognitive) composite *action* with the aim of getting more sensory data, compensating for their equivocality, and obtaining cognitive feedback (confirmatory and disconfirmatory), and/or with the aim of manipulating them (manipulative abduction), also exploiting *cognitive delegations* to the environment and to *artifacts* (cf. section 1.3).

   [Indeed these high-level inferential processes affect perception in various ways, like it is shown by the evidence coming from the studies illustrated above].

Thus affordances emerge through perceptions that are semi-encapsulated and affected by the two processes above which in turn grant the final (pragmatic) *action*.

In this light the Brunswikian concept of ecological validity can be seen in terms of abductive plausibility of the inference at play, given the available data/cues. The quality of the inferential abductive performance measures the degree of adaptation between an organism's behavior and the structure of the environment, i.e. the fitness of the behavior based on the particular adopted inference. When the cues are object of simple and immediate perceptual appraisal the situation reflects what Magnani has illustrated in the case of the so-called "visual abduction" [8].[20] On the contrary, in the other cases, organisms inferentially abduce a "hypothesis/judgment" about distal structure of the environment. It is worthy to quote again Gibson intuition, which can better be grasped in this perspective "Perceiving is the simplest and best kind of knowing. But there are other kinds, of which three were suggested. Knowing by means of instruments extends perceiving into the realm of the very distant and the

---

[20] In this last case we could say there is a one-to-one mapping between proximal and distal structure [58, p. 261].

very small; [...] Knowing by means of language makes knowing explicit instead of tacit" [1, p. 263]. An example can be a forecast – usually probabilistic – about the behavior of wind based on the current wind speed measured at a ground station as illustrated in a computer screen – the "cue". It is important to say that in this last case the proximal perception affords the day-after action of wearing clothes suitable to the weather.

We have said above that in the mediating inferential process also cognitive delegations to the environment (for example to automated artifacts) can also play a role, for example in facilitating reliable action/decisions.[21] In both human and non-human animals artifacts can reduce the uncertainty of the relationship between organism and environment. Even in these cases, for example – in the case of humans – when technology is directly designed to respond with greater precision than people can do, a person's abductive judgment can still fail to correspond with the distal environment. Recent research based on Brunswikian tradition has emphasized the essential ecological character of the cognitive engineering enterprise in the framework of systems composed of the interaction among humans, mediating technologies, and tasks environments. Many results have shown in various interesting ways how many technological devices help humans to fulfill their adaptation to the environment by enhancing hypothesis generation/judgment and, consequently, decision-making. Often the technology itself fails to provide the correct judgment about a given situation, in other cases the gap in the proximal/distal relationship is embedded in the interaction with the user [84].

For example [85] study the airline pilot's performance in landing and taxiing to the gate in foggy conditions by showing, thanks to an ecological analysis, how the cockpit artifacts (and their interfaces which provide proximal cues/chances for action) affect the overall cognitive performance. Often the problem is related to the fact that technologies exhibit a discrete ecology that does not sufficiently involve the approximation and convergence performed by continuous ecologies of natural environmental structures. This fact for example explains why it has been recently shown that current design of cockpit automation leaves pilots less supported in special uncertain (and more challenging) – unsafe – situations: control systems proximal to pilots are discrete, whereas the behavior of the distal controlled system – the aircraft – is continuous [86]. Finally, in the mediating abductive inferences occurring through artifacts, time (and the so-called "time-stress" and "time-pressure" effects and their relationship to knowledge deficits and task simplification) is still a key element which strongly affects information incompleteness and the degree of success of various heuristic schemes.

1.5.3 Attunement, Affordances, and Cognitive Artifacts: Extracting and Creating Affordances

Organisms need to become *attuned* to the relevant offered features and much of the cognitive tools built to reach this target are the result of evolution and of merely pre-wired and embodied perceptual capacities like imagistic, empathetic, trial and error,

---

[21] The case of computer interfaces and time pressure is treated in [83].

and analogical devices. These wired capabilities, that in our epistemological perspective have to be considered "cognitive" even if instinctual, can be seen as devices of organisms that provide potential implicit abductive powers, (cf. the following section): they can provide an immediate and overall appraisal of the situation at hand and so orient action, and can be seen as forms of pseudo-explanation of what is occurring over there, as emerging in that material contact with the environment granted in the perceptual interplay. It is through this embodied process that affordances can arise both in wild and artificially modified niches. Peirce had already contended more than one hundred years ago that abduction even takes place when a new born chick picks up the right sort of corn. This is an example, so to say, of spontaneous abduction – analogous to the case of some unconscious/embodied abductive processes in humans (cf. section 1.3):

> When a chicken first emerges from the shell, it does not try fifty random ways of appeasing its hunger, but within five minutes is picking up food, choosing as it picks, and picking what it aims to pick. That is not reasoning, because it is not done deliberately; but in every respect but that, it is just like abductive inference (Cf. the article "The proper treatment of hypotheses: a preliminary chapter, toward and examination of Hume's argument against miracles, in its logic and in its history" [1901] [87, p. 692]).

Animals can act on a goal that they cannot perceive – the predator that waits for the prey for example – so organism's appraisal of the situation includes factors that cannot be immediately perceived, the tamarins quickly learn to select the tool that can grant more food in varied situations among different tools offered. Animals occupy different environmental niches that "afford" their possibility to act, like Gibson's theory of affordances teaches, but in this interplay cognitive niches built and created by them also provide affordances: higher-level cognitive endowments either shaped by the evolution or plastically learnt are at play. Well-known dishabituation experiments have shown how for example infants use high-level physical principles to relate themselves to the environment and to be afforded by it: they look longer at the facts that they find surprising showing what expectations they have; animals like dolphins respond to structured complex gestural signs in ways, so it is in the case of monkeys that perform complicated technical manipulations of objects and in birds that build artifacts to house beings that have not yet been born. The problem here is that organisms already have affordances available because of their instinctive endowments, but also they can *dinamically* abductively "extract" natural affordances through "affecting" and modifying perception (which becomes semi-encapsulated). Organisms can also "create" affordances by building artifacts and cognitive niches. These last affordances were not previously available taking advantage of both their instinctual and cognitive capacities. Like Gibson says: "[...] The human young must learn to perceive these affordances, in some degree at least, but the young of some animals do not have time to learn the ones that are crucial for survival" [1, p. 406].

It can be hypothesized that in many organisms the perceptual world is the only possible model of itself (cf. [62]) and they can be accounted for in terms of a merely reflex-based organisms, and so no other internal representations are available. In the case of affordances in other sensitive organisms the coupling with the environment can be more flexible because it is important in coupling with the niche to determine what environmental dynamics are currently the most relevant, among the several that afford and that are available. An individual that is looking for its prey and at the same time for a mate (which both immediately afford it without any ambiguity) is contemporarily in front of two different affordances and has to abductively select (see the following sections) the most suitable one by weighting them. Both affordances and the more or less higher-level plastic cognitive processes of their selection in specific situations can be "stabilized", but they can also be modified, increased, and substituted with new ones.[22]

It is important to note that recent research based on Schrödinger's focusing on energy, matter and thermodynamic imbalances provided by the environment, draws the attention to the fact that all organisms, including bacteria, are able to perform elementary *cognitive functions* because they "sense" (i.e. they are afforded by) the environment and process internal information for "thriving on latent information embedded in the complexity of their environment" (Jacob, Shapira, and Tauber [89, p. 496]. Indeed Schrödinger maintained that life requires the consumption of negative entropy, i.e. the use of thermodynamic imbalances in the environment. As a member of a complex superorganism – the colony, a multi-cellular community – each bacterium possesses the ability to sense and communicate with the other units comprising the collective and performs its work within a distribution task so, bacterial communication entails collective sensing and cooperativity through interpretation of chemical messages, distinction between internal and external information, and a sort of self vs. non-self distinction (peers and cheaters are both active).

In this perspective "biotic machines" are *meaning*-based forms of intelligence to be contrasted with the *information*-based forms of artificial intelligence: biotic machines generate new information, assigning contextual meaning to gathered information: self-organizing organisms like bacteria are afforded – through a real cognitive act – and by "relevant" information that they subsequently couple with the regulating, restructuring, and *plastic* activity of the contextual information (intrinsic meaning) already internally stored, which reflects the intra-cellular state of the cells. Of course the "meaning production" involved in the processes above refers to structural aspects of communication that cannot be related to the specific sentential and model-based cognitive skills of humans, primates, and other simpler animals, but still shares basic functions with these like sensing, information processing, and collective abductive contextual production of meaning. As stressed by Ben Jacob, Shapira, and Tauber

> In short, bacteria continuously sense their milieu and store the relevant information and thus exhibit "cognition" by their ability to process information and responding accordingly. From those fundamental sensing faculties, bacterial information processing has evolved communication capabilities that allow the creation of cooperative structures among individuals to form super-organisms [89, p. 504].

---

[22] It has to be said that in animals, still at the higher level of non-merely reflex-based cognitive abilities, no representational internal states need always be hypothesized [88].

1.6 Affordances and Abduction: The Plasticity of Environmental Situatedness

We have quoted above Gibson conviction that "The hypothesis that things have affordances, and that we perceive or learn to perceive them, is very promising, radical, but not yet elaborated" [1, p. 403]. Let us deepen this issue: we can say that the fact that a chair affords sitting means we can perceive some clues (robustness, rigidity, flatness) from which a person can easily say "I can sit down". Now, suppose the same person has another object O. In this case, the person can only perceive its flatness. He/she does not know if it is rigid and robust, for instance. Anyway, he/she decides to sit down on it and he/she does that successfully. Again, we are faced with the problem of direct and indirect visual perception. It is thanks to the effect of action that we can detect and stabilize the new affordances.

Now, our point is that we should distinguish between the two cases: in the first one, the cues we come up with (flatness, robustness, rigidity) are *highly diagnostic* to know whether or not we can sit down on it, whereas in the second case we eventually decide to sit down, but we do not have any precise clue about. How many things are there that are flat, but one cannot sit down on? A nail head is flat, but it is not useful for sitting. This example further clarifies two important elements: firstly, finding/constructing affordances certainly deals with a (semiotic) inferential activity (cf. [90]); secondly, it stresses the relationship between an affordance and the information that specify it that only arise in the *eco-cognitive interaction* between environment and organisms. In this last case the information is reached through a simple action, in other cases through action and complex manipulations. We maintain that the notion of abduction can further clarify this puzzling problem (cf. below section 1.7).

The term "highly diagnostic" explicitly refers to the abductive framework. In section 1.3 we have said that abduction is a process of *inferring* certain facts and/or laws and hypotheses that render some sentences plausible, that *explain* or *discover* some (eventually new) phenomenon or observation. The distinction between theoretical and manipulative abduction extends the application of that concept beyond the internal dimension. From Peirce's philosophical point of view, all thinking is in signs, and signs can be icons, indices or symbols. Moreover, all inference is a form of sign activity, where the word sign includes "feeling, image, conception, and other representation" [40, 5.283], and, in Kantian words, all synthetic forms of cognition. That is, a considerable part of the thinking activity is "model-based" and consequently non sentential. Of course model-based reasoning acquires its peculiar creative relevance when embedded in abductive processes, so that we can individuate a *model-based abduction*. In the case of diagnostic reasoning in medicine, a physician detects various symptoms (that are signs or clues), for instance, cough, chest pain, and fever, *then* he/she may infer that it is a case of pneumonia.

The original Gibsonian notion of affordance (cf. section 1.4) especially deals with those situations in which the "perceptual" signs and clues we can detect prompt or suggest a certain action rather than others.[23]

They are already available and belong to the normality of the adaptation of an organism to a given ecological niche. Nevertheless, if we acknowledge that environments and organisms' instinctual and cognitive plastic endowments change, we may argue that affordances can be related to the variable (degree of) *abductivity* of a configuration of signs: *a chair affords sitting* in the sense that the action of sitting is a result of a sign activity in which we perceive some physical properties (flatness, rigidity, etc.), and therefore we can ordinarily "infer" (in Peircean sense) that a possible way to cope with a chair is sitting on it. So to say, in most cases it is a spontaneous abduction to find affordances because this chance is already present in the perceptual and cognitive endowments of human and non-human animals.

We maintain that describing affordances that way may clarify some puzzling themes proposed by Gibson, especially the claim concerning the fact that we directly perceive affordances and that the value and meaning of a thing is clear at first glance. As we have just said, organisms have at their disposal a standard endowment of affordances (for instance through their wired sensory system), but at the same time they can extend and modify the range of what can afford them through the appropriate cognitive abductive skills (more or less sophisticated). As maintained by several authors (cf. Rock [91], Thagard [92], Hoffman [93], Magnani [8]), what we see is a result of an embodied cognitive abductive process. For example, people are adept at imposing order on various, even ambiguous, stimuli (Magnani [8, p. 107]). Roughly speaking, we may say that what we *see* is what our visual apparatus can, so to say, "explain". It is worth noting that this process happens almost simultaneously without any further mediation. Perceiving affordances has something in common with it. Visual perception is a more automatic and "instinctual" semi-encapsulated activity, we have already said that Peirce claimed to be essentially abductive, even if not propositional. Indeed he considers inferential any cognitive activity whatever, not only conscious abstract thought: he also includes perceptual knowledge and subconscious cognitive activity. For instance he says that in subconscious mental activities visual representations play an immediate role [94].

We also have to remember that environments change and so the perceptive capacities when enriched through new or higher-level cognitive skills, which go beyond the ones granted by the merely instinctual levels. This dynamics explains the fact that if affordances are usually stabilized this does not mean they cannot be modified and changed and that new ones can be formed.

First of all, affordances appear durable in human and animal behavior, like kinds of habits, as Peirce would say [40, 2.170]. For instance, that a chair affords sitting is a fair example of what we are talking about. This deals with what we may call

---

[23] In the original Gibsonian view the notion of affordance is mainly referred to proximal and immediate perceptual chances, which are merely "picked up" by a stationary or moving observer. We maintain that perceiving affordances also involves evolutionary changes and the role of sophisticated and plastic cognitive capacities.

stabilized affordances. That is, affordances that we have experienced as highly successful. Once evolutionarily formed, or created/discovered through cognition, they are stored in embodied or explicit cognitive libraries and retrieved upon occasion. Not only, they can be a suitable source of new chances, through analogy. We may have very different objects that equally afford sitting. For instance, a chair has four legs, a back, and it also stands on its own. The affordances exhibited by a traditional chair may be an analogical source and transferred to a different new artifact that presents the affordance of a chair for sitting down (and that to some extent can still be described as a chair). Consider, for instance, the variety of objects that afford sitting without having four legs or even a back. Let us consider a stool: it does not have even a back or, in some cases, it has only one leg or just a pedestal, but it affords sitting as well as a chair.

Second, affordances are also subjected to changes and modifications. Some of them can be discarded, because new configurations of the cognitive environmental niche (for example new artifacts) are invented with more powerful offered affordances. Consider, for instance, the case of blackboards. Progressively, teachers and instructors have partly replaced them with new artifacts which exhibit affordances brought about by various tools, for example, slide presentations. In some cases, the affordances of blackboards have been totally re-directed or re-used to more specific purposes and actions. For instance, one may say that a logical theorem is still easier to be explained and understood by using a blackboard, because of its affordances that give a temporal, sequential, and at the same time global perceptual depiction to the matter.

Of course – in the case of humans – objects can afford different persons in different ways. This is also the case of experts: they take advantage of their advanced knowledge within a specific domain to detect signs and clues that ordinary people cannot detect. For instance, a patient affected by pneumonia affords a physician in a completely different way compared with that of any other uncultured person. Being abductive, the process of perceiving affordances mostly relies on a continuous activity of hypothesizing which is cognition-related. That A affords B to C can be also considered from a semiotic perspective as follows: A signifies B to C. A is a sign, B the object signified, and C the interpretant. Having cognitive skills (for example knowledge contents and inferential capacities but also suitable pre-wired sensory endowments) about a certain domain enables the interpretant to perform certain abductive inferences from signs (namely, perceiving affordances) that are not available to those who do not possess those apparatuses. To ordinarily people a cough or chest pain are not diagnostic, because they do not know what the symptoms of pneumonia or other diseases related to cough and chest pain are. Thus, they cannot make any abductive inference of this kind and so perform subsequent appropriate medical actions.

1.7 Innovating through Affordance Creation

Consider, for instance, a huge stone and a chair. Indeed, both these objects afford sitting. The difference is that in the case of a stone affordances are simply already, so to say, *given*: we find a stone and we ordinarily "infer" (in Peircean sense) that it can be useful for sitting. In contrast, those of a chair are somehow *manufactured* from scratch. In the case of a chair, we have entirely created an object that displays a set of affordances. This process of building affordances can be made clearer taking advantage of the abductive framework we have introduced above.

That an object affords a certain action means that it embeds those *signs* from which we "infer" – through various cognitive endowments, both instinctual and learnt – that we can interact with it in a certain way. As already said, in the case of a stone, humans exploit a pre-existing configuration or structure of meaningful sign data that are in some sense already established in the interaction organism/environment shaped by past evolution of the human body (and in part "material cultural" evolution, for example when hominids exploited a stone/chair to sit down in front of a rudimentary altar). In the case of a chair, this configuration is invented. If this perspective is correct, we may argue that building "artificial" affordances means configuring signs in an external environment expressly to trigger new proper inferences of affordability. In doing this, we perform smart manipulations and actions that – we conjecture – can produce new (and sometimes "unexpected") affordances. Accordingly, creating affordances is at the same time making new ways of inferring them feasible.[24]

1.7.1 Latent Constraints

In section 1.5.2 we have said that the organism's "end stage", in the Brunswikian interplay proximal/distal, not only involves perception, but also a more complicated *inferential processing* which in turn occurs through either

1. a mere *internal* cognitive processing (expressed by model-based abduction and, in the case of human beings, also by sentential abduction, cf. section 1.3) or

2. a (cognitive) composite *action* with the aim of getting more sensory data, compensating for their equivocality, and obtaining cognitive feedback (confirmatory and disconfirmatory), and/or with the aim of manipulating them (manipulative abduction), also exploiting *cognitive delegations* to the environment and to *artifacts* (cf. section 1.3).

   [Indeed these high-level inferential processes affect perception in various ways, like its is shown by the evidence coming from the studies illustrated above].

We have consequently said that affordances emerge through perceptions that in humans are basically semi-encapsulated, and also affected by the two processes above and that grants the subsequent (pragmatic) *action*.

---

[24] On this note, see [95] about the role of abduction in designing computer interfaces.

Consequently, two kinds of "actions", cognitive and pragmatic (performatory) are at play. Kirlik [96], also resorting to a distinction between pragmatic and epistemic action offered by Kirsh and Maglio [51],[25] offers an analysis of the problem that is extremely interesting to further clarify the dichotomy. In Kirlik's words, the first type of action we have just indicated plays an "epistemic" or "knowledge-granting" role and the second a "performatory role in the execution of interactive tasks" [96, p. 214]. It is well-known that traditionally, cognitive scientists have seen action systems in a purely performatory/pragmatic light, thus lacking any immediate "cognitive" features other than ability to execute commands. In the new epistemological perspective expert performers use action in everyday life to create an external model of task dynamics that can be used in lieu of an internal model: for example a child shaking a birthday present to guess its contents is dithering, a common human response to perceptually impoverished conditions.

Not only a way for moving the world to desirable states, action performs an epistemic and not merely performatory role: people structure their worlds to simplify cognitive tasks but also in the presence of incomplete information or faced with a diminished capacity to act upon the world when they have less opportunities to gain knowledge. Epistemic action can also be described as resulting from the exploitation of latent constraint in the human-environment system. This additional constraint grants additional information: in the example of the child shaking a birthday present she is taking action that will cause variables relating to the contents of the package to covary with perceptible acoustic and kinesthetic variables. Epistemic actions result from exploiting *latent constraints* in the human-environment system as chances for further inferences. "Prior to shaking, there is no active constraint between these hidden and overt variables causing them to carry information about each other". Similarly, "one must put a rental car 'through its paces' because the constraints active during normal, more reserved driving do not produce the perceptual information necessary to specify the dynamics of the automobile when driven under more forceful conditions" [98, p. 24]. Moreover, a very interesting experiment is reported concerning short-order cooking at a restaurant grill in Atlanta: the example shows how cooks with various levels of expertise use external models in the dynamic structure of the grill surface to get new information otherwise inaccessible.

In Brunswikian terms some variable values are proximal because they can be perceived and others cannot because they are distal. This distinction, Kirlik observes, is relative:

> A particular variable (e.g. the velocity of an automobile) could be described as proximal if the purpose of a study was to understand how a driver might use velocity to infer less readily available information relating to a vehicular control task (e.g. whether he or she can take a given turn safely). In other cases, velocity could be considered a distal variable if the goal of the effort was to understand how velocity itself was estimated on the basis

---

[25] Kirsh describes everyday situations (e. g. grocery bagging, salad preparation, where people adopt actions to simplify choice, perception, and reduce the amount of internal computation through the use of suitable cognitive delegations to the environment and to artifacts [97].

of even more immediate perceptual information (e.g. optical flow patterns, speedometer readings, etc.) [96, p. 216].

Of course from the perspective of action the values of proximal variables can directly be manipulated. On the contrary distal variables can be changed (and so inferred) only by manipulating proximal variables. On this basis Kirlik describes a generalized framework that provides an ontology for describing an environment of a performer who is undertaking an interactive activity. Variables exhibited to an agent in a world of surfaces, objects, events, and artifacts present various values:

1. Type 1 [PP.PA]: variables are proximal from the perspective of both perception and action. "The location of the book you are now reading can most likely be represented by variables of this type: most of us can directly manipulate the values of these variables by hand and arm movements, and we can also perceive the location of the book directly. One can think of these variables as representing Gibson's directly perceptible affordances" [96, p. 216].
2. Type 1 [PP.DA]: variables proximal for perception but distal for action. Variables are directly available to perception but we can change their values by manipulating proximal variables that cause changes in the values of the distal action variable. You feel it is cold in the room but you need to manipulate the thermostat to change the temperature and consequently your feeling, but you cannot directly change the temperature.
3. Type 1 [DP.PA]: variables distal for perception and proximal for action. It is the case that many of my actions through for instance an artifact can change the values of environment, which I cannot directly perceive. A common outcome of the manipulation of many artifacts is that they for example have moral consequences, perhaps even unexpected ones, for human beings and objects very distant and remote from us, which might last forever. "When posing for a photograph I change the location of my image in the viewfinder of a camera without being able to perceive how it has been changed" [96, pp. 217–218].[26]
4. Type 1 [PP.PA]: variables that are distal from both perception and action, very common in the case of technology. "We infer the values of these variables from interface displays, and we change the value of these variables by using interface controls" (*ibid.*). D. Norman [3] "concepts of the gulf of evaluation and gulf of execution were invented to describe the challenges we sometimes face in using the proximal action variables we have immediately available to infer and alter the values of the distal variables which are typically the true target of our interaction" [96, p. 218].

---

[26] A related issue is illustrated by Magnani in the recent [36]: many artifacts play the role of "moral mediators". This happens when macroscopic and growing phenomenon of global moral actions and collective responsibilities result from the "invisible hand" of manipulations of artifacts and interactions among agents at a local level, like for example in the case of the effect of the Internet on privacy: for example my manipulations on the net can affect the privacy of people with effects that I cannot perceive.

In the framework above, which stresses the importance of inner-outer inter-action in proximal-distal dynamics, it is very easy to interpret artifacts built by humans as ways of adaptation that, through the construction of suitable cognitive feasible niches, mediate and augment our interaction with the distal world. They aim at enhancing intellectual functioning by offering suitable differentiated affordances/proximal-cues, which are easier to perceptually detect and present new opportunities for action. This strategy, which modifies previous available human cognitive ecology, can offload some cognitive demands to the world through a better exploitation of artifacts.[27]

1.7.2 Creating Chances through Manipulating

It is now clear that the history of culture, artifacts, and various tools can be viewed as a continuous process of building and crafting new affordances upon pre-existing ones or even from scratch. From cave art to modern computers, there has been a co-evolution between humans and the environment they have built and they live in. Indeed, what a computer can afford embraces an amazing variety of opportunities and chances comparing with the ones exhibited by other tools and devices. More precisely, a computer as a Practical Universal Turing Machine (cf. [100] can mimet-ically reproduce a considerable part of the most complex operations that the human brain-mind systems carry out (cf. [35]). For instance, computers even result in many respects more powerful than humans in memory capacity and in mathematical rea-soning. From a semiotic perspective, computers bring into existence new artifacts that present "signs" (in Peircean sense) – for exploring, expanding, and manipulat-ing our own brain cognitive processing (and so they contribute to "extend the mind beyond the brain"), that is, by offering and creating new affordances. As just argued above in section 1.2, building affordances deals with a semiotic activity, mainly ab-ductive, in which signs are appropriately scattered all around in order to prompt (or suggest) a certain interaction rather than others.

In order to clarify this point, consider the simple diagrammatic demonstration that the sum of the internal angles of any triangle is 180°.

This is an example of construction of affordances taken from the field of elemen-tary geometry. In this case a simple manipulation of the – suitably externally depicted – triangle in Figure 1.1(a) gives rise to another external configuration (Figure 1.1(b)) that uncovers new visual affordances. The process occurs through construction and

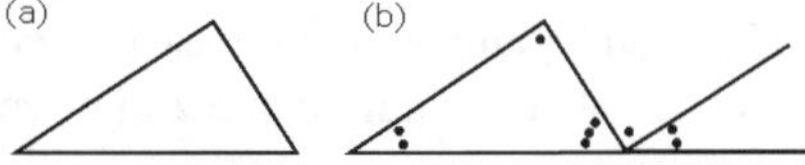

Fig. 1.1. Diagrammatic demonstration that the sum of the internal angles of any triangle is 180°. (a) Triangle. (b) Diagrammatic manipulations.

---

[27] Kirlik [99] describes various interesting examples on the use of artifacts to enhance cogni-tive skills in technological disciplines like architecture and aeronautics, but also in expert everyday performances like cooking.

modification of the initial external diagram, so as we can easily arrive at the generalized result that the sum of the internal angles of any triangle is 180°. The process
happens in a distributed interplay between a continuous externalization through cognitive acts, its manipulation, and re-internalization that acknowledges what has been
discovered outside, picking up the result and re-internalizing it. In the action new
affordances arise and lead to the result whereby, the sum of the internal angle is
found.

From the epistemological point of view this example is a typical example of the
so-called manipulative abduction (cf. [8] chapter three, and [101]). In terms of affordances this is a cognitive manipulation (entirely abductive, the aim is to find an
explanatory hypothesis) in which an agent, when faced with merely "internal" representational geometrical "thoughts", from which it is hard or impossible to extract
new meaningful features of a concept, organizes epistemic action that structures the
environment in such a way that unearths new affordances as chances which favor new
outcomes. As already mentioned, the detection of affordances is hypothesis-driven:
it is not said that just anybody can detect it. In the example of the triangle, only
a person that has been taught about geometry can infer the affordances within the
manipulated construction built upon the original triangle. In this sense, "artificial"
affordances are intimately connected to culture and knowledge available in specific
human cognitive niches but also to their availability to the suitable individuals involved in the epistemic process.

We can say that in a given geometrical diagram various affordances still exist
regardless of correct interpretation or even perception by the agent: indeed the diagram embeds geometrical knowledge that potentially means something to someone.
To correctly grasp the affordances

1. the agent has to "know" geometry,
2. the particularly expert agent – through suitable cognitive manipulations in the
   internal/external representations interplay – can also extract/find/discover "new"
   affordances like chances offered to possibly increase of knowledge.

The diagram offers nested affordances:[28]

1. it is a simple picture, which almost everybody (and many animals) can see and
   understand as a perceptual picture that afford colors and shapes, depending on
   the perceptive wired endowments of the organism in front of it – for example
   expert and uncultured persons but also cats – (strict Gibsonian case);
2. it is a picture which can be seen and understood as a geometrical diagram with
   all its technical properties and features (in this case higher cognitive endowments
   are needed in the organism at stake: a human being of course, other animals do
   not understand geometry),

---

[28] Turner [102] clearly acknowledges the character of affordances as nested. They are seen as
"complex" affordances expressed by "significances" delegated to the natural and artificial
environment – in the perspective of an interesting semiotic and Heideggerian approach to
distributed cognition.

3. it is an artifact which can offer – through even higher creative cognitive manipulations – new affordances to be picked up and possibly inserted in the available encyclopedia of geometrical knowledge. Imagine a child who must "demonstrate" a theorem of the elementary geometry (the sum of the internal angles of a triangle). The child does not have to demonstrate this theorem for the first time, in the sense that it has already been discovered (demonstrated) historically and reported in every manual of Euclidean geometry. However, excluding the case when he passively repeats by rote, he can achieve this demonstration by using the series of elementary extracted affordances illustrated above, based on the appropriate use of basic geometric concepts which are already available to him. We can also say that the child used a heuristic, that is an advanced procedure of discovery. And this heuristic, naturally, considered from the viewpoint of an already available manual of geometry (as an abstract and static system of knowledge) is a true "demonstration", and obviously does not lead to a discovery. Rather, it is a kind of re-discovery. Also from the viewpoint of the child-subject, it is a re-discovery, because he re-discovers a property that is already given at the beginning). Instead the inferences that were employed at the moment of the first historical discovery (maybe Greek) of that property of the triangles (and the assessment of the respective theorem) is a kind of creative achievement (as we have said, a creative manipulative abduction). Moreover, both types of reasoning are mainly model-based, considering the fact they are performed with "hybrid" forms of representation, including considerable non-verbal devices (like the geometrical diagrams).[29]

The case indicated in item 1) also explains why an affordance can be grasped at the same time by animals, infants, and adult human beings, despite their cognitive differences: all of them can perceive "the brink of a cliff as *fall-off-able* according to a common perceptual process" (Scarantino [55, p. 960]), even if they have different perceptual endowments. "This is much the same as we would describe a piano as having an affordance of music playability. Nested within this affordance, the piano keys have the affordance of depressability" (McGrenere and Ho [103, p. 340]). We can add that the piano also offers the chance, in the cognitive interplay artifact/agent, of providing new affordances of *new* good melodies, not previously invented in a merely internal way, in the agent's mind, but found over-there, in the hybrid interplay with the artifact. Of course both artifacts, the diagram and the piano, offer in themselves various constrained conditions for affordances, depending on their properties, quality, materiality, and design, and of course various degrees of affordances. From Kirlik's perspectives in terms of variables and proximal/distal distinctions the example above can be illustrated in this way: the agent builds a diagram in which

---

[29] Of course the agent can modify and change in a more or less creative way the features of the artifacts just to make the available affordances visible or more exploitable, or to build new ones that are offered as options. This is for example the case of a user who customizes a computational interface making an alias for a long command string. He/she can gain easier use of the tool by simply invoking a single key or by the simultaneous pressing of multiple keys instead of writing a long string of characters (cf. Mcgrenere and Ho [103]).

he/she can work on the surface by taking advantage of the constraints that guarantee that latent variables intrinsic to the materiality of the artifact at hand "take care of themselves, so to speak" [96, p. 221].

The need of having a rich internal model of the depicted triangle is weakened because various aspects are discharged from the agent and delegated to the external representation, which offers a proximal perceptual and a manipulative environment containing all the resources needed to effectively perform the creative task of finding new properties of a triangle. The result is immediately perceived and consequently can be picked up and internally ri-represented. The diagram itself is a model of the dynamics of the geometrical reasoning and expresses an example of situated cognitive action completely intertwined with perception.

From a semiotic perspective, at first glance we do not have sufficient cognitive capacities to internally infer what the sum of internal angles is. Manipulating the externalized configuration (the external diagram of a triangle) we come up with a new configuration of perceptual signs that adds properties not contained neither in the initial internal representation of a triangle nor in the initial external depiction. This new configuration of signs gives birth to a new set of affordances that allow us to find the solution. In a Euclidean sense it is a way of "demonstrating" a new theorem. This example furnishes an epistemological example of the nature of the cognitive interplay between internal neuronal semiotic configurations that permit the representational thought about a triangle (together with the help of various embodied "cognitive" kinesthetic abilities) and external representations: also for Peirce, more than a century before the new ideas derived from the field of distributed reasoning, the two aspects are pragmatically intertwined.

Indeed, the human possibility of creating affordances is constrained in this interplay. More precisely, this possibility depends on:

- the specific cognitive delegation to externalities suitably artificially modified and
- the particular intrinsic constraints of the manipulated materiality at play, which usefully provide to us new – so to say – "insights".

The first aspect is closely related to the idea of humans as chance-seekers illustrated above. As extensively argued throughout this paper, securing meaning to the environment relies on smart manipulations that are always hypothetical and therefore withdrawable. That is, it depends on our ability to exploit latent chances. Regarding the second aspect, the environment has constraints of its own. When designing new affordances, we do not work on a blank slate. Conversely, we are in a network of pre-existing ones that constrain our possibilities. On one hand, the possibility of uncovering new latent chances depends on human agency. On the other hand, this agency is mediated/constrained by what the environment offers in terms of pre-existing affordances and its potentiality to provide new ones, as maintained.

As already mentioned, this "hypothetical" status of affordances reminds us that it is not necessarily the case that just anybody can detect it. Affordances are a mere potentiality for organisms. First of all perceiving affordances results from an abductive activity in which we infer possible ways to cope with an object from the signs and cues available to us. Some of them are more or less stableand in some cases they are neurally pre-wired (pre-specified) in the perceptual system – they are "invariants", to exploit a term derived from physics, also used by Gibson. This is especially true when dealing with affordances that have a high cognitive valence.[30] The differences that we can appreciate are mostly *inter-species* – so to speak. A chair affords a child as well as an adult. But this is not the case of a cat. The body of a cat – actually, the cat can sit down on a chair, but also it can sleep on it – has been shaped by evolution quite differently from us.

In the high-level cognitive performance on triangles there is something different, since *intra-species* differences seem to be strongly involved. For instance, only a person that has been taught about geometry can infer (and so "perceive") the affordances "inside" the new manipulated construction built on the original triangle we have illustrated above. This concerns the problem of "expertise" we already noted. First of all, artificial affordances are intimately connected to culture and the social settings in which they arise and the suitable availability of knowledge of the individual(s) in question. Secondly, affordances deal with learning. Perceiving some affordances like those of a triangle is not a *built-in* activity, indeed, once manufactured, they can be learnt and taught. Of course acknowledging this last fact places much more emphasis upon the dynamic character of affordances in organisms's plastic cognitive life, beyond the their evolutionary character.

Hence, the capacity to perform smart manipulations is concerned with the activity of creating external representations. Humans are continuously engaged in cognitive mimetic processes (cf. [104] and [35]) in which they represent thoughts, private ideas, solutions, into external structures and suitably constructed artifacts. In doing so they create external counterparts of some internal, already available propositional and model-based representations, suitably stored in their brains. In some cases these external counterparts, which initially just simply mirror ideas or thoughts already present in the mind, later on can be creatively used to find room for finding new concepts and new ways of inferring that cannot be exhibited by the mere "internal" representation alone [35]. In building these external representations (that can be hold as merely mimetic, but that can become "creative"), people manipulate the world in such a way that new cognitive *chances* are uncovered. In doing this, new affordances are thus manufactured and made "socially" available. More generally, we can conclude that abduction also deals with the continuous activity of manipulating the environment in order to create new chances and opportunities to action, namely, affordances.

1.8 Conclusion: the Eco-Cognitive Inheritance

In this paper we have argued that cognitive niche construction is one of the most distinctive traits of human and animal cognition. As a matter of fact, humans and

---

[30] Perceiving the affordances of a chair is indeed rooted and "stabilized" in our cultural evolution because for human beings it is easy – and *possible*, given our cognitive-biological configuration – to gain the corresponding cognitive ability as a "current" and "reliable" ability (cf. [55, p. 959].

many other organisms continuously manipulate the environment in order to exploit its offerings. In doing this, they are engaged in a process of altering or even creating external structures to lessen and overcome their limits. New ways of coping with the environment, through both evolution and cultural evolution (i.e "cognitive niche construction") are thus created.

In depicting the intricate relationship between humans (and other animals) and the environment we have argued that the concept of affordance may provide new valuable insights. In our view, affordances can be considered as chances that can be either already present and available or manufactured upon occasion via niche construction. We have pointed out that the notion of affordance is to some extent ambiguous. One of the most puzzling questions is whether the process of affordance in visual perception is or is not mediated. To address this problem we have taken advantage of some recent neurocognitive results and stressed and deepened the attention on the evolutionary aspects of affordances.

First of all the relationship between humans and environment is highly dynamic and evolution provided humans and other animals with a set of various hard-wired ways to be afforded; nevertheless, if humans and other animals have at their disposal a standard endowment of hard-wired capacities to be afforded by the environment, they can also enrich, manipulate, and transform it through new and plastic cognitive skills by enhancing their *cognitive niches*. In the case of humans it is easy to see that they can create, modify, and often stabilize affordances that, in turn, form a great part of what it is called the *eco-cognitive inheritance system*: once made available, they can be learnt, as Gibson himself acknowledged, and passed on through generations. This is clearly captured by the idea that the perception of affordances is "semi-encapsulated", i.e. it is not insulated from plastic cognitive processes that lean on learning and experience. We contend this perspective develops some Gibson's intuitions which are clearly stated in the following passage "Perceiving is the simplest and best kind of knowing. But there are other kinds [...]. Knowing by means of instruments extends perceiving into the realm of the very distant and the very small; [...] knowing by means of language makes knowing explicit instead of tacit" [1, p. 263].

We have also argued that abduction can fruitfully shed light on the process of detecting and building new affordances as fundamental cognitive "chances". We have maintained that abduction can clarify the process through which humans design external representations and artifacts, so providing new ways of being afforded. Also taking advantage of Peircean insights on abductive reasoning, we have contended that both finding and constructing affordances refers to a (semiotic) inferential activity: we come up with an affordance insofar as an object exhibits those signs from which we infer a possible way to interact with it and to perform suitable actions. This inferential process relies on various cognitive endowments, both instinctual and learnt, and occurs in the eco-cognitive interaction between environment and organisms.

Finally, it is worth noting that the changes and modifications made upon the environment can be passed on, and they become socially available to other organisms. This introduces an interesting issue for future development. As argued above, the evolutionary impact of a niche depends on its persistence and stability. In turn, its persistence is related to the possibility of transmitting the various resources made ecologically available through cognitive niche construction. Now, given the tremendous success of humans as ecological engineers, it follows that humans should be equipped with a mechanism that facilitates the exploitation and transmission of those ecological resources. That mechanism is, indeed, connected to information-gaining ontogenetic processes. Consequently, from an evolutionary perspective cognitive niche construction cannot be separated from various brain-based learning processes, which also are prominently social. This social dimension is intimately connected to the capacity of people to exploit the construction of a "social" medium as the basis for creating chances, solving problems or making decisions.

References

1. Gibson, J.J.: The Ecological Approach to Visual Perception. Houghton Mifflin, Boston, MA (1979)
2. Brunswik, E.: The Conceptual Framework of Psychology. University of Chicago Press, Chicago (1952)
3. Norman, D.: The Psychology of Everyday Things. Basic Books, New York (1988)
4. Tucker, M., Ellis, R.: On the relations between seen objects and components of potential actions. Journal of Experimental Psychology **24(3)** (2006) 830–846
5. Knoblich, G., Flach, R.: Predicting the fefects of actions: interactions of perception and action. Psychological Science **12(6)** (2001) 467–472
6. Derbyshire, N., Ellis, R., Tucker, M.: The potentiation of two components of the reach-to-grasp action during object categorisation in visual memory. Acta Psychologica **122** (2006) 74–98
7. Simon, H.: A behavioral model of rational choice. The Quarterly Journal of Economics **69** (1955) 99–118
8. Magnani, L.: Abduction, Reason, and Science. Processes of Discovery and Explanation. Kluwer Academic/Plenum Publishers, New York (2001)
9. Tooby, J., DeVore, I.: The reconstruction of hominid behavioral evolution through strategic modeling. In Kinzey, W.G., ed.: Primate Models of Hominid Behavior. Suny Press, Albany (1987) 183–237
10. Pinker, S.: How the Mind Works. Norton, New York (1997)
11. Pinker, S.: Language as an adaptation to the cognitive niche. In Christiansen, M.H., Kirby, S., eds.: Language Evolution. Oxford University Press, Oxford (2003)
12. Clancey, W.J.: Situated Cognition: on Human Knowledge and Computer Representations. Cambridge University Press, Cambridge (1997)
13. Magnani, L.: Chance discovery and the disembodiment of mind. In Oehlmann, R., Abe, A., Ohsawa, Y., eds.: Proceedings of the Workshop on Chance Discovery: from Data Interaction to Scenario Creation, International Conference on Machine Learning (ICML 2005). (2005) 53–59
14. Ellis, R.D.: Questioning Consciousness: the Interplay of Imagery, Cognition, and Emotion in the Human Brain. John Benjamins, Amsterdam (1995)
15. Ramachandran, V.S., Hirstein, W.: Three laws of qualia: what neurology tells us about the biological functions of consciousness. Journal of Consciousness Studies **4** (1997) 429–457

16. Thomas, N.J.T.: Are theories of imagery theories of imagination? An active perception approach to conscious mental content. Cognitive Science **23** (1999) 207–245
17. Godfrey-Smith, P.: Complexity and the Function of Mind in Nature. Cambridge University Press, Cambridge (1998)
18. Laland, K., Odling-Smee, J., Feldman, M.: Niche construction, biological evolution and cultural change. Behavioral and Brain Sciences **23(1)** (2000) 131–175.
19. Laland, K.N., Odling-Smee, F.J., Feldman, M.W.: Cultural niche construction and human evolution. Journal of Evolutionary Biology **14** (2001) 22–33
20. Odling-Smee, F., Laland, K., Feldman, M.: Niche Construction. A Neglected Process in Evolution. Princeton University Press, New York, NJ (2003)
21. Odling-Smee, J.J.: The Role of Behavior in Evolution. Cambridge University Press, Cambridge (1988)
22. Sterelny, K.: Made by each other: organism and their environment. Biology and Philosophy **20** (2005) 21–36
23. Dawkins, R.: Extended phenotype - but not extended. a reply to laland, turner and jablonka. Biology and Philosophy **19** (2004) 377–397
24. Laland, K., Odling-Smee, J., Feldman, M.: On the breath and significance of niche construction: a reply to grittiths, okasha and sterelny. Biology and Philosophy **20** (2005) 37–55
25. Day, R.L., Laland, K., Odling-Smee, J.: Rethinking adaptation. The niche-construction perspective. Perspectives in Biology and Medicine **46(1)** (2003) 80–95
26. Magnani, L.: Animal abduction. From mindless organisms to artifactual mediators. In Magnani, L., Li, P., eds.: Model-Based Reasoning in Science, Technology, and Medicine. Springer, Berlin (2007) Forthcoming.
27. Godrey-Smith, P.: Environmental complexity and the evolution of cognition. In Sternberg, R., Kaufman, K., eds.: The Evolution of Intelligence. Lawrence Erlbaum Associates, Mawhah, NJ (2002) 233–249
28. Jung, C.G.: On psychic energy. In: The Collected Works of C. G. Jung. Routledge & Kegan Paul, London (1960) 3–91 Vol. 8, 'The Structure and Dynamics of the Psyche', edited by H. Read, M. Fordham, and G. Adler.
29. Aunger, R.: The Eletric Meme. A New Theory of How We Think. The Free Press, New York (2002)
30. Sinha, C.: Epigenetics, semiotics, and the mysteries of the organism. Biological Theory **1(2)** (2006) 112–115
31. Zhang, J.: The nature of external representations in problem-solving. Cognitive Science **21**(2) (1997) 179–217
32. Hutchins, E.: Cognition in the Wild. The MIT Press, Cambridge, MA (1995)
33. Clark, A., Chalmers, D.J.: The extended mind. Analysis **58** (1998) 10–23
34. Wilson, R.A.: Boundaries of the Mind. Cambridge University Press, Cambridge (2004)
35. Magnani, L.: Mimetic minds. Meaning formation through epistemic mediators and external representations. In Loula, A., Gudwin, R., Queiroz, J., eds.: Artificial Cognition Systems. Idea Group Publishers, Hershey, PA (2006) 327–357
36. Magnani, L.: Morality in a Technological World. Knowledge as Duty. Cambridge University Press, Cambridge (2007)
37. Bardone, E., Secchi, D.: The distributed cognition approach to rationality: getting the framework (2006) Presented at the 2006 Academy of Management Annual Meeting, Atlanta
38. Gatti, A., Magnani, L.: On the representational role of the environment and on the cognitive nature of manipulations. In Magnani, L., Dossena, R., eds.: Computing, Philosophy and Cognition, London, College Pubblications (2005) 227–242

39. Knuuttila, T., Honkela, T.: Questioning external and internal representation: the case of scientific models. In Magnani, L., Dossena, R., eds.: Computing, Philosophy and Cognition, London, College Pubblications (2005) 209–226

40. Peirce, C.S.: Collected Papers of Charles Sanders Peirce. Harvard University Press, Cambridge, MA (1931-1958) vols. 1-6, Hartshorne, C. and Weiss, P., eds.; vols. 7-8, Burks, A. W., ed.

41. Magnani, L.: Epistemic mediators and model-based discovery in science. In Magnani, L., Nersessian, N., eds.: Model-Based Reasoning: Scientific Discovery, Technological Innovation, Values. Kluwer, New York, NY (2002) 325–329

42. Magnani, L.: Abductive reasoning: philosophical and educational perspectives in medicine. In Evans, D.A., Patel, V.L., eds.: Advanced Models of Cognition for Medical Training and Practice. Springer, Berlin (1992) 21–41

43. Nersessian, N.: Model-based reasoning in conceptual change. In Nersessian, N.J., Magnani, L., Thagard, P., eds.: Model-based Reasoning in Scientific Discovery, New York, Kluwer Academic/Plenum Publishers (1999) 5–22

44. Gooding, D.: Experiment and the Making of Meaning. Kluwer, Dordrecht (1990)

45. Piaget, J.: Adaption and Intelligence. University of Chicago Press, Chicago (1974)

46. Anderson, M.L.: Embodied cognition: a field guide. Artificial Intelligence **149(1)** (2003) 91–130

47. Clancey, W.J.: Simulating activities: relating motives, deliberation, and attentive coordination. Cognitive Systems Research **3**(1-4) (2002) 471–500

48. Agree, P., Chapman, D.: What are plans for? In Maes, P., ed.: Designing Autonomous Agents. The MIT Press, Cambridge, MA (1990) 17–34

49. Brooks, R.A., Stein, L.: Building brains for bodies. Autonomous Robots **1** (1994) 7–25

50. Hutchins, E.: Cognitive artifacts. In Wilson, R.A., Keil, F.C., eds.: Encyclopedia of the Cognitive Sciences. The MIT Press, Cambridge, MA (1999) 126–7

51. Kirsh, D., Maglio, P.: On distinguishing epistemic from pragmatic action. Cognitive Science **18** (1994) 513–549

52. Wells, A.J.: Gibson's affordances and Turing's theory of computation. Ecological Psychology **14(3)** (2002) 141–180

53. Greeno, J.G.: Gibson's affordances. Psychological Review **101(2)** (1994) 336–342

54. Stoffregen, T.A.: Affordances as properties of the animal-environment system. Ecological Psychology **15(3)** (2003) 115–134

55. Scarantino, A.: Affordances explained. Philosophy of Science **70** (2003) 949–961

56. Chemero, A.: An outline of a theory of affordances. Ecological Psychology **15(2)** (2003) 181–195

57. Natsoulas, T.: To see is to perceive what they afford: James J. Gibon's concept of affordance. Mind and Behaviour **2(4)** (2004) 323–348

58. Vicente, K.J.: Beyond the lens model and direct perception: toward a broader ecological psychology. Ecological Psychology **15(3)** (2003) 241–267

59. Zhang, J., Patel, V.L.: Distributed cognition, representation, and affordance. Cognition & Pragmatics **2** (14) 333–341

60. Mithen, S.: The Prehistory of the Mind. A Search for the Origins of Art, Religion and Science. Thames and Hudson, London (1996)

61. Bermùdez, J.L.: Thinking without Words. Oxford University Press, Oxford (2003)

62. Brooks, R.A.: Intelligence without representation. Artificial Intelligence **47** (1991) 139–159

63. Domenella, R.G., Plebe, A.: Can vision be computational? In Magnani, L., Dossena, R., eds.: Computing, Philosophy and Cognition, London, College Pubblications (2005) 227–242

64. Beers, R.: Expressions of mind in animal behavior. In Mitchell, W., Thomson, N.S., Miles, H.L., eds.: Anthropomorphism, Anecdotes, and Animals. State University of New York Press, Albany, NY (1997) 198–209
65. Wilson, R.: Collective memory, group minds, and the extended mind thesis. Cognitive Processing **6** (2005) 227–236
66. Zhang, J.: The nature of external representations in problem solving. Cognitive Science **21**(2) (1997) 179–217
67. Clark, A.: *Being There: Putting Brain, Body, and World Together Again.* The MIT Press, Cambridge, MA (1997)
68. Wheeler, M.: Is language and ultimate artifact? *Language Sciences* **26** (2004) 693–715
69. Brunswik, E.: Representative design and probabilistic theory in a functional psychology. Psychological Review **62** (1955) 193–217
70. Brunswik, E.: Oranismic achievement and envornmental probability. Psychological Review **50** (1943) 255–272
71. Hammond, K.R., Steward, T.R., eds.: The Essential Brunswik. Beginnings, Explications, Applications, Oxford/New York, Oxford University Press (2001)
72. Rothrock, L., Kirlik, A.: Inferring fast and frugal heuristics from human judgment data. In Kirlik, A., ed.: Human-Technology Interaction. Methods and Models for Cognitive Engineering and Human-Computer Interaction. Oxford University Press, Oxford/New York (2006) 131–148
73. Goldstein, W.M.: Introduction to bruswikian theory and mehod. In Kirlik, A., ed.: Human-Technology Interaction. Methods and Models for Cognitive Engineering and Human-Computer Interaction. Oxford University Press, Oxford/New York (2006) 10–26
74. Magnani, L.: Multimodal abduction. External semiotic anchors and hybrid representations. Logic Journal of the IGPS **14**(1) (2006) 107–136
75. Pipers, J.R., Oudejans, R.R.D., Bakker, F.C., Beek, P.J.: The role of anxiety in perceiving and realizing affordances. Ecological Pshychology **18(3)** (2006) 131–161
76. Witt, J.K., Proffitt, D.R., Epstein, W.: Tool use affects perceived distance, but only when you intend to us. Journal of Experimental Psychology **32(6)** (2006) 1405–1421
77. Godges, B.H., Lindheim, O.: Carrying babies and groceries: the effect of moral and social weight on caring. Ecological Pshychology **18(2)** (2006) 93–111
78. Balcetis, E., Dunning, D.: See what you want to see: motivational influences on visual perception. Journal of Personality and Social Psychology **91(4)** (2006) 612–625
79. Lozano, S., Hard, B.M., Tversky, B.: Perspective taking promotes action understanding and learning. Journal of Experimental Psychology **32(6)** (2006) 1405–1421
80. Anquetil, T., Jeannerod, M.: Simulated actions in the first and in the third person perspective. Brain Research **1130** (2007) 125–129
81. Paccalin, C., Jeannerod, M.: Changes in breathing during observation of effortful actions. Brain in Research **862** (2000) 194–200
82. Decety, J., Grèzes, J.: The power of simulation: imagining one's own and other behavior. Brain in Research **1079** (2006) 4–14
83. Adelman, L., Yeo, C., Miller, S.: Understanding the effects of computer displays and time pressure on the performance of distributed teams. In Kirlik, A., ed.: Human-Technology Interaction. Methods and Models for Cognitive Engineering and Human-Computer Interaction. Oxford University Press, Oxford/New York (2006) 43–54
84. Kirlik, A., ed.: Human-Technology Interaction. Methods and Models for Cognitive Engineering and Human-Computer Interaction, Oxford/New York, Oxford University Press (2006)
85. Byrne, D., Kirlik, A., Fick, C.S.: Kilograms matter: rational analysis, ecological rationality, and closed-loop modeling of interactive cognition and behavior. In Kirlik, A., ed.:

Human-Technology Interaction. Methods and Models for Cognitive Engineering and Human-Computer Interaction. Oxford University Press, Oxford/New York (2006) 267–286

86. Degani, A., Shafto, M., Kirlik, A.: What makes vicarious functioning work? Exploring the geometry of human-technology interaction. In Kirlik, A., ed.: Human-Technology Interaction. Methods and Models for Cognitive Engineering and Human-Computer Interaction. Oxford University Press, Oxford/New York (2006) 179–196

87. Peirce, C.S.: The Charles S. Peirce Papers: Manuscript Collection in the Houghton Library. The University of Massachusetts Press, Worcester, MA (1967) Annotated Catalogue of the Papers of Charles S. Peirce. Numbered according to Richard S. Robin. Available in the Peirce Microfilm edition. Pagination: CSP = Peirce / ISP = Institute for Studies in Pragmaticism.

88. Tirassa, M., Carassa, A., Geminiani, G.: Describers and explorers: a method for investigating cognitive maps. In Nualláin, S.Ó., ed.: Spatial Cognition. Foundations and Applications, Amsterdam/Philadelphia, John Benjamins (1998) 19–31

89. Ben Jacob, E., Shapira, Y., Tauber, A.I.: Seeking the foundation of cognition in bacteria. From Schrödinger's negative entropy to latent information. Physica A **359** (2006) 495–524

90. Windsor, W.L.: An ecological approach to semiotics. Journal for the Theory of Social Behavior **34(2)** (2004) 179–198

91. Rock, I.: Inference in perception. PSA: Proceedings of the Biennial Meeting of the Philosophy of Science Association **2** (1982) 525–540

92. Thagard, P.: Computational Philosophy of Science. The MIT Press, Cambridge, MA (1988)

93. Hoffman, D.D.: Visual Intelligence: How We Create What We See. Norton, New York (1998)

94. Magnani, L.: Disembodying minds, externalizing minds: how brains make up creative scientific reasoning. In Magnani, L., ed.: Model-Based Reasoning in Science and Engineering, Cognitive Science, Epistemology, Logic, London, College Publications (2006) 185–202

95. Magnani, L., Bardone, E.: Designing human interfaces. The role of abduction. In Magnani, L., Dossena, R., eds.: Computing, Philosophy and Cognition, London, College Pubblications (2005) 131–146

96. Kirlik, A.: Abstracting situated action: implications for cognitive modeling and interface design. In Kirlik, A., ed.: Human-Technology Interaction. Methods and Models for Cognitive Engineering and Human-Computer Interaction. Oxford University Press, Oxford/New York (2006) 212–226

97. Kirsh, A.: The intelligent use of space. Artificial Intelligence **73** (1995) 31–68

98. Kirlik, A.: The Ecological Expert: acting to create information to guide action. In: Human Interaction with Complex Systems, (HICS'98) Proceedings. IEEE Computer Society Press (2006) 22–25. Fourth Annual Symposium, Dayton, OH.

99. Kirlik, A.: Reiventing intelligence for an invented world. In Sternberg, R.J., Preiss, D.D., eds.: Intelligence and technology: the impact of tools on the nature and development of human abilities. Lawrence Erlbaum Associates, Mawhah, NJ (2007) 105–134

100. Turing, A.M.: Collected Works of Alan Turing, Mechanical Intelligence. Elsevier, Amsterdam (1992) Ed. by D.C. Ince.

101. Magnani, L.: Semiotic brains and artificial minds. How brains make up material cognitive systems. In Gudwin, R., Queiroz, J., eds.: Semiotics and Intelligent Systems Development. Idea Group Inc., Hershey, PA (2007) 1–41

102. Turner, P.: Affordances as context. Interacting with Computers **17** (2005) 787–800
103. McGrenere, J., Ho, W.: Affordances: clarifying and evolving a concept. In: Proceedings of Graphics Interface. (2000) 179–186 May 15-17, 2000, Montreal, Quebec, Canada.
104. Donald, M.: A Mind So Rare. The Evolution of Human Consciousness. Norton, London (2001)

Discovering and Communicating through Multimodal Abduction

The Role of External Semiotic Anchors and Hybrid Representations

Lorenzo Magnani

University of Pavia, Department of Philosophy, and Computational Philosophy Laboratory, 27100 Pavia, Italy,
and Sun Yat-sen University, Department of Philosophy, 510275 Guangzhou (Canton), P.R. China
lmagnani@unipv.it

Summary. Our brains make up a series of signs and are engaged in making or manifesting or reacting to a series of signs: through this semiotic activity they are at the same time engaged in "being minds" and so in thinking intelligently, in communicating and in extracting chances from the einvironment. An important effect of this semiotic activity of brains is a continuous process of "externalization of the mind" that exhibits a new cognitive perspective on the mechanisms underling the semiotic emergence of abductive processes of meaning formation. To illustrate this process I will take advantage of the analysis of some aspects of the cognitive interplay between internal and external representations and communications. I consider this interplay critical in analyzing the relation between meaningful semiotic internal resources and devices and their dynamical interactions with the externalized semiotic materiality suitably stocked in the environment. Hence, minds are material, "extended" and artificial in themselves. A considerable part of human abductive thinking is occurring through an activity consisting in a kind of reification in the external environment (that originates what I call *semiotic anchors*) and a subsequent re–projection and reinterpretation through new configurations of neural networks and chemical processes. I also illustrate how this activity takes advantage of hybrid representations and how it can nicely account for various processes of creative and selective abduction, central to communications processes and chance/risk extraction, bringing up the question of how *multimodal* aspects involving a full range of sensory modalities are important in hypothetical multidisciplinary reasoning.

2.1 The Centrality of Abduction in Multidisciplinary Hypothetical Reasoning

If we decide to increase knowledge on both cognitive and semiotic aspects of multidisciplinary hypothetical reasoning it is necessary to develop a cognitive model of creativity able to represent not only "novelty" and "unconventionality", but also some features commonly referred to as the entire creative process, such as the hybrid

L. Magnani: *Discovering and Communicating through Multimodal Abduction*, Studies in Computational Intelligence (SCI) **123**, 41–62 (2008)
www.springerlink.com

modeling activity developed in the communicative interplay between internal and external representations. The philosophical concept of *abduction* may be a candidate to solve this problem, and offers an approach to model creative processes of meaning generation and communication in a completely explicit and formal way.

A hundred years ago, C. S. Peirce [8] coined the concept of abduction in order to illustrate that the process of scientific discovery is not irrational and that a methodology of discovery is possible. Peirce interpreted abduction essentially as an "inferential" *creative process* of generating a new hypothesis. Abduction has a logical form – fallacious, if we model abduction by using classical syllogistic logic – distinct from deduction and induction. Reasoning which starts from reasons and looks for consequences is called *deduction*; that which starts from consequences and looks for reasons is called *abduction*.

Abduction – a distinct form of reasoning – is the process of *inferring* certain facts and/or laws and hypotheses that render some sentences plausible, that *explain* or *discover* some (eventually new) phenomenon or observation; it is the process of reasoning in which explanatory hypotheses are formed and evaluated. There are two main epistemological meanings of the word abduction [20]: 1) abduction that only generates "plausible" hypotheses ("selective" or "creative") and 2) abduction considered as inference "to the best explanation", which also evaluates hypotheses (cf. Figure 2.1). An illustration from the field of medical knowledge is represented by the discovery of a new disease and the manifestations it causes which can be considered as the result of a creative abductive inference. Therefore, "creative" abduction deals with the whole field of the growth of scientific knowledge. This is irrelevant in medical diagnosis where instead the task is to "select" from an encyclopedia of pre-stored diagnostic entities. We can call both inferences ampliative, selective and creative, because in both cases the reasoning involved amplifies, or goes beyond, the information incorporated in the premises.

I have introduced [20] the concept of *theoretical abduction* as a form of neural and basically internal processing. I maintain that there are two kinds of theoretical abduction, "sentential", related to logic and to verbal/symbolic inferences, and

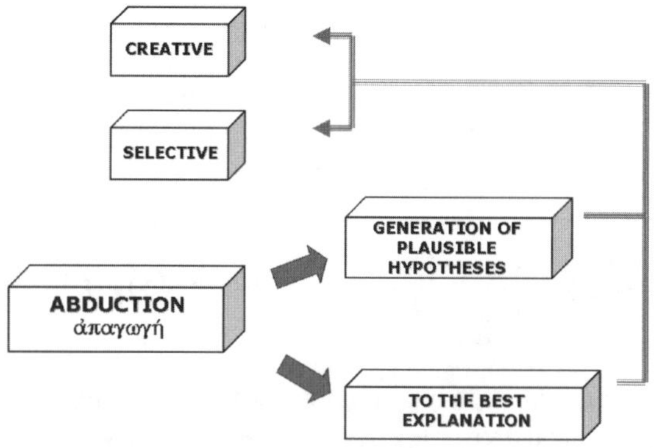

Fig. 2.1. Creative and selective abduction.

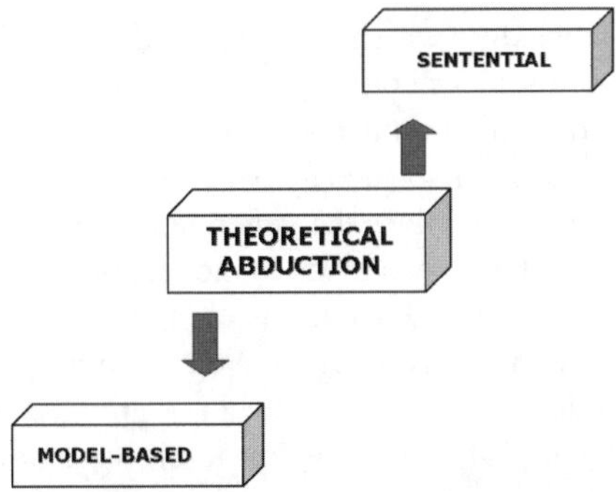

Fig. 2.2. Theoretical abduction.

"model-based", related to the exploitation of models such as diagrams, pictures, etc, cf. below in this paper, section 2.1.2 and subsection 2.3 (cf. Figure 2.2).

Theoretical abduction certainly illustrates much of what is important in creative abductive reasoning, in humans and in computational programs, but fails to account for many cases of explanations occurring in science when the exploitation of environment is crucial. It fails to account for those cases in which there is a kind of "discovering through doing", cases in which new and still unexpressed information is codified by means of manipulations of some external objects I have called *epistemic mediators* [20]. The concept of *manipulative abduction*[1] captures a large part of everyday and scientific thinking where the role of action is central, and where the features of this action are implicit and hard to be elicited: action can provide otherwise unavailable information that enables the agent to solve problems by starting and by performing a suitable abductive process of generation or selection of hypotheses.

In section 2.4 I will describe how manipulative abduction can nicely account for communication and risk/chance extraction in the relationship between meaningful behavior and dynamical interactions with the environment. The following sections illustrate that at the roots of the creation of new meanings there is a process of externalization that exhibits a new cognitive description of the mechanisms underling the emergence of meaning processes through semiotic delegations to the environment. Hence, interesting (and new) information and knowledge packages are generated and stored over there in the external human niches, in various supports more or less accessible that can be picked up in further communicative and chance/risk discovering processes.

2.1.1 The "Internal" Side of Creative Reasoning

Throughout his career Peirce defended the thesis that, besides deduction and induction[2], there is a third mode of inference that constitutes the only method for really

---

[1] Manipulative abduction and epistemic mediators are introduced and illustrated in [21] and [20].

[2] Peirce clearly contrasted abduction with induction and deduction, by using the famous syllogistic model. More details on the differences between abductive and inductive/deductive inferences can be found in [22] and [20].

improving scientific knowledge, which he called *abduction*. Science improves and grows continuously, but this continuous enrichment cannot be due to deduction, nor to induction: deduction does not produce any new idea, whereas induction produces very simple ideas. New ideas in science are due to *abduction*, a particular kind of non-deductive[3] inference that involves the generation and evaluation of explanatory hypotheses. Many attempts have been made to model abduction by developing some formal/sentential tools in order to illustrate its computational properties and the relationships with the different forms of deductive reasoning [see, for example, [23]. Some of the formal models of abductive reasoning are based on the theory of the *epistemic state* of an agent [24], where the epistemic state of an individual is modeled as a consistent set of beliefs that can change by expansion and contraction (*belief revision framework*).

2.1.2 Model-Based Abduction and its External Dimension

We do not have to limit ourselves to the *formal/sentential* view of theoretical abduction but we have to consider a broader *inferential* one: the *model-based* sides of creative abduction (cf. below).

From Peirce's philosophical point of view, all thinking is in signs, and signs can be icons, indices or symbols. Moreover, all inference is a form of sign activity, where the word sign includes "feeling, image, conception, and other representation" [8, 5.283], and, in Kantian words, all synthetic forms of cognition. That is, a considerable part of the thinking activity is model-based. Of course model-based reasoning acquires its peculiar creative relevance when embedded in abductive processes, so that we can individuate a *model-based abduction*. Hence, we must think in terms of model-based abduction (and not in terms of sentential abduction) to explain complex processes like scientific conceptual change. Different varieties of *model-based abductions* [25] are related to the high-level types of scientific conceptual change [see, for instance, [26].

Following Nersessian [9, 27], the term "model-based reasoning" is used to indicate the construction and manipulation of various kinds of representations, not mainly sentential and/or formal, but mental and/or related to external mediators. Obvious examples of model-based reasoning are constructing and manipulating visual representations, thought experiment, analogical reasoning, but also for example the so-called "tunnel effect" [28], occurring when models are built at the intersection of some operational interpretation domain – with its interpretation capabilities – and a new ill-known domain.

Manipulative abduction [20] - contrasted with theoretical abduction - happens when we are thinking through doing and not only, in a pragmatic sense, about doing. So the idea of manipulative abduction goes beyond the well-known role of experiments as capable of forming new scientific laws by means of the results (nature's answers to the investigator's question) they present, or of merely playing a predictive

---

[3] Non-deductive if we use the attribute "deductive" as designated by classical logic.

role (in confirmation and in falsification). Manipulative abduction refers to an extra-theoretical behavior that aims at creating communicable accounts of new experiences to integrate them into previously existing systems of experimental and linguistic (theoretical) practices. The existence of this kind of extra-theoretical cognitive behavior is also testified by the many everyday situations in which humans are perfectly able to perform very efficacious (and habitual) tasks without the immediate possibility of realizing their conceptual explanation. In the following sections manipulative abduction will be considered from the perspective of the relationship between internal and external representations.

2.2 Mimetic and Creative Representations

Human brains organize themselves through a semiotic activity that is reified in the external environment and then re-projected and reinterpreted through new configurations of neural networks and chemical processes. I also think the externalization of mind can nicely account for low-level semiotic processes of meaning creation, bringing up the question of how could higher-level processes be comprised and how would they interact with lower-level ones.

2.2.1 External and Internal Representations

I have illustrated in a previous paper [7] dealing with some paleoanthropological issues that through the mediation of the material culture the modern human mind for example can arrive to *internally* "think" the new complicated abstract meaning of animals and people at the same time. We can account for this process of externalization from an impressive cognitive point of view.

I maintain that representations are external and internal. We can say that

- *external representations* are formed by external materials that express (through reification) concepts and problems already stored in the brain or that do not have a *natural home* in it;
- *internalized representations* are internal re-projections, a kind of recapitulations, (learning) of external representations in terms of neural patterns of activation in the brain. They can sometimes be "internally" manipulated like external objects and can originate new internal reconstructed representations through the neural activity of *transformation* and *integration*.

This process explains why human beings seem to perform both computations of a *connectionist* type[4] such as the ones involving representations as

---

[4] Here the reference to the word "connectionism" is used on the plausible assumption that all mental representations are brain structures: verbal and the full range of sensory representations are neural structures endowed with their chemical functioning (neurotransmitters and hormones) and electrical activity (neurons fire and provide electrical inputs to other neurons). In this sense we can reconceptualize cognition neurologically: for example the

- (I Level) *patterns of neural activation* that arise as the result of the interaction between body and environment (and suitably shaped by the evolution and the individual history): pattern completion or image recognition,

and computations that use representations as

- (II Level) *derived combinatorial syntax and semantics* dynamically shaped by the various external representations and reasoning devices found or constructed in the environment (for example geometrical diagrams); they are neurologically represented contingently as pattern of neural activations that "sometimes" tend to become stabilized structures and to fix and so *to permanently belong to the I Level* above.

The I Level originates those *sensations* (they constitute a kind of "face" we think the world has), that provide room for the II Level to reflect the structure of the environment, and, most important, that can follow the computations suggested by these external structures. It is clear we can now conclude that the growth of the brain and especially the synaptic and dendritic growth are profoundly determined by the environment.

When the fixation is reached the patterns of neural activation no longer need a direct stimulus from the environment for their construction. In a certain sense they can be viewed as *fixed internal records* of *external structures* that *can exist* also in the absence of such external structures. These patterns of neural activation that constitute the I Level Representations always keep record of the experience that generated them and, thus, always carry the II Level Representation associated to them, even if in a different form, the form of *memory* and not the form of a vivid sensorial experience. Now, the human agent, via neural mechanisms, can retrieve these II Level Representations and use them as *internal* representations or use parts of them to construct new internal representations very different from the ones stored in memory [29].[5]

I think there are two basic kinds of external representations active in this process of externalization of the mind: *creative* and *mimetic*. Mimetic external representations mirror concepts and problems that are already represented in the brain and need to be enhanced, solved, further complicated, etc. so they sometimes can creatively give rise to new concepts and meanings. In the examples I will illustrate in the following sections it will be clear how for instance a mimetic geometric representation can become creative and give rise to new meanings and ideas in the hybrid interplay between brains and suitable "cognitive niches"[6] that consequently are appropriately reshaped.

---

solution of a problem can be seen as a process in which one neural structure representing an explanatory target generates another neural structure that constitutes a hypothesis for the solution.

[5] The role of external representations has already been stressed in some central traditions of cognitive science and artificial intelligence, from the area of distributed and embodied cognition and of robotics [17, 18, 30] to the area of active vision and perception [1, 31].

[6] This expression, used in the different framework of the problem of language as biological adaptation to the environment appears very appropriate also in this context [3, 4].

In the following section I will illustrate some fundamental aspects of the interplay above in the light of basic semiotic and thus communicative aspects of abductive reasoning.

2.3 Model-Based Abduction, Semiosis, Communication

What exactly is model-based abduction from a philosophical and cognitive point of view? I have already said that Peirce stated that all thinking is in signs, and signs can be icons, indices, or symbols and that all *inference* is a form of sign activity, where the word sign includes "feeling, image, conception, and other representation" [8, 5.283]. In this light it can be maintained that a considerable part of the creative meaning processes is *model-based*. Moreover, a considerable part of meaning creation processes (not only in science) occurs in the middle of a relationship between brains and external objects and tools that have received cognitive and/or epistemological delegations (cf. the previous and the following subsection). Let me address some philosophical issues.

Following this Peircian perspective about inference I think it is extremely useful from a cognitive point of view to consider the concept of reasoning in a very broad way (cf. also [32, p. 8]). We have three cases:

1. reasoning can be fully conscious and typical of high-level worked-out ways of inferring, like in the case of scientists' and professionals' performances;
2. reasoning can be "acritical" [8, 5.108], which includes every day inferences in conversation and in various ordinary patterns of thinking;
3. reasoning can resort to "operations of the mind which are logically analogous to inference excepting only that they are unconscious and therefore uncontrollable and therefore not subject to logical criticism" [8, 5.108].

Immediately Peirce adds a note to the third case "But that makes all the difference in the world; for inference is essentially deliberate, and self-controlled. Any operation which cannot be controlled, any conclusion which is not abandoned, not merely as soon as criticism has pronounced against it, but in the very act of pronouncing that decree, is not of the nature of rational inference – is not reasoning" (*ibid.*).

As Colapietro clearly states [33, p. 140], it seems that for Peirce human beings semiotically involve unwitting trials and unconscious processes. Moreover, it seems clear that unconscious thought can be in some sense considered "inference", even if not rational; indeed, Peirce says, it is not reasoning. Peirce further indicates that there are in human beings multiple trains of thought at once but only a small fraction of them is conscious, nevertheless the prominence in consciousness of one train of thought is not to be interpreted an interruption of other ones.

In this Peircian perspective, which I adopt in this essay, where inferential aspects of thinking dominate, there is no intuition, in an anti-Cartesian way. We know all important facts about ourselves in an *inferential* abductive way:

> [...] we first form a definite idea of ourselves as a hypothesis to provide a
> place in which our errors and other people's perceptions of us can happen.
> Furthermore, this hypothesis is constructed from our knowledge of "out-
> ward" physical facts, such things as the sounds we speak and the bodily
> movements we make, that Peirce calls signs [32, p. 8].

Recognizing in a series of *material*, physical events, that they make up a series
of signs, is to know the existence of a "mind" (or of a group of minds) and to be
absorbed in making, manifesting, or reacting to a series of signs is to be absorbed
in "being a mind". "[...] all thinking is dialogic in form" [8, 6.338], both at the
intrasubjective[7] and intersubjective level, so that we see ourselves exactly as others
see us, or see them exactly as they see themselves, and we see ourselves through our
own speech and other interpretable behaviors, just others see us and themselves in
the same way, in the commonality of the whole process [32, p. 10].

As I will better explain later on in the following sections, in this perspective
minds are material like brains, in so far as they consist in intertwined internal and external semiotic processes: "[...] the psychologists undertake to locate various mental
powers in the brain; and above all consider it as quite certain that the faculty of language resides in a certain lobe; but I believe it comes decidedly nearer the truth
(though not really true) that language resides in the tongue. In my opinion it is much
more true that the thoughts of a living writer are in any printed copy of his book than
they are in his brain" [8, 7.364].

2.3.1 Man is an External Sign

Peirce's semiotic motto "man is an external sign" is very clear about the materiality
of mind and about the fact that the conscious self[8] is a cluster actively embodied of
flowing intelligible signs:

> It is sufficient to say that there is no element whatever of man's conscious-
> ness which has not something corresponding to it in the word; and the reason
> is obvious. It is that the word or sign which man uses *is* the man himself.
> For, as the fact that every thought is a sign, taken in conjunction with the
> fact that life is a train of thoughts, proves that man is a sign; so, that every
> thought is an *external sign*, proves that man is an external sign. That is to
> say, the man and the *external sign* are identical, in the same *sense* in which
> the words *homo* and *man* are identical. Thus my language is the sum total of
> myself; for the man is the thought [8, 5.314].

It is by way of signs that we ourselves *are* semiotic processes – for example a
more or less coherent cluster of narratives. If all thinking is in signs it is not true that
thoughts are in us because we are in thoughts.

---

[7] "One's thoughts are what he is 'saying to himself', that is saying to that other self that
is just coming to life in the flow of time. When one reasons, it that critical self that one
is trying to persuade: and all thought whatsoever is a sign, and is mostly in the nature of
language" [8, 5.421].

[8] Consciousness arises as "a sort of public spirit among the nerve cells" [8, 1.354].

I think it is at this point clearer what I meant in section 2.1.2, when I explained the concept of model-based abduction and said, adopting a Peircian perspective, that all thinking is in signs, and signs can be icons, indices, or symbols and that, moreover, all *inference* is a form of sign activity, where the word sign includes feeling, image, conception, and other representation. The model-based aspects of human cognition are central, given the central role played for example by signs like images and feeling in the inferential activity "[...] man is a sign developing according to the laws of inference. [...] the entire phenomenal manifestation of mind is a sign resulting from inference" [8, 5.312 and 5.313].

Moreover, the "person-sign" is future-conditional, that is not fully formed in the present but depending on the future destiny of the concrete semiotic activity (future thoughts and experience of the community) in which she will be involved. If Peirce maintains that when we think we appear as a sign [8, 5.283] and, moreover, that everything is present to us is a phenomenal manifestation of ourselves, then feelings, images, diagrams, conceptions, schemata, and other representations are phenomenal manifestations that become available for interpretations and thus are guiding our actions in a positive or negative way. They become *signs* when we think and interpret them. It is well-known that for Peirce all semiotic experience – and thus abduction - is also providing a guide for action. Indeed the whole function of thought is to produce habits of action.[9]

In the following sections I will describe how the interplay of signs, objects, and interpretations is working in important aspects of abductive reasoning. Of course model-based cognition acquires its peculiar creative relevance when embedded in abductive processes. I will show some examples of model-based inferences. It is well known the importance Peirce ascribed to diagrammatic thinking (a kind of iconic thinking), as shown by his discovery of the powerful system of predicate logic based on diagrams or "existential graphs". As we have already stressed, Peirce considers inferential any cognitive activity whatever, not only conscious abstract thought; he also includes perceptual knowledge and subconscious cognitive activity. For instance in subconscious mental activities visual representations play an immediate role [34].

Many commentators always criticized the Peircian ambiguity in treating abduction in the same time as inference and perception. It is important to clarify this problem, because perception and imagery are kinds of that model-based cognition which we are exploiting to explain abduction: in [7] I conclude we can render consistent the two views, beyond Peirce, but perhaps also within the Peircian texts, taking advantage of the concept of *multimodal* abduction, which depicts hybrid aspects of abductive reasoning.

Thagard [35, 36] observes, that abductive inference can be visual as well as verbal, and consequently acknowledges the sentential, model–based, and manipulative nature of abduction I have illustrated above. Moreover, both data and hypotheses can be visually represented:

---

[9] On this issue cf. for example the contributions contained in recent special issue of the journal *Semiotica* devoted to abduction [34].

For example, when I see a scratch along the side of my car, I can generate the mental image of grocery cart sliding into the car and producing the scratch. In this case both the target (the scratch) and the hypothesis (the collision) are visually represented. [...] It is an interesting question whether hypotheses can be represented using all sensory modalities. For vision the answer is obvious, as images and diagrams can clearly be used to represent events and structures that have causal effects [36].

Indeed hypotheses can be also represented using other sensory modalities:

[...] I may recoil because something I touch feels slimy, or jump because of a loud noise, or frown because of a rotten smell, or gag because something tastes too salty. Hence in explaining my own behavior my mental image of the full range of examples of sensory experiences may have causal signif-icance. Applying such explanations of the behavior of others requires pro-jecting onto them the possession of sensory experiences that I think are like the ones that I have in similar situations. [...] Empathy works the same way, when I explain people's behavior in a particular situation by inferring that they are having the same kind of emotional experience that I have in similar situations [36].

Thagard illustrates the case in which a professor with a recently rejected manu-script is frowning: another colleagues can empathizes by remembering how annoying she felt in the same circumstances, projecting a mental image onto the colleague that is a non-verbal representation able to explain the frown. Of course a verbal explana-tion can be added, but this just complements the empathetic one. It is in this sense that Thagard concludes that abduction can be fully multimodal, in that both data and hypotheses can have a full range of verbal and sensory representations. Some basic aspects of this constitutive hybrid (and thus intrinsically multidisciplinary) nature of abduction – involving words, sights, images, smells, etc. but also kinesthetic experi-ences and other feelings such as pain – will be investigated in the following sections.

2.4 Constructing and Communicating Meaning through Mimetic and Creative External Objects

2.4.1 Constructing Meaning through Manipulative Abduction

Manipulative abduction occurs when many external things, usually inert from the semiotic (and so for example epistemic) point of view, can be transformed into what I have called, in the case of scientific reasoning, "epistemic mediators" [20] that give rise to new signs, new chances for interpretations, and new interpretations.

We can cognitively account for this process of externalization[10] taking advan-tage of the concept of *manipulative* abduction (cf. Figure 2.3). It happens when we

---

[10] A significant contribution to the comprehension of this process in terms of the so–called "disembodiment of the mind" derives from some studies in the field of cognitive

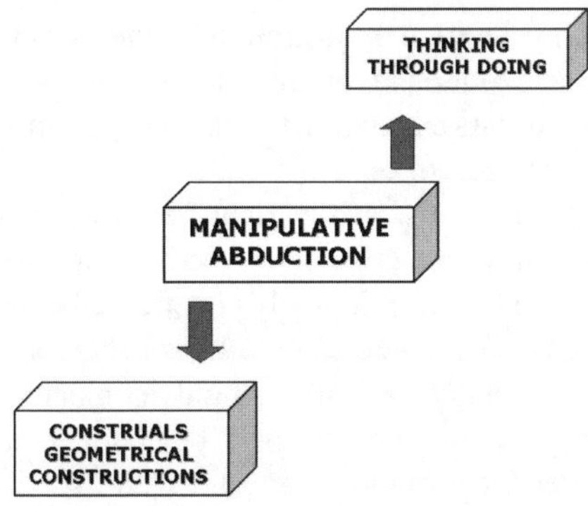

Fig. 2.3. Manipulative abduction.

are thinking *through* doing and not only, in a pragmatic sense, about doing. It happens, for instance, when we are creating geometry constructing and manipulating an external suitably realized icon like a triangle looking for new meaningful features of it, like in the case given by Kant in the "Transcendental Doctrine of Method" ([21] and the following section). It refers to an extra–theoretical behavior that aims at creating communicable accounts of new experiences to integrate them into previously existing systems of experimental and linguistic (semantic) practices.

Gooding [10] refers to this kind of concrete manipulative reasoning when he illustrates the role in science of the so-called "construals" that embody tacit inferences in procedures that are often apparatus and machine based. The embodiment is of course an expert manipulation of meaningful semiotic objects in a highly constrained experimental environment, and is directed by abductive movements that imply the strategic application of old and new *templates* of behavior mainly connected with extra-rational components, for instance emotional, esthetical, ethical, and economic.

The hypothetical character of construals is clear: they can be developed to examine or discard further chances, they are provisional creative organization of experience and some of them become in their turn hypothetical *interpretations* of experience, that is more theory-oriented, their reference/meaning is gradually stabilized in terms of established observational practices. Step by step the new interpretation - that at the beginning is completely "practice-laden" - relates to more "theoretical" modes of understanding (narrative, visual, diagrammatic, symbolic, conceptual, simulative), closer to the constructive effects of theoretical abduction. When the reference/meaning is stabilized the effects of incommensurability with other established observations can become evident. But it is just the construal of certain phenomena that can be shared by the sustainers of rival theories. Gooding [10] shows how Davy and Faraday could see the same attractive and repulsive actions at work in the phenomena they respectively produced; their discourse and practice as to the role of their

---

paleoanthropology that describe various related aspects of the birth of the material "culture". In [7] I have illustrated this issue relating it to the Turing ideas on "unorganized" and "organized" brains.

construals of phenomena clearly demonstrate they did not inhabit different, incommensurable worlds in some cases. Moreover, the experience is constructed, reconstructed, and distributed across a social network of negotiations among the different scientists by means of construals.

It is difficult to establish a list of invariant behaviors that are able to describe manipulative abduction in science. As illustrated above, certainly the expert manipulation of objects in a highly semiotically constrained experimental environment implies the application of old and new *templates* of behavior that exhibit some regularities. The activity of building construals is highly conjectural and not immediately explanatory: these templates are hypotheses of behavior (creative or already cognitively present in the scientist's mind-body system, and sometimes already applied) that abductively enable a kind of epistemic "doing": for example it allows us to find epistemica chances which in some cases can reflect concrete risks in the studied situation. Hence, some templates of action and manipulation can be *selected* in the set of the ones available and pre-stored, others have to be *created* for the first time to perform the most interesting creative cognitive accomplishments of manipulative abduction.

Moreover, I think that a better understanding of manipulative abduction at the level of scientific experiment could improve our knowledge of induction, and its distinction from abduction: manipulative abduction could be considered as a kind of basis for further meaningful inductive generalizations. Different generated construals can give rise to different inductive generalizations.

Some common features of these tacit templates that enable us to manipulate things and experiments in science to favor meaning formation are related to: 1. sensibility towards the aspects of the phenomenon which can be regarded as *curious* or *anomalous*; manipulations have to be able to introduce potential inconsistencies in the received knowledge (Oersted's report of his well-known experiment about electromagnetism is devoted to describe some anomalous aspects that did not depend on any particular theory of the nature of electricity and magnetism; Ampère's construal of experiment on electromagnetism - exploiting an artifactual apparatus to produce a static equilibrium of a suspended helix that clearly shows the role of the "unexpected"); 2. preliminary sensibility towards the *dynamical* character of the phenomenon, and not to entities and their properties, common aim of manipulations is to practically reorder the dynamic sequence of events in a static spatial one that should promote a subsequent bird's-eye view (narrative or visual-diagrammatic); 3. referral to experimental manipulations that exploit *artificial apparatus* to free new possibly stable and repeatable sources of information about hidden knowledge and constraints (Davy well-known set-up in terms of an artifactual tower of needles showed that magnetization was related to orientation and does not require physical contact). Of course this information is not artificially made by us: the fact that phenomena are made and manipulated does not render them to be idealistically and subjectively determined; 4. various contingent ways of epistemic acting: *looking* from different perspectives, *checking* the different information available, *comparing* subsequent events, *choosing, discarding, imaging* further manipulations, *re-ordering* and *changing relationships* in the world by implicitly *evaluating* the usefulness of a new order (for instance, to help memory).

From the general point of view of everyday situations manipulative abductive reasoning exhibits other very interesting templates: 5. action elaborates a *simplification* of the reasoning task and a redistribution of effort across time when we "need to manipulate concrete things in order to understand structures which are otherwise too abstract" [11], or when we are in presence of *redundant* and unmanageable information; 6. action can be useful in presence of *incomplete* or *inconsistent* information - not only from the "perceptual" point of view - or of a diminished capacity to act upon the world: it is used to get more data to restore coherence and to improve deficient knowledge; 7. action as a *control of sense data* illustrates how we can change the position of our body (and/or of the external objects) and how to exploit various kinds of prostheses (Galileo's telescope, technological instruments and interfaces) to get various new kinds of stimulation: action provides some tactile and visual information (e.g., in surgery), otherwise unavailable; 8. action enables us to build *external artifactual models* of task mechanisms instead of the corresponding internal ones, that are adequate to adapt the environment to the agent's needs: experimental manipulations exploit *artificial apparatus* to free new possible stable and repeatable sources of information about hidden knowledge and constraints.

The whole activity of manipulation is devoted to build various external *epistemic mediators*[11] that function as versatile semiotic tools able to provide an enormous new source of data, information, and knowledge. Therefore, manipulative abduction represents a kind of redistribution of the epistemic and cognitive effort to manage objects and information that cannot be immediately represented or found internally (for example exploiting the resources of visual imagery).[12]

If we see scientific discovery like a kind of opportunistic ability of integrating information from many kinds of simultaneous constraints to produce explanatory hypotheses that account for them all, then manipulative abduction will play the role of eliciting possible hidden constraints by building external suitable experimental structures.

2.4.2 Manipulating and Communicating Meanings through External Semiotic Anchors

If the structures of the environment play such an important role in shaping our semiotic representations and, hence, our cognitive processes, we can expect that physical manipulations of the environment receive a great cognitive relevance.

Several authors have pointed out the role that physical actions can have at a cognitive level. In this sense Kirsh and Maglio [16] distinguish actions into two categories,

---

[11] I derive this expression from the cognitive anthropologist Hutchins, that coins the expression "mediating structure" to refer to various external tools that can be built to cognitively help the activity of navigating in modern but also in "primitive" settings [6, 15].

[12] It is difficult to preserve precise spatial relationships using mental imagery, especially when one set of them has to be moved relative to another.

namely *pragmatic actions* and *epistemic actions*. Pragmatic actions are the actions that an agent performs in the environment in order to bring itself physically closer to a goal. In this case the action modifies the environment so that the latter acquires a configuration that helps the agent to reach a goal which is understood as physical, that is, as a desired state of affairs. Epistemic actions are the actions that an agent performs in a semiotic environment in order to discharge the mind of a cognitive load or to extract information that is hidden or that would be very hard to obtain only by internal computation.

In this section I want to focus specifically on the relationship that can exist between manipulations of the environment and representations. In particular, I want to examine whether external manipulations can be considered as means to construct external representations.

If a manipulative action performed upon the environment is devoted to create a configuration of signs that carries relevant information, that action will well be able to be considered as a cognitive semiotic process and the configuration of elements it creates will well be able to be considered an external representation. In this case, we can really speak of an embodied cognitive process in which an action constructs an external representation by means of manipulation. We define *cognitive manipulating* as any manipulation of the environment devoted to construct external configurations that can count as representations.

An example of cognitive manipulating is use of diagrams in mathematical reasoning. In this case diagrams carry relevant semiotic information about the internal angles of a triangle "anchoring" new meanings.

The entire process through which an agent arrives at a physical action that can count as cognitive manipulating can be understood by means of the concept of manipulative abduction [20]. Manipulative abduction is a specific case of cognitive manipulating in which an agent, when faced with an external situation from which it is hard or impossible to extract new meaningful features of an object, selects or creates an action that structures the environment in such a way that it gives information which would be otherwise unavailable and which is used specifically to infer explanatory hypotheses.

In this way the semiotic result is achieved on *external* representations used in lieu of the internal ones. Here action performs an *epistemic* and not a merely performatory role, for example relevant to abductive reasoning.

2.4.3 Communication and Semiosis through Re-Embodiment

Some interesting semiotic aspects of the above illustrated communicative process can be nicely analyzed. Imagine that a suitable *fixed internal record* exists – deriving from the cognitive exploitation of the previous suitable interplay with *external structures* - at the level of neural activation and that for instance it embeds an abstract concept endowed with all its features, for example the concept of triangle. Now, the human agent, via neural mechanisms and bodily actions, can "re-embody" that concept by making an external perceivable *sign*, for instance available to the attention (and so potentially communicable) of other human or animal senses and brains. For instance that human agent can use what in semiotics is called a *symbol* (with its conventional character: *ABC*, for example), but also an *icon* of relations (a suitable diagram of a triangle), or a *hybrid representation* that will take advantage of both. In Peircian terms:

> A representation of an idea is nothing but a sign that calls up another idea. When one mind desires to communicate an idea to another, he embodies his idea by making an outward perceptible image which directly calls up a like idea; and another mind perceiving that image gets a like idea. Two persons may agree upon a conventional sign which shall call up to them an idea it would not call up to anybody else. But in framing the convention they must have resorted to the primitive diagrammatic method of embodying the idea in an outward form, a picture. Remembering what likeness consists in, namely, in the natural attraction of ideas apart from habitual outward associations, I call those signs which stand for their likeness to them *icons*. Accordingly, I say that the only way of directly communicating an idea is by mean of an icon; and every indirect method of communicating an idea must depend for its establishment upon the use of an icon [19, 787, 26–28].[13]

It is well-known that for Peirce every picture is a icon and thus every diagram, even if it lacks a sensuous similarity with the object, but just exhibits an analogy between the relations of the part of it and of the object:

> Particularly deserving of notice are icons in which the likeness is aided by conventional rules. Thus, an algebraic formula is an icon, rendered such by the rules of commutation, association, and distribution of the symbols; that it might as well, or better, be regarded as a compound conventional sign. It may seem at first glance that it is an arbitrary classification to call an algebraic expression an icon; that it might as well, or better, be regarded as a compound of conventional sign. But it is not so. For a great distinguishing property of the icon is that by direct observation of it other truths concerning its object can de discovered than those which suffice to determine its construction. Thus, by means of two photographs a map can be drawn, etc. Given a conventional or other general sign of an object, to deduce any other truth than which it explicitly signifies, it is necessary, in all cases, to replace that sign by an icon. This capacity of revealing unexpected truth is precisely that wherein the utility of algebraic formulae consists, so that the icon in character is the prevailing one [19, 787, CSP 26–28].

Stressing the role of iconic dimensions of semiosis[14] in the meantime celebrates the virtues in communication of analogy, as a kind of "association by resemblance",

---

[13] We have to note that for Peirce an idea "[…] is not properly a conception, because a *conception* is not an idea at all, but a *habit*. But the repeated occurrence of a general idea and the experience of its *utility*, results in the formation or strengthening of that habit which is the conception" [8, 7.498].

[14] We have to remember that in this perspective any proposition is a diagram as well, because it represents a certain relation of symbols and indices.

as contrasted to "association by contiguity". The emphasis on iconic and analogical aspects of both everyday and scientific reasoning clearly favors the role of interdisciplinarity in communication and production of multiple clusters of data.

Human beings delegate cognitive (for example communicative and epistemic) features to external representations through semiotic attributions because for example in many problem solving situations the internal computation would be impossible or it would involve a very great effort because of human mind's limited capacity. First a kind of "alienation" is performed, second a recapitulation is accomplished at the neuronal level by re-representing internally that which was "discovered" outside. Consequently only later on we perform cognitive operations on the structure of data that synaptic patterns have "picked up" in an analogical way from the environment. We can maintain that internal representations used in cognitive processes like many events of *meaning creation* and *communication* have a deep origin in the experience lived in the semiotic environment.

I already illustrated in section 2.2 that I think there are two kinds of artifacts that play the role of *external objects* (representations) active in this process of externalization of the mind: *creative* and *mimetic*. Mimetic external representations mirror concepts and problems that are already represented in the brain and need to be enhanced, solved, further complicated, etc. so they sometimes can creatively give rise to new concepts and meanings. Hence, interesting (and new) information and knowledge packages are generated and stored over there in the external human niches, in various supports more or less accessible that can be picked up in further data communication and chance/risk discovering processes.

Following my perspective it is at this point evident that the "mind" transcends the boundary of the individual and includes parts of that individual's environment. It is in this sense that the mind is semiotic and artificial.

2.4.4 Delegated and Intrinsic Constraints in External Agents

We have said that through the cognitive interplay with external representations the human agent is able to pick up and use what suggested by the constraints and features intrinsic to their external materiality and to their relative established conventionality: data, artificial languages, proofs, examples, etc. At the beginning of this kind of process the human agent embodies a sign in the external world that for example in classical geometry is an icon endowed with "intentional" delegated cognitive conventional and public features – meanings - that resort to some already known properties of the Euclidean geometry: a certain language and a certain notation, the definition of a triangle, the properties of parallel lines that also hold in case of new elements and "auxiliary" constructions obtained through manipulation, etc. Then she looks, through diagram manipulations, for possible necessary consequences that occur over there, in the diagram/icon and that obey both

- the conventional *delegated* properties and
- the properties *intrinsic* to the materiality of the model.

This external model is a kind of autonomous cognitive *agent* offered to new interpretations of the problem/object in question. The model can be picked up later and acknowledged by the human agent through fixation of a new neural configuration – a new "thought". This operation can be imagined as acting in other epistemic settings, for example in manipulations of multiple data of a specific multidisciplinary field, as a way for extracting/discovering new chances/risks implicity embedded in the case/circumstances under analysis.

The distinction above between delegated and intrinsic and immanent properties is also clear if we adopt the Peircian semiotic perspective. Peirce – speaking about the case of syllogistic logic, and not of geometry or algebra - deals with this problem by making an important distinction between what is going on in the brain of the logical human agent and the autonomous power of the chosen external system of representation or diagrammatization [37]. The presence of this "autonomous power" explains why I attribute to the system of representation a status of cognitive agency similar to the one of a human person, even if of course lacking aspects like direct intention and responsibility. Imagine for instance, the powerful autonomous agency which is represented by a Practical Universal Turing Machine (a PC with its software). In the case of diagrams, Peirce says, it makes use

> [...] of a particular system of symbols - a perfectly regular and very limited kind of language. It may be a part of a logician's duty to show how ordinary ways of speaking and of thinking are to be translated into that symbolism of formal logic; but it is no part of syllogistic itself. Logical principles of inference are merely rules for the illative transformation of the symbols of the particular system employed. If the system is essentially changed, they will be quite different [8, 2.599].

Of course the argumentation above also holds for the case of iconic geometric representation and can be extended in many other epistemic setting where for example the external support is a computational tool. This distinction integrates the one I have introduced above in the two levels of representations, and in some sense blurs it by showing how the *hybrid* character of the system composed by the two levels themselves, where the whole package of sensorial and kinesthetic abilities are involved.

The construction of the diagram also depends on those delegated semiotic properties that are embedded in what Peirce calls "precept" as he says in the passage we have already quoted above and not only on the constraints expressed by the materiality of the model itself.[15] A diagram has various semiotic properties just like a computation tool presents a lot of constraints but also of knowledge chances, when suitably exploited in the external/internal interplay I have described in section 2.2.

Pickering [39] depicts the role of some externalities (representations, artifacts, tools, etc.) in terms of a kind of non-human agency that interactively stabilizes with

---

[15] It is worth noting that this process is obviously completely related to the Peircian idea of pragmatism [38], that he simply considers "the experimental method" which is the procedure of all science.

human agency in a dialectic of resistance and accommodation [39, p. 17 and p. 22]. The two agencies, for example in scientific reasoning, originate a co-production of cognition the results of which cannot be presented and identified in advance: the outcome of the co-production is intrinsically "unpredictable". Latour's notions of the de-humanizing effect of technologies are based on the so-called "actor network theory",[16] which also stresses the semiotic role of externalities like the so-called non human agents. The actor network theory basically maintains that we should think of science, technology, and society as a field of human and non-human (material) agency. Human and non-human agents are associated with one another in networks, and they evolve together within these networks. Because the two aspects are equally important, neither can be reduced to the other: "An actor network is simultaneously an actor whose activity is networking heterogeneous elements and a network that is able to redefine and transform what is it made of [...]. The actor network is reducible neither to an actor alone nor to a network" [41, p. 93].

The operation on a diagram has reduced complexity enabling concentration on essential relations and has revealed new data. Moreover, through manipulations of the diagram new perspectives are offered to the observation, or interesting anomalies with respect the internal expectations are discovered. In the case of mathematicians, Peirce maintains, the diagram "puts before him an icon by the observation of which he detects relations between parts of the diagram other than those which were used in its construction" [46, III, p. 749]: "unnoticed and hidden relations among the parts" are discovered [8, 3.363]. This activity is a kind of "thinking through doing": "In geometry, subsidiary lines are drawn. In algebra permissible transformations are made. Thereupon, the faculty of observation is called into play. [...] Theorematic reasoning invariably depends upon experimentation with individual schemata" [8, 4.233].

We have said that firstly the human agent embodies a sign in the external world that is in this geometrical case an icon endowed with "intentional" delegated cognitive conventional and public features – meanings - that resort to some already known properties of the Euclidean geometry: these features can be considered a kind of immanent rationality and regularity [38] that establishes a disciplinary field to envisage conclusions.[17] The system remains relative to the chosen conventional framework. They are real as long as there is no serious doubt in their adequacy: "The 'real,' for Peirce, is part of an evolutionary process and while 'pragmatic belief' and unconscious habits might be doubled from a scientific point a view, such a science might also formulate serious doubts in its own representational systems" [38, p. 295].

Let us imagine we choose a different representational system still exploiting material and external diagrams. Through the manipulation of the new symbols and

---

[16] This theory has been proposed by Callon, Latour himself, and Law [40–45].

[17] Paavola, Hakkarainen, and Sintonen [47] consider the interplay between internal and external aspects of abductive reasoning in the framework of the interrogative model of the so-called "explanation-seeking why-questions". They emphasize the interaction with the "environment" and show the importance of the heuristic strategies and of their trialogic nature (inquirer and fellow inquirers, object of inquiry, mediating artefacts and processes), also taking advantage of Davidson's ideas concerning triangulation.

diagrams we expect very different conclusions. An example is the one of the non-Euclidean discoveries. In Euclidean geometry, by adopting the postulate of parallels we necessarily arrive to the ineluctable conclusion that the sum of internal angles of a triangle is 180°, but this does not occur in the case of the non-Euclidean geometry where a different selected representational system - that still uses Euclidean icons - determines quite different possibilities of constructions, and thus different results from iconic experimenting.[18]

2.5 Conclusion

The main thesis of this paper is that the process of externalization of mind is a significant cognitive perspective able to unveil some basic features of abductive reasoning in both everyday and epistemic settings. Its fertility in explaining the semiotic communicative interplay between internal and external levels of cognition is evident and stressed its obvious interdisciplinary character. I maintain that various aspects of creative meaning formation and communication could take advantage of the research on this interplay: for instance study on external mediators can provide a better understanding of the processes of explanation and discovery (an chance discovery) in science and in some areas of artificial intelligence related to mechanizing discovery processes, where the aim at discovering chances and risks in the related studied situation is central.[19]

The cognitive referral to the central role of the relation between meaningful behavior and dynamical interactions with the environment becomes critical to the problem of meaning formation and communication. The perspective above, resorting to the exploitation of a very interdisciplinary interplay will further shed light on how concrete manipulations of external objects influence the abductive generation of hypotheses and so on the characters of what I call manipulative abduction showing how we can find methods of constructivity – and their computational counterparts – in scientific and everyday reasoning based on external models and "epistemic mediators" [50], as tools that can enhance in many ways risk/chance construction or extraction/elicitation.

References

1. Gibson, J.J.: *The Ecological Approach to Visual Perception*. Houghton Mifflin, Boston, MA (1979)
2. Magnani, L.: *Abduction, Reason, and Science. Processes of Discovery and Explanation*. Kluwer Academic/Plenum Publishers, New York (2001)
3. Tooby, J., DeVore, I.: The reconstruction of hominid behavioral evolution through strategic modeling. In Kinzey, W.G., ed.: *Primate Models of Hominid Behavior*. Suny Press, Albany (1987) 183–237

---

[18] I have illustrated this problem in detail in [48].

[19] On the recent achievements in the area of the machine discovery simulations of model-based creative tasks cf. [49].

4. Pinker, S.: Language as an adaptation to the cognitive niche. In Christiansen, M.H., Kirby, S., eds.: *Language Evolution: The States of the Art*. Oxford University Press (2003)

5. Zhang, J.: The nature of external representations in problem-solving. *Cognitive Science* **21**(2) (1997) 179–217

6. Hutchins, E.: *Cognition in the Wild*. MIT Press, Cambridge, MA (1995)

7. Magnani, L.: Mimetic minds. meaning formation through epistemic mediators and external representations. In Loula, A., Gudwin, R., Queiroz, J., eds.: *Artificial Cognition Systems*. Idea Group Publishers, Hershey, PA (2006) 327–357

8. Peirce, C.S.: *Collected Papers of Charles Sanders Peirce*. Harvard University Press, Cambridge, MA (1931-1958) vols. 1-6, Hartshorne, C. and Weiss, P., eds.; vols. 7-8, Burks, A. W., ed.

9. Nersessian, N.J.: Model-based reasoning in conceptual change. In Magnani, L., Nersessian, N.J., Thagard, P., eds.: *Model-based Reasoning in Scientific Discovery*, New York, Kluwer Academic/Plenum Publishers (1999) 5–22

10. Gooding, D.: *Experiment and the Making of Meaning*. Kluwer, Dordrecht (1990)

11. Piaget, J.: *Adaption and Intelligence*. University of Chicago Press, Chicago (1974)

12. Anderson, M.L.: Embodied cognition: a field guide. *Artificial Intelligence* **149**(1) (2003) 91–130

13. Agree, P., Chapman, D.: What are plans for? In Maes, P., ed.: *Designing Autonomous Agents*. The MIT Press, Cambridge, MA (1990) 17–34

14. Brooks, R.A., Stein, L.: Building brains for bodies. *Autonomous Robots* **1** (1994) 7–25

15. Hutchins, E.: Cognitive artifacts. In Wilson, R.A., Keil, F.C., eds.: *Encyclopedia of the Cognitive Sciences*. The MIT Press, Cambridge, MA (1999)

16. Kirsh, D., Maglio, P.: On distinguishing epistemic from pragmatic action. *Cognitive Science* **18** (1994) 513–549

17. Brooks, R.A.: Intelligence without representation. *Artificial Intelligence* **47** (1991) 139–159

18. Zhang, J.: The nature of external representations in problem solving. *Cognitive Science* **21**(2) (1997) 179–217

19. Peirce, C.S.: *The Charles S. Peirce Papers: Manuscript Collection in the Houghton Library*. The University of Massachusetts Press, Worcester, MA (1967) Annotated Catalogue of the Papers of Charles S. Peirce. Numbered according to Richard S. Robin. Available in the Peirce Microfilm edition. Pagination: CSP = Peirce / ISP = Institute for Studies in Pragmaticism.

20. Magnani, L.: *Abduction, Reason, and Science. Processes of Discovery and Explanation*. Kluwer Academic/Plenum Publishers, New York (2001)

21. Magnani, L.: *Philosophy and Geometry. Theoretical and Historical Issues*. Kluwer Academic Publisher, Dordrecht (2001)

22. Flach, P., Kakas, A.: *Abductive and Inductive Reasoning: Essays on Their Relation and Integration* (2000) edited book.

23. Bylander, T., Allemang, D., Tanner, M.C., Josephson, J.R.: The computational complexity of abduction. *Artificial Intelligence* **49** (1991) 25–60

24. Boutilier, C., Becher, V.: Abduction as belief revision. *Artificial Intelligence* **77** (1995) 43–94

25. Magnani, L.: Inconsistencies and creative abduction in science. In: *AI and Scientific Creativity. Proceedings of the AISB99 Symposium on Scientific Creativity*, Edinburgh, Society for the Study of Artificial Intelligence and Simulation of Behaviour, University of Edinburgh (1999) 1–8

26. Thagard, P.: *Conceptual Revolutions*. Princeton University Press, Princeton (1992)

27. Nersessian, N.J.: Should physicists preach what they practice? Constructive modeling in doing and learning physics. *Science and Education* **4** (1995) 203–226

28. Cornuéjols, A., Tiberghien, A., Collet, G.: A new mechanism for transfer between conceptual domains in scientific discovery and education. *Foundations of Science* **5**(2) (2000) 129–155 Special Issue on "Model-based Reasoning in Science: Learning and Discovery", ed. by L. Magnani and N. J. Nersessian and P. Thagard.

29. Gatti, A., Magnani, L.: On the representational role of the environment and on the cognitive nature of manipulations. In Magnani, L., ed.: *Computing, Philosophy and Cognition*, London, King's College Pubblications (2006) 227–242

30. Clark, A.: *Natural-Born Cyborgs. Minds, Technologies, and the Future of Human Intelligence*. Oxford University Press, Oxford (2003)

31. Thomas, H.J.: Are theories of imagery theories of imagination? An active perception ap-proach to conscious mental content. *Cognitive Science* **23**(2) (1999) 207–245

32. Brent, J.: A brief introduction to the life and thought of Charles Sanders Peirce. In Muller, J., Brent, J., eds.: *Peirce, Semiosis, and Psychoanalysis*. John Hopkins, Baltimore and London (2000) 1–14

33. Colapietro, V.: Futher consequences of a singular capacity. In Muller, J., Brent, J., eds.: *Peirce, Semiosis, and Psychoanalysis*. John Hopkins, Baltimore and London (2000) 136–58

34. Queiroz, J., Merrell, F.: Abduction: between subjectivity and objectivity (2005) Special Issue of the Journal *Semiotica*.

35. Thagard, P.: How does the brain form hypotheses? Towards a neurologically realistic computational model of explanation. In Thagard, P., Langley, P., Magnani, L., Shunn, C., eds.: *Symposium "Generating explanatory hypotheses: mind, computer, brain, and world"*, Stresa, Italy, Cognitive Science Society, CD-Rom (2005) Proceedings of the 27th International Cognitive Science Conference.

36. Thagard, P.: Abductive inference: from philosophical analysis to neural mechanisms. In Feeney, A., Heit, E., eds.: *Inductive Reasoning: Cognitive, Mathematical, and Neuroscientific Approaches*, Cambridge, Cambridge University Press (2007) Forthcoming.

37. Hoffmann, M.H.G.: Peirce's diagrammatic reasoning as a solution of the learning paradox. In Debrock, G., ed.: *Process Pragmatism: Essays on a Quiet Philosophical Revolution*. Rodopi Press, Amsterdam (2003) 121–143

38. Hoffmann, M.H.G.: How to get it. Diagrammatic reasoning as a tool for knowledge development and its pragmatic dimension. *Foundations of Science* **9** (2004) 285–305

39. Pickering, A.: *The Mangle of Practice. Time, Agency, and Science*. The University of Chicago Press, Chicago and London (1995)

40. Callon, M.: Four models for the dynamics of science. In Jasanoff, S., Markle, G.E., Petersen, J.C., Pinch, T.J., eds.: *Handbook of Science and Technology Studies*, Los Angeles, Sage (1994) 29–63

41. Callon, M.: Society in the making: the study of technology as a tool for sociological analysis. In Bjiker, W.E., Hughes, T.P., Pinch, T., eds.: *The Social Construction of Technological Systems*, Cambridge, MA, MIT Press (1997) 83–106

42. Latour, J.: *Science in Action: How to follow Scientists and Engineers through Society*. Harvard University Press, Cambridge, MA (1987)

43. Latour, J.: *The Pasteurization of France*. Harvard University Press, Cambridge, MA (1988)

44. Callon, M., Latour, B.: Don't throw the baby out with the bath school! A reply to collins and yearley. In Pickering, A., ed.: *Science as Practice and Culture*, Chicago and London, The University of Chicago Press (1992) 343–368

45. Law, J.: *Modernity, Myth, and Materialism.* Blackwell, Oxford (1993)
46. Peirce, C.S.: *The New Elements of Mathematics by Charles Sanders Peirce.* Mouton/ Humanities Press, The Hague-Paris/Atlantic Higlands, NJ (1976) vols I-IV, edited by C. Eisele.
47. Paavola, S., Hakkarainen, K., Sintonen, M.: Abduction with dialogical and trialogical means. *Logic Journal of the IGPL* **14(1)** (2006) 137–150
48. Magnani, L.: Semiotic brains and artificial minds. How brains make up material cognitive systems. In Gudwin, R., Queiroz, J., eds.: *Semiotics and Intelligent Systems Development.* Idea Group Inc., Hershey, PA (2007) 1–41
49. Magnani, L., Nersessian, N., Pizzi, C.: *Logical and Computational Aspects of Model-Based Reasoning* (2002) edited book.
50. Magnani, L.: Conjectures and manipulations. Computational modeling and the extra-theoretical dimension of scientific discovery. *Minds and Machines* **14** (2004) 507–537

Part II

Creative Community Working on Multidisciplinary Data

3

Augmented Analytical Exploitation of a Scientific Forum

Xijin Tang[1], Yijun Liu[2], and Wen Zhang[1]

[1] Institute of Systems Science, Academy of Mathematics and Systems Science Chinese Academy of Sciences, Beijing 100080 P.R.China

[2] Institute of Policy and Management, Chinese Academy of Sciences, Beijing 100080 P.R.China

`{ xjtang, yijunliu, zhangwen } @amss.ac.cn`

Summary. In reality, group work, especially various meetings, exists as a feasible way for people to communicate and collaborate to deal with problems. Various academic meetings and conferences serve as an important part of social process toward scientific knowledge growth. It is significant to understand more about the outcome of those meetings for maintaining the scale of those dialogues and facilitating policy making. In this paper, some augmented analytical methods are applied to a famous scientific forum on frontiers of science and technology in China, Xiangshan Science Conference, to expose some ignored information which is eagerly required by conference organizers, policy makers and researchers. Those methods, such as visualization of expert opinion structure, augmented information support by Web text-mining, clustering of contributed ideas and various analysis about individual's participation, etc. are integrated into a group argumentation environment (GAE), which aims to support divergent group thinking process for emergence of a *ba* for knowledge creation and provide a variety of perspectives towards the concerned topics by those addressed conferencing mining techniques.

3.1 Introduction

Whatever efficiency or effectiveness of those meetings held in daily life, that kind of group work exists as a feasible way for people to share ideas, interests, understandings and achievements about some focused topics and to search and find solutions toward a variety of problems. For scientific researchers, weekly seminars, academic conferences, scientific forums, etc. are usual ways for collaboration, discussion and exchange of experience and practice on the related issues or topics. By P. Thagard's view, scientific knowledge growth consists of the psychological processes of discovery and acceptance, the physical processes involving instruments and experiments, and the social processes of collaboration, communication, and consensus that brought about transformations in knowledge [1]. Thus it is necessary for governmental departments, research organizations, societies, etc. to maintain those group

X. Tang et al.: *Augmented Analytical Exploitation of a Scientific Forum*, Studies in Computational Intelligence (SCI) **123**, 65–79 (2008)
`www.springerlink.com`

activities to facilitate sociological approaches to the study of science. Lots of efforts had been invested. However, how to evaluate the outcome of those activities regarding those visible or invisible efforts has not been comprehensively studied by organizers in comparison to those direct outputs, such as budgets, publications, participation, etc. after those group activities were completed.

In China, XiangShan Science Conference (XSSC) is the most famous platform for scientific discussions and debates. Similar to Gordon Research Conferences in USA, XSSC is fully supported by government since 1993 and serves as a scientific forum which consists of a series of small-scale academic workshops where a group of scientists working at the frontier of a particular area meet to discuss in depth all aspects of the most advanced topics in the relevant fields and then new plans for research may be incubated. With its excellent academic contents, broad scientific visions and featured operating mechanism, XSSC has gained renowned reputation in open-mindedness and innovation, made important contributions to national science development and exerted a profound impact on the decision making process of the various government departments concerned. Even with a variety of statistic figures about outputs of XSSC, and aggregation of various records about those workshops including lectures, discussions and debates, comments and summaries posted on the conference web site, few studies have been undertaken toward those records to detect more hidden information by quantitative methods.

In this paper, a suite of analytical methods is applied to some explorations from such a knowledge repository where stored active scientists' understandings, wisdoms in scientific research at a context of economic and social reforms in China in recent 20 years. Versatile analyses are undertaken, such as visualization of expert opinion structure, various clustering of contributed opinions, augmented information support by Web text-mining, various measures about participants' contribution and roles in those series workshops, etc. All those analytical tools have been integrated into our developed group argumentation environment (GAE), which aims to support divergent group process for the emergence of a *ba* for idea generation and knowledge creation and provide a variety of perspectives towards the concerned topics by those conferencing mining techniques.

3.2 Exploitation of XSSC as Group Argumentation

Till now, almost 300 workshops across multiple disciplines and with over 10,000 participants of different ages had been held under the name of XSSC. Due to its features in facilitating interdisciplinary discussions, each workshop could be regarded as a group thinking process toward one or some scientific problems. Some hot topics, such as brain and consciousness, complexity, etc. have been discussed at many workshops. Basic information of each workshop is published at the conference web site www.xssc.ac.cn. Then people who did not attend the workshop can acquire some information from related web pages, such as a summary page of each workshop which includes all primary talks, some typical questions and discussions during the workshop together with a list of all participants.

For the sake of processing, here we consider that a group discussion is composed by a set of utterance records with a structure as

$$< topic, participant, utterance, keywordsset, time >$$

Such a record indicates a *participant* submits an *utterance* with a set of *keywords* at the *time* point along a discussing process about the *topic*. Here *topic, participant, utterance* and *time* can be directly fixed as the event happens, while *keywords set*, which are manually selected and assigned to each utterance according the meaning and context of discussion. For example, the record

<E4_CoData,
xjTang,
Augmented Analytical Exploitation of a Scientific Forum,
{conference mining, Xiangshan Science Conference},
2006-10-24>

indicates that participant *xjTang* gave a talk *Augmented Analytical Exploitation of a Scientific Forum* about the topic *E4_CoData* on October 24, 2006. *Conference mining* and *Xiangshan Science Conference* are two representative keywords of the utterance. Keywords are usually indicated clearly together with the abstract by the authors during submission to general academic conferences.

Based on such a conceptual model about group discussion, two kinds of matrices are generated.

1) Frequency matrix

Two frequency matrices F_p and F_u can be acquired. Each element of matrix F_p denotes the frequency of keyword i referred by participant j during the whole discussing process. Each element of matrix F_u denotes the frequency of keyword i referred by the utterance j as shown in Table 3.1. The keywords are articulated as attributes of participants or utterances.

Given frequency matrix F_u, dual-scaling method is employed to analyze the correspondence relations between utterances and keywords. Proposed by Nishisato, dual scaling is a multi-variant statistical method that is of similar characteristics with correspondence analysis and exploratory factor analysis [2]. In Table 3.1, each element in $X = (x_1, x_2, \ldots, x_m)^T$ refers to the weight of the corresponding keyword while vector $Y = (y_1, y_2, \ldots, y_n)^T$ refers to the sum of weighted scores about the corresponding utterance. With the principal components for given relations between keywords and utterances acquired by dual scaling, both the utterances and keywords can be mapped into 2-dimensional space. As a result, a pair of utterances with more common keywords may locate closer in the 2-dimention space. Such a process may also apply to spatially mapping with relations between participants and keywords set with frequency matrix F_p.

In the computerized support tool, group argumentation environment (GAE), for group divergent work, an electronic brainstorming room (BAR) is a basic module for

Table 3.1. Frequency Matrix: Utterance sets and keyword sets

X \ Y		keyword$_1$ x_1	keyword$_2$ x_2	$\cdots$	keyword$_m$ x_m	
utterance$_1$	y_1	a_{11}	a_{12}	$\cdots$	a_{1i}	$y_1 = \sum_{i=1}^{m} a_{1i}x_i$
utterance$_2$	y_2	a_{21}	a_{22}	$\cdots$	a_{2i}	$y_2 = \sum_{i=1}^{m} a_{2i}x_i$
$\vdots$		$\vdots$	$\vdots$	$\ddots$	$\vdots$	$\vdots$
utterance$_n$	y_n	a_{n1}	a_{n2}	$\cdots$	a_{ni}	$y_n = \sum_{i=1}^{m} a_{ni}x_i$

(a_{ji} denotes the frequency keyword$_i$ appeared in the utterance$_j$.)

diverse idea publishing and serves as the virtual space for participants for communication and information sharing, similar to a general BBS. Besides general functions as BBS, GAE-BAR provides a visualized area to exhibit the dynamic process of discussing for one topic using the results of dual scaling method for both frequency matrices. The results for matrix F_p processed by dual scaling method are displayed as a common view; while the results for matrix F_u are displayed as a personal view. Both views serve as a visualized shared memory space where displayed the global structure of participants' joint thought about the concerned topic. Moreover, with selected participants or selected utterances, a visualized group thinking map cab be accessed via a retrospective view. Retrospect analysis in GAE-BAR mainly help users to "drill down" into the discussing process with visualized snapshots about pieces of discussion such as selected intervals of discussion or selected participants, and detect the micro-community forming, which may be useful in understanding about participants' thinking structure, story-telling or group thinking context awareness, or case studies for other problem solving.

Furthermore, based on the spatial relations in common view or personal view, clustering of keywords or utterances can be done. In GAE-BAR, centroid-based K-means method is applied to keywords clustering, while KJ method is applied to utterances grouping.

2) Matrix of Agreement or Discrepancy
The second kind of matrix concerns relationships between participants. If there are n participants at one period of time in the course, let U_i is the keywords set referred by the participant i, $i = (1, 2, \ldots n)$, then two matrices are acquired:

- Matrix of agreement or similarity, denoted as A_1 where $a_{ij}^1 = |U_i \cap U_j|$. The element is number of the keywords shared between participants i and j. Obviously, $a_{ii}^1 = |U_i|$.
- Matrix of dissimilarity or discrepancy, denoted as A_2 where $a_{ij}^2 = |(U_i \cap \bar{U}_j) \cup (\bar{U}_i \cap U_j)|$. The element is the number of different keywords between two participants i and j. Obviously, $a_{ii}^2 = 0$.

Both matrices are symmetrical. According to the eigenvector corresponding with the maximum eigenvalue for each matrix, we get rank of participants which may reflect participants' contributions in the course. It is estimated that eigenvector of Matrix A_1 reflects who holds more common concerns in the course, and that of Matrix A_2 reflects who is of more diverse perspectives than others. If we pay more attention to the extent of consensus or agreement, characteristics of Matrix A_1 may give some hints; if we focus on how diverse of ideas in the course, more information could be acquired from A_2. Both measures are firstly discussed in Ref. [3].

Since both matrices are non-negative, in order to ensure the existence of a unique maximum eigenvalue for each matrix, a small value, eg. 0.001, may be added to each element and the original non-negative matrix is transformed into a positive matrix. Such a change can be explained from a practical view. For matrix of agreement, those participants attending the same discussion of the concerned topic may underlie that they at least share same interests in the topic. For matrix of discrepancy, each participant is different, and even a participant himself may change ideas along the discussing process, then differences always exist even to one participant along a discussion process.

Such kind of measures may be helpful to selection of appropriate experts for relevant workshop or even problem solving later. Next, an example is given to show how those analytical methods are applied to mining of XSSC.

3.3 Augmented Analyses toward Topics on Complexity of XSSC by View of Group Argumentation

As complex systems and complexity research has becoming very hot since the 21st century, here we select some workshops, whose principal topics concentrate on complexity or complex system, to study how relevant research has been noticed and undertaken by Chinese scientists. Till May of 2004, seven relevant workshops have been held as listed in Table 3.2.

Initially the script of group argumentation includes all talks given by 17 selected scientists, whose group thinking space via common viewer in GAE-BAR is as shown in Figure 3.1.

The rectangular icon refers to the user ID of participants in the discussing process, and the oval icon refers to keyword as articulated as attributes of participants. If the mouse locates at a rectangular, then all utterances given by the corresponding participant are popped up. The more shared keywords between participants, the higher mutual relevance between them, which is reflected by the distances between participants in the map. As discussion goes on, the structure of the diagram will be changing. Then as appended all talks given by those selected scientists who attended the 7th workshop held in May of 2004, a big change happened in the common view. The final common view is displayed at Figure 3.2(a) which also shows the layout of the client window of GAE-BAR. As to details of functions or framework of GAE-BAR, please refer Ref. [3, 4]. Figure 3.2(b) shows a retrospective view with 11 selected scientists who are closely located in Figure 3.2(a). Then as indicated in the

Table 3.2. Topics about Complexity

Workshop No.	Title	ConveningTime
20	Open Complex Giant System Methodology	June 20-23, 1994
29	Theoretical method and some major Scientific Problems relevant to Natural	March 29-31, 1995
68	Theory and Practice of Open Complex Giant System Methodology	January 6-9,1997
110	Cybernetics and Revolutions in Science and Technology	December 22-23,1998
112	Complexity Science	March 18-20, 1999
190	Complex Systemin Process Engineering	September 17-19, 2002
227	System, Control and Complexity Science	May 25-27, 2004

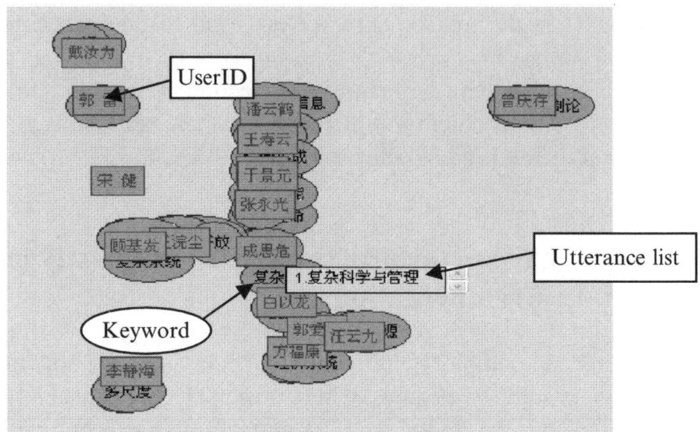

Fig 3.1. Original spatial map about "complexity" (6 workshops selected, till 2002)

middle of Figure 3.2(b), another micro community is detected where "control", "cybernetics" and "artificial intelligence" are foci. Retrospective analysis can also used by observers as a readily accessible record of the topics when facing similar issues during problem-solving process.

Next some feature functions of GAE-BAR as shown in Figure 3.2 are addressed.

(1). Recording Original Idea Provider

For all keywords proposed during a discussing process, GAE-BAR can tell users who firstly propose each keyword, when and how often that keyword is referred later, as shown in Figure 3.2(c). Such a mechanism is to check the originality of participants. In her 2-space transformation model, Boden claims the idea as P-creative if the person in whose mind the idea arises haven't had it before, no matter how many others may have had the same idea already [5]. Then the record of original idea provider is the record of individual's P-creativity among the discussing group under the same theme.

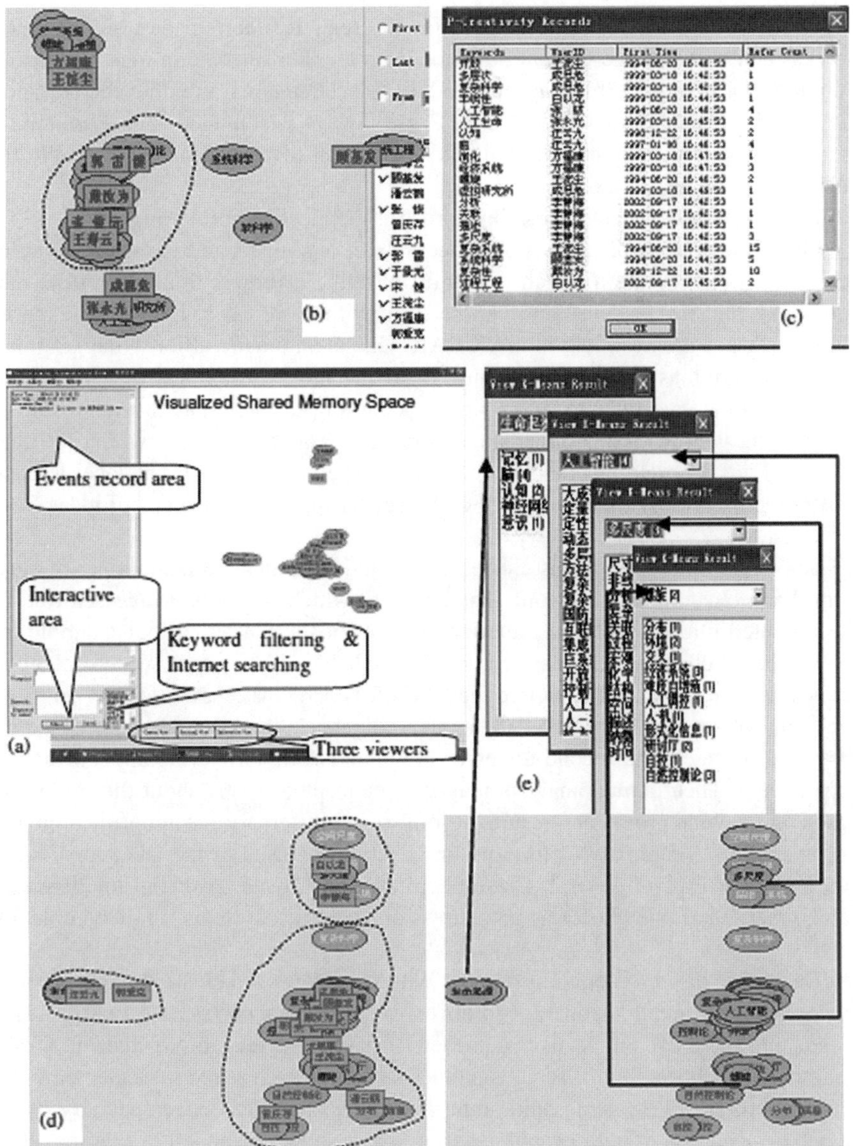

Fig 3.2. Functions of GAE-BAR
(a) Main client window; (b) Retrospective view; (c) Original keyword provider;
(d) Clustering analysis (k =4) with participants; (e) List of a cluster of keywords by (d).

The highest frequency of the referred keywords may indicate the foci of the argumentation. As shown in Figure 3.2(c), two frequently referred keywords are "complex system" (15 times) and "complexity" (9 times). Less referred keywords may also be noticed by active participants. Such a record then help users aware of those which have not been much noticed so far (P-creativity) during the argumentation process and may change current attention with another interesting idea for further thinking. For divergent thinking, more ideas as scattered across either the common view or the personal view at GAE-BAR are always encouraged. The P-creativity recorder is then a supplement to help users aware of diverse ideas, instead only focusing those highlighted ones.

(2). Clustering of Utterances/Keywords for Perspective Formatting

Given the records of discussion process, both organizers and participants prefer somewhat summarization which is helpful to refine divergent thinking results into something that makes sense and can be dealt with more easily by human experts. Now two ways are available in GAE-BAR. One is automatic affinity diagramming (usually referred as KJ method because of its inventor, Kawakita Jiro) which maps the whole utterance set into 16×16 cells according to their spatial relationship at the persona view with a 2-dimensional structure. Those utterances which fall into same cell are regarded as one cluster. Human experts may assign one label to each cluster themselves by their judgment and discover meaningful groups of ideas from a raw list.

Another way uses k-means clustering method (k is an assumed number of clusters). Each cluster has a centroid. The keyword which is closest to the centroid of the affiliated cluster could be regarded as label of the cluster. In our example, as $k = 4$, then 4 clusters are generated (see Fig. 3.2(d)). The keyword set of each cluster is as shown in Fig. 3.2(e) where the labels for each cluster are 生命起源 "the origin of life" (the bottom window), 人工智能 "artificial intelligence" (the second bottom window), 多尺度 "multi-scale" (the second upper window) and 螺旋 "spiral" (the upper window) respectively. Then human analysts may give conceptual terms about those clusters based on machine processing results or join some clusters for meaningful summary. For example, "complexity of life and brain" could be used as the label for Cluster "the origin of life" in general. Cluster "spiral" and Cluster "artificial intelligence" can be merged as one cluster as both stress on "complex systems" and "complexity science".

Regarding the corresponding experts with those keyword clusters, expert grouping is acquired simultaneously. Moreover, such kind of correspondence provides some hints about the perspectives and even roles of those related scientists in XSSC. For example, Cluster "multi-scale" includes keywords such as spatial scale, temporal scale, micro scale, etc. and could somewhat reflect the knowledge specialty or research perspectives of both experts 白以龙 and 李静海 as shown in the left of Figure 3.2(d). This point is in accord with the reality. Similarly, Cluster "the origin of life" may indicate knowledge specialty of both expert 汪云九 and 郭爱克 who are active in the field of complexity of life and brain. Such an association may be very helpful together with measures of participation as addressed next.

(3). Measures of participants' contributions to the discussions

Table 3.3 lists the evaluation of 17 participants' involvement to the discussion of the topic of complexity based on agreement and discrepancy matrices.

It is shown that the participant 戴汝为 holds highest rank at both agreement and disagreement measures, which may be justified by his active role as one of chairpersons

Table 3.3. Measures of 17 experts' participation to complexity related workshops at XSSC

The eigenvector of maximum eigen value of agreement matrix:	(0.0952, 0.5795, 0.0979, 0.3114, 0.0363, 0.0712, 0.1447, 0.2049, 0.3348, 0.4596, 0.0964, 0.0412, 0.0446, 0.3449, 0.0896, 0.0408, 0.1196)
Rank of the top five participants:	戴汝为 > 于景元 > 张钹 > 宋健 > 成思危
Meaning of the indicator:	Expert with higher rank may hold more common concerns during the brainstorming session
The eigenvector of maximum eigenvalue of discrepancy matrix:	(0.3075, 0.3288, 0.2724, 0.2888, 0.1848, 0.1882, 0.1905, 0.2184, 0.2381, 0.2594, 0.1630, 0.2679, 0.2347, 0.2229, 0.2464, 0.2557, 0.1813)
Rank of the top five participants:	戴汝为 > 李静海 > 成思危 > 白以龙 > 汪云九
Meaning of the indicator:	Expert with higher rank may be of more diverse perspectives during the brainstorming session

or plenary speech contributors among those 7 workshops, which furthermore exposes his big influence in complex system field in China.

Above indicators of agreement or discrepancy are just one kind of measures. Higher agreement for one participant may indicate he share more foci than those of lower ranks. If some participants always follow or response other ideas, his agreement rank may also be higher. At that moment it is better to check original idea provider (P-creativity recorder) to exclude those followers. The topics and whole discussing process should be considered when undertaking practical analyses.

3.4 Augmented Information Support for XSSC

Users of GAE-BAR can acquire external information along a discussing process via any searching engines (such as Google, Baidu, etc.) based on keywords. To help participants' concentration on discussion, an augmented information support (AIS) tool specifically designed for XSSC is implemented based on Web content mining technologies. AIS-GAE includes four functional modules, web crawler, indexing, automatic summarization and user interface for searching, to implement general Web mining tasks. Web crawler collects the web documents given the seed websites and a defined exploring depth. Different kinds of useful information, such as workshop report, etc. are extracted from those web documents according to their structure characteristics and stored into a database with the index for each page. User interface is the entry for people to search the relevant information. Figure 3.3 shows that the 137 Web pages are found and listed below by order of relevance in searching keyword "complex". Each item includes the original URL, the sentence with the highest

Fig 3.3. Main page of user interface of AIS-GAE

relevance rank at the corresponding page referred by URL, the link to abstract of that page, the link to participants list and the link to related conferences.

The result lists really expose a rough scenario that how Chinese scientists approach complexity or complex system from different disciplines and how those scientists interact across different disciplines via the platform provided by XSSC. For example, the theme of workshop indicated by the first item in Figure 3.3 is about system and control, Item 2 is about complexity in the brain, Item 3 is from the perspective of complex system modeling and system engineering practice, and Item 4 is about a workshop on medical sciences and life, etc.

According to the specific structure of XSSC web page, a list of participants is extracted from each summary page of one workshop. Then a full list of all participants is acquired. Therefore besides an abstract of related introduction or overview of the workshop, a list of participants together with their affiliations is also provided by AIS-GAE. Figure 3.4 displays the summarization of the original web document listed as the 3rd item of all 137 results as shown in Figure 3.3. Figure 3.5 shows part participants of that workshop. Figure 3.6 provides a list of other XSSC workshops at the same cluster as the referred workshop. Clustering of XSSC Web texts is discussed at Chapter 9 in this book.

AIS-GAE could provide help for three kinds of people relevant to XSSC, i) the workshop organizers who can search past relevant workshops information and

参加讨论会的有钱学森、许国志、曾庆存、陈能宽、周干峙、张钹、汪成为、赵玉芬等10位院士和来自系统科学、数学、物理、生物、化学、计算机、软科学、军事、经济、气象、石油、化工、建筑、材料、认知科学、人工智能、社会科学、哲学等领域的近50名专家学者。1992年初，钱学森院士提出建立从定性到定量综合集成研讨厅体系，这就使得综合集成法有了一个可操作的具体系统。1992年底进一步提出"要把人的思维、思维的成果、人的知识、智慧以及各种情报、资料统统集成起来，可以叫大成智慧工程"。一、开放的复杂巨系统的一般理论及其方法论进展钱学森院士在他的书面发言中再次从科学方法论的高度论证了开放的复杂巨系统及其方法论的有效性，他说：关于开放的复杂巨系统，由于其开放性和复杂性，我们不能用还原论的办法来处理它，不能象经典统计物理以及由此派生的处理开放的简单巨系统的方法那样来处理，我们必须用依靠宏观观察，只求解决一定时期的发展变化的方法。他强调处理开放的复杂巨系统中的问题，需要用从定性到定量的综合集成方法论。戴汝为院士作了题为"大成智慧工程(metasynthetic engineering)"的评述报告，从一个更加宏大的范围、更加深刻的层次高度上论述了开放的复杂巨系统以及从定性到定量的综合集成方法论。对复杂系统的描述可以采用计算机建模的方法，也就是说，复杂系统的模型可以是程序表达的模型，而不局限于简单系统那样采用数学的方法进行建模。建模是综合集成方法的关键性环节，建立什么样的模型，以及参数如何调节都是以人为主，计算机为辅，是人机结合的产物，它的直接表现就是计算机程序。他在分析了人工智能的发展历程之后指出，现在人工智能的发展已经从传统ai转向非传统ai的研究。

Fig 3.4. Abstract of the original Web page with the searching keyword "complex" highlighted in red

Fig 3.5. Participant list of a XSSC workshop

Fig 3.6. Related workshops

acquire a whole vision toward XSSC; ii) the reviewers of a workshop application who can compare with past workshops regarding available similar themes and give their judgments about necessity and originality of applications; and iii) the invited participants of a workshop who may prepare their talks with more visions and engage actively during the workshop.

As time goes on, it is natural that the changes in position and even organization will happen to some participants. Such kind of additional information about participants will be pushed to the users as they try to search a participant using AIS-GAE. For example, we type a name such as 于景元 instead of a general keyword and start

Fig 3.7. Search a participant by AIS-GAE

Fig 3.8. Push information by AIS-GAE

searching. The searching result is as shown in Fig. 3.7. Here 6 relevant Web pages are found.

Moreover, as the inputted word is detected as a name based on participants list, AIS-GAE then pushes relevant information of that participant to the users. Due to popup blocks, push information may need to be released (see a button of release in Fig. 3.7). As shown in Fig. 3.8, the popup information includes the latest position and affiliation of the relevant participant together with all XSSC workshops he has attended (here, Professor 于景元 had participated 5 workshops).

Most participants of general XSSC workshops are nominated by workshop presidents and then invited by the XSSC organizers. In spite of a tenet of open-mindedness and cross disciplines, there still exists a tendency to recommend popular scientists with higher position, better reputation or from famous organizations among many XSSC workshops. Based on participant list, a human network can then be constructed by AIS-GAE. In the network, the vertex refers to a participant. If two

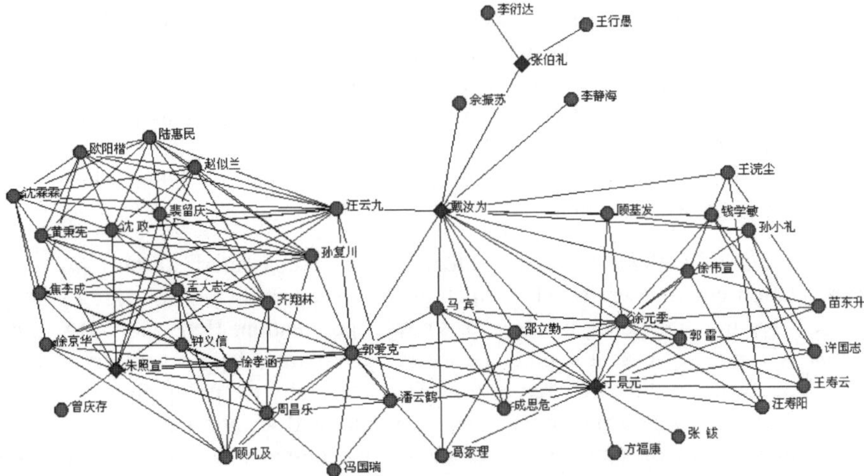

Fig 3.9. Human network composed by participants who simultaneously attend at least twice workshops (: cutpoints)

participants i and j attend a workshop simultaneously, then there is an edge between vertex i and vertex j. The strength of the edge is the frequency of co-occurrence of both participants along all concerned workshops. Fig. 3.9 shows a human activity network where the strength of the edge is at least greater than 2, which means both participants sharing one edge had been simultaneously attended those complexity workshops at least twice.

Here social network analysis is naturally applied to detect some features from that network. For example, 4 cutpoints are found in Fig. 3.9. Those corresponding scientists are 戴汝为 (middle), 于景元 (right), 张伯礼 (upper) and 朱照宣 (left), which reflects their important roles in the complexity related workshops, such as the former two scientists had been served as presidents for several times. In consideration of their academic background, the cutpoint scientists may also be regarded as gatekeeper of different school about complexity research in China. If combined with those measures of participations and participants group provided by GAE-BAR, more information about participants may be acquired.

3.5 Concluding Remarks

In this paper, we focus on mining information from a knowledge repository, Xiangshan Science Conference, a famous scientific forum on frontiers of science and technology in China. Unlike traditional simple statistics and qualitative evaluation toward conference, here adopted some augmented analytical methods, such as

- Visualized thinking space or structure of a group of scientists toward specified topics;

- Various clustering about the workshop discussion for effective summarization;
- Measuring the contributions of participants, which aims to provide more help to organizers in inviting or selecting participants;
- Augmented information support by Web content mining, especially push information about participants;
- Participants network analysis to detect influential scientists or find gatekeepers about some research streams.

Those versatile methods have been integrated into a computerized tool, group argumentation environment (GAE), which aims to support divergent group thinking process for the emergence of a *ba* for knowledge creation. Here, GAE-BAR and AIS-GAE may serve as a conference assistant system to help organizers and researchers aware of those ignored information, especially the productive process of series workshops where burgeoning new disciplines and scientists. Not withstanding complaints never fade about the low efficiency of group meetings in reality, they are still feasible and effective ways for communication and information sharing, opinion collection and expert knowledge acquisition. Therefore computerized support for group work, especially toward the awareness of group working context, are of continuous research.

With GAE, in-depth studies toward more facets about XSSC may help to expose more precious information from such a scientific think tank, and then serve as reasons to maintain the scale of those dialogues and facilitate policy making toward fundamental research in China.

As a matter of course, GAE may be of advantages in supporting small-scale group argumentation at current stage. With practice of GAE, especially on XSSC, there emerged new issues which are of in-depth research. For example, about clustering of participants' ideas, similar with constructing human activity network, a keyword network can also be constructed where the vertex refers to a keyword, and if both keyword i and keyword j occur simultaneously in one utterance, then an edge exists between two vertexes while the weight of the edge refers to the frequency of co-occurrence of both keywords along all the concerned workshops [6]. With the community (subgroup) of keywords detected from this network, more perspectives toward those concerned topics may be acquired together with clustering of keywords taken in GAE-BAR. Moreover, analysis of structure about those communities (keywords or participants) detected from either network can be studied, for example using relevant algorithms proposed by Newman and Girvan [7].

It is worth indicating that all analyses toward XSSC introduced here were undertaken after the finish of those workshops. It is our motivation to apply GAE to those on-going workshops which may be especially better to facilitate the construction of a context of show-and-tell for both participants and other interested people with things to "show" who want to "hear" tell, and for people who give tell to things shown. Augmented support may facilitate active interactions between participants by think-and-play and enable productive work for both scientists and organizers. Then more diverse and visualized association techniques about the discussion topics will be under further exploration or adopted and integrated into GAE. For example, ideas and technologies of chance discovery proposed by Japanese scientists [8] to detect possible networks to show "islands and bridges" (hidden chance) is worth some synthesizing with current GAE conferencing mining methods. Whatever's about next research, our original idea is to explore more structuring approaches toward unstructured problem solving.

Acknowledgments

The authors are grateful to faculty members of Xiangshan Science Conference Office for their encouragement to this study, which is supported by Natural Sciences Foundation of China under Grant No. 70571078 and 70221001.

References

1. Thagard, P.: How Scientists Explain Disease, Princeton University Press, Princeton, New Jersey (1999)
2. Nishisato, S.: Analysis of Categorical Datadual scaling and its applications. University of Toronto Press, Toronto, (1980) 1-53
3. Tang, X. J., Liu, Y. J.: Exploring computerized support for group argumentation for idea generation. In: Nakamori, Y. et al. (eds.): Proceedings of the 5th International Symposium on Knowledge and Systems Sciences, Japan (2004) 296-302
4. Tang, X. J., Liu, Y. J.: Computerized Support for Idea Generation during Knowledge Creating Process. In Cao, C. G., Sui, Y. F. (eds.): Knowledge Economy Meets Science and Technology (KEST'2004), Tsinghua University Press, Beijing, (2004) 81-88
5. Stefik, M., Smoliar, S.: The Creative Mind: Myths and Mechanisms: six reviews and a response, Artificial Intelligence, Vol.79, No. 1 (1995) 65-67
6. Tang, X. J., Liu, Y. J.: Computerized Support for Qualitative Meta-synthesis as Perspective Development for Complex Problem Solving. In Adam, F. et al. (eds.): Creativity and Innovation in Decision Making and Decision Support (proceedings of IFIP WG 8.3 International Conference on Creativity and Innovation in Decision Making and Decision Support), Vol.1, Decision Support Press, London, (2006) 432-448
7. Newman, M E J, Girvan, M.: Finding and Evaluating Community Structure in Networks, Physical Review E (2004) 69:026113.
8. Ohsawa, Y., Tsumoto, S. (eds) Chance Discoveries in Real World Decision Making, Springer-Verlag, Berlin (2006)

Multi-Data Mining for Understanding Leadership Behavior

Naohiro Matsumura and Yoshihiro Sasaki

[1] Graduate School of Economics, Osaka University, 1-7 Machikaneyana, Toyonaka, Osaka,
560-0043 Japan
`matumura@econ.osaka-u.ac.jp`
[2] School of Economics, Osaka University, 1-7 Machikaneyana, Toyonaka, Osaka, 560-0043
Japan
`cdg36390@par.odn.ne.jp`

In this chapter, we propose an approach for understanding leadership behavior in
dot-jp, a non-profit organizations in Japan, by analyzing heterogeneous multi-data
composed of questionnaire and e-mail archives. From questionnaire, we obtain qualitative relations between staff members. From e-mail archives, we obtain quantitative
relations between staff members. By integrating these different kind of relations, we
discovered that a leader should give supportive messages as well as directive messages to staff members to constract his/her reliable relation with staff members.

4.1 Introduction

After the non-profit organization (NPO) law was established in December 1998 in
Japan, specified non-profit activities have been promoted by the Japanese government, and as a result, more NPOs have been established than before. According to
a government web site, the number of authorized NPOs has increased as shown in
Figure 4.1. This increase reflects the growing needs of people to take part in social
activities. Staff members in NPOs engage in social activities. However, the motivation also causes managing NPOs to be difficult. Specifically, the difficulty comes
from the missions of NPOs, the voluntarism of staff members, and the scale of the
organizations:

- The primary mission of an NPO is public benefit, not profitability. Changing or
  correcting this mission is justified in profit-driven organizations if doing so would
  increase profits, but not in NPOs.
- The commitment of people to an NPO depends on their voluntarism, not on oblig-
  ation. Therefore, leaders cannot force staff members to engage in tasks against
  their will.
- As the number of staff members increases, sharing missions becomes more challenging and weakens the voluntarism of staff members.

N. Matsumura and Y. Sasaki: *Multi-Data Mining for Understanding Leadership Behavior*, Studies in Computational
Intelligence (SCI) **123**, 81–94 (2008)
`www.springerlink.com`

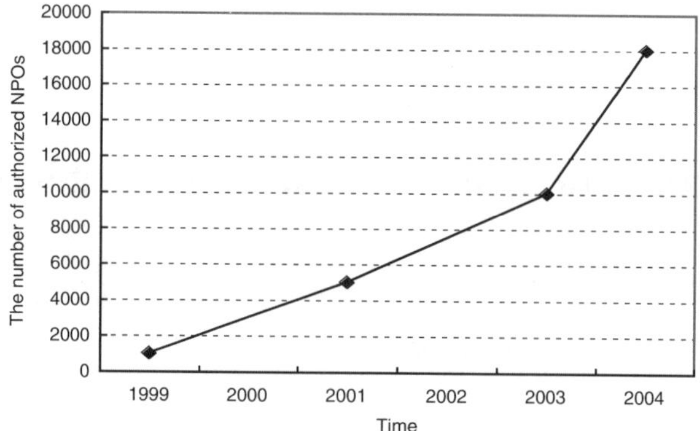

Fig 4.1. The number of NPOs since Nov. 1999 to Sep. 2004

Much research on NPOs has showed that reliable relationships between staff members are crucial properties for making the most of human and knowledge capital [2, 3, 11]. Leadership behaviors also play an important role in determining the atmosphere or culture of an organization [9] and the ability to create knowledge from the experiences of staff members [4]. In an ongoing project about NPOs [8], the aspects of capital, scale of operation, human resources, and partnerships with governments and organizations are being studied. However, the relationships between staff members are not being investigated, although they are quite relevant for sharing missions and motivating voluntarism.

In this chapter, we propose an approach for understanding leadership behavior in organizations by analyzing heterogeneous multi-data composed of questionnaire and e-mail archives. We first determined the results of the questionnaire for 97 staff members working in dot-jp[1], an NPO in Japan, to understand the degrees of satisfaction of staff members with leaders. Then, we extracted human influence networks from the archives of e-mail used at dot-jp to understand the relationships between leaders and other staff members. Finally, we integrated the results of the questionnaire with human influence networks and determined reliable leadership behaviors.

4.2 Overview of Dot-jp

We studied dot-jp, an NPO in Japan that organizes seminars and internship programs to give university students the opportunity to participate in political activities with diet members. Through the internship program, students learn how diet members engage in political activities.

The unit of active period for dot-jp is six months, and about half of the staff switches positions with newcomers when a new period starts. At the beginning of each period, staff members have to advertise the activities of dot-jp to university students about seminars. Also, staff members have to contact diet members, explain the

---

[1] http://www.dot-jp.or.jp/

activities of dot-jp, and ask diet members to take students for the internship program. Through the seminars, staff members find internship programs for each student who is interested in the political activities of a particular diet member. Then, the internship programs start, and students experience political activities as assistants to the diet members.

The headquarters of dot-jp is in Osaka, and seven branch offices are distributed all over Japan (Branches A, B, C, D, E, F, and G[2]). Each branch office has nine to twenty one staff members, and three of the staff members are appointed to be different managers.

- **Area manager:** The person responsible for managing all staff members (The area manager is the leader in a branch office.)
- **Seminar manager:** The person responsible for bringing students to seminars
- **Internship Manager:** The person responsible for contacting diet members for internship programs

Most staff members in dot-jp are university students, and having face-to-face meetings with each other is difficult because of their distant residences from each other and classes. For these reasons, staff members mainly use e-mail to exchange information, plan and arrange events, and discuss matters. Communication via e-mail creates a virtual office and complements real office communication.

Yoshihiro Sasaki, the second author of this chapter was an area manager for and currently is a marketing director in dot-jp. In addition, dot-jp cooperated fully with our project. Therefore, we did in-depth research by distributing our questionnaire, having face-to-face interviews, and extracting human influence networks from the archives of e-mail used in dot-jp.

4.3 Questionnaire

Understanding the degree of satisfaction of staff members in an organization can be a key to discovering specific problems related to faults in human resource management [1]. We sent questionnaires to all staff members (104) working for dot-jp to understand their degree of satisfaction, specifically about their managers and their branch offices. Then, by comparing the results of questionnaires received from 97 staff members (correction rate: 93%) with the achievement rate of activity, explained later, we determined factors that lead to successful activities at dot-jp.

4.3.1 Questions

To investigate the degree of satisfaction of staff members with their branch offices, and area, seminar, and internship managers, we sent questionnaires to 104 staff members working in the seven branch offices in March 2005, the last month of the 14th period. The questions in the questionnaire were as follows:

---

[2] We used fake names to maintain secrecy.

- **Q1.** Please rate your degree of satisfaction with your branch office.
- **Q2.** Please rate your degree of satisfaction with the area manager in your branch office.
- **Q3.** Please list up to three substantive leaders in your branch office.

Answers were selected from five alternatives (1: Very satisfied, 2: Satisfied, 3: Neutral, 4: Dissatisfied 5: Very dissatisfied).

4.3.2 Achievement Rate

The activity for each branch office was numerically evaluated by comparing the total number of students coming to seminars and diet members accepting student interns with the desired number for both groups set at an early stage of each period. The achievement rate was calculated using the following equation.

$$Achievement\ rate(\%) = \frac{x}{y} \times 100 \tag{4.1}$$

$where$

$x : actual\ number\ of\ students\ and\ diet\ members$

$y : desired\ number\ of\ students\ and\ diet\ members$

4.3.3 Assumption for Leadership Behavior

The results of the questionnaire (Q1, Q2, and Q3) and the achievement rate for each branch office are shown in Table 4.1. To understand the relationships between the degrees of satisfaction and the achievement rates, we classified the branch offices into three groups based on their achievement rates.

Table 4.1. Achievement rate and averaged degree of satisfaction for each branch office and area manager

Branch	Achievement rate (%)	Averaged degree of satisfaction†	
		Branch	Area manager
A	135	4.78	5.00
B	109	4.37	4.37
C	106	2.69	3.00
D	97	3.58	4.50
E	99	2.69	2.75
F	79	2.87	3.69
G	69	2.82	4.50

† Answers were selected from five alternatives
(1: Very dissatisfied, 2: Dissatisfied, 3: Neutral, 4: Satisfied
5: Very satisfied)

- **Group-1:** The high-achievement-rate group, including Branches A, B, and C. Staff members in Branches A and B had high degrees of satisfaction with their branch offices and area managers. Staff members in Branch C had a high degree of satisfaction with their branch office but a low degree of satisfaction with their area manager.
- **Group-2:** The middle-achievement-rate group, including Branches D and E. The degree of satisfaction with the area manager in Branch D was quite high compared to that in Branch E.
- **Group-3:** The low-achievement-rate group, including Branches F and G. The degree of satisfaction with the area manager in Branch G was much higher than that in Branch F.

We did not consider that these groups reflect the management ability of area managers because the achievement rates affected the degrees of satisfaction with area managers (i.e., high achievement rate positively affects the degree of satisfaction). However, the degrees of satisfaction within the same group reflected the management ability of area managers since they excluded effects related to the achievement rate. We made an assumption regarding the management ability of area managers by comparing branch offices in the same groups with different degrees of satisfaction with area managers.

[**Assumption**] The high degrees of satisfaction with area managers come from a high level of management ability.

We interviewed 10 staff members individually, face-to-face, and made sure that the assumption was empirically getting to the point, (i.e., area managers of Branches C, E, and F were not trusted regarding their management ability, whereas area managers of other branches had the esteem of their staff members). In the following Sections, we further investigated leadership behavior by comparing human influence networks obtained from the archives of e-mail with this assumption.

4.4 IDM

The IDM was originally an algorithm for measuring values of influence of messages, senders, and terms from threaded messages [5]. Recently, the algorithm was revised and expanded to measure values of influence to extract human influence networks [6, 7].

In the IDM, the influence between a pair of senders is defined as the number of terms propagating human-to-human via messages. Here, let a message chain be a series of threaded messages, and the influence of a message x on a message y (x precedes y) in the same message chain be $i_{x \to y}$. Then, $i_{x \to y}$ is defined as

$$i_{x \to y} = |w_x \cap \cdots \cap w_y|, \tag{4.2}$$

where w_x and w_y are the set of terms in x and y, respectively, and $|w_x \cap \cdots \cap w_y|$ is the number of terms propagating from x to y via other messages. If x and y are not in the same message chain, $i_{x \to y}$ is defined as 0 because the terms in x and y are used in a different context and no relationships exist between them[3].

Based on the influence between messages, the influence of a sender p on a sender q can be measured as the total influence of p's messages on the messages of others through q's messages replying to p's messages. Let the set of p's messages be α, the set of q's messages replying to any of α be β, and the message chains starting from a message z be ξ_z. The influence from p to q, $j_{p \to q}$, is then defined as

$$j_{p \to q} = \sum_{x \in \alpha} \sum_{z \in \beta} \sum_{y \in \xi_z} i_{x \to y}. \tag{4.3}$$

The influence of p on q is regarded as q's contribution toward the spread of p's messages.

The influence of each sender is also measurable using $j_{p \to q}$. Let the influence of p on others be $I_p^{<out>}$, the influence of others on p be $I_p^{<in>}$, and the set of all staff members except p be γ. Then, $I_p^{<out>}$ and $I_p^{<in>}$ are defined as

$$I_p^{<out>} = \sum_{q \in \gamma} j_{p \to q} \tag{4.4}$$

$$I_p^{<in>} = \sum_{q \in \gamma} j_{q \to p}. \tag{4.5}$$

As an example of measuring influence, we used a simple message chain, as shown in Figure 4.2, where Message2 and Message4 are posted as replies to Message1 and Message3 is posted as a reply to Message2.

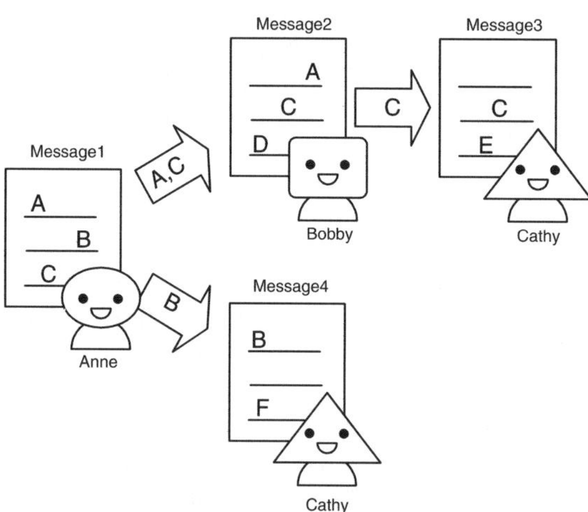

Fig 4.2. Message chain of four messages sent by three individuals

[3] We recognize that the terms in different message chains sometimes have some relationships with each other. To deal with this, IDM software has an option that sets temporal relationships to the messages with some similarities regardless of message chains, but we did not use the option because of the accuracy of temporal relationships.

Here, the influence between a pair of senders is measured as follows.

- The influence of *Anne* on *Bobby* is 3 (i.e., $j_{Anne \to Bobby} = 3$), because two terms (A and C) were propagated from *Anne* to *Bobby*, and one term (C) was propagated from *Anne* to *Cathy* via *Bobby*.
- The influence of *Anne* on *Cathy* is 1 (i.e., $j_{Anne \to Cathy} = 1$), because one term (B) was propagated from *Anne* to *Cathy*.
- The influence of *Bobby* on *Cathy* is 1 (i.e., $j_{Bobby \to Cathy} = 1$), because one term (C) was propagated from *Bobby* to *Cathy*.
- The influence of *Bobby* on *Anne* and of *Cathy* on *Anne* is 0 (i.e., $j_{Bobby \to Anne} = 0$ and $j_{Cathy \to Anne} = 0$), because no terms were propagated to *Anne* from either *Bobby* or *Cathy*.

From this influence, the value of influence ($I_p^{<out>}$) and influenced value ($I_p^{<in>}$) of each staff member was calculated as follows.

- The influence of *Anne* on *Bobby* is 3 (i.e., $j_{Anne \to Bobby} = 3$), because two terms (A and C) were propagating from *Anne* to *Bobby*, and one term (C) was propagating from *Anne* to *Cathy* via *Bobby*.
- The influence of *Anne* on *Cathy* is 1 (i.e., $j_{Anne \to Cathy} = 1$), because one term (B) was propagating from *Anne* to *Cathy*.
- The influence of *Bobby* on *Cathy* is 1 (i.e., $j_{Bobby \to Cathy} = 1$), because one term (C) was propagating from *Bobby* to *Cathy*.
- The influence of *Bobby* on *Anne* and of *Cathy* on *Anne* is 0 (i.e., $j_{Bobby \to Anne} = 0$ and $j_{Cathy \to Anne} = 0$), because no term was propagating to *Anne* from either *Bobby* or *Cathy*.

The influence of *Anne* on *Cathy* is ignored even though one term (C) was propagated from *Anne* to *Cathy* via *Bobby* because we wanted to measure the direct influence between individuals. Instead, the indirect influence of *Anne* on *Cathy* via *Bobby* is considered as a contribution of *Bobby*, and we added it to the influence of *Anne* on *Bobby*.

By mapping the propagating terms between senders, a human influence network can be obtained, as shown in Figure 4.3, where the number of outgoing terms corresponds to the values of influence ($I_p^{<out>}$) and the number of incoming terms corresponds to the influenced values ($I_p^{<in>}$). From the figure, the relationships of influence between them can be visually understood.

4.5 Human Influence Network

4.5.1 E-mail Archives

Activities, such as seminars and events, are held in the field, but many important things, such as information sharing, decision-making, event planning, and consensus building, are carried out over e-mail. Area managers are in charge of the e-mail, as

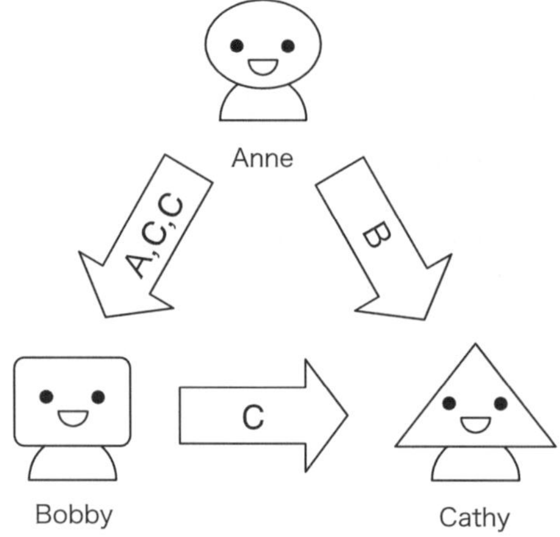

Fig 4.3. Human influence network extracted from message chain in Figure 4.2

Table 4.2. Number of staff members and e-mails exchanged during the 14th period (October 2004 – March 2005)

	Branch A	Branch B	Branch C	Branch D	Branch E	Branch F	Branch G
# of staffs	21	9	16	14	16	16	12
# of e-mails	2297	1198	2465	2076	3258	1309	1717

opposed to seminar and intern managers operating in the field. Therefore, leadership behaviors of area manager are reflected in the human influence networks extracted from the archives of e-mail.

An overview of the e-mail archives in the seven branch offices of the 14th period (from October 2004 to March 2005) that we analyzed is shown in Table 4.2. Thousands of e-mails were exchanged during the period.

The e-mails were written in Japanese. Before applying the IDM to the archives, we first used JUMAN[4], a morphological analysis system, to obtain terms of nouns, verbs, and adjectives, and then removed noise words (or "stop words" [10]) to measure the influence and influenced values accurately. A list of noise words was made in advance from the results of pre-testing before the analysis.

4.5.2 Verification of Assumption

We applied the IDM to the archives of e-mail in each branch office and obtained $I_p^{<out>}$ and $I_p^{<in>}$ as influence and influenced values of each staff member. The top six staff members by $I_p^{<out>}$ for each branch office are shown with $I_p^{<in>}$ in Table 4.3, where the same names across branch offices represent distinct staff

---

[4] http://www.kc.t.u-tokyo.ac.jp/nl-resource/juman.html (In Japanese)

Table 4.3. For each branch office, the top six staffs by $I_p^{<out>}$ are shown with $I_p^{<in>}$. Note that the same staff names in different branches represent distinct staffs, and the staff names with "*"(asterisk mark) represent area managers

Branch	Staff	$I_p^{<out>}$	$I_p^{<in>}$	Branch	Staff	$I_p^{<out>}$	$I_p^{<in>}$
A	p2*	541	827	E	p4	1022	8
	p39	518	0		p3	1019	1032
	p18	506	177		p11	754	563
	p4	470	244		p7	737	664
	p41	454	910		p1*	686	849
	p5	417	235		p2	560	491
B	p1*	841	385	F	p11	578	524
	p9	772	1051		p20	506	871
	p5	542	317		p18	432	52
	p2	484	666		p1*	404	263
	p11	304	198		p16	355	98
	p17	251	537		p28	325	0
C	p6	1113	403	G	p3	779	1270
	p1*	836	142		p4*	559	623
	p8	515	1257		p6	537	36
	p18	490	95		p8	521	466
	p2	470	163		p1	513	480
	p17	405	261		p28	445	0
D	p1*	1495	717				
	p4	923	68				
	p13	893	1880				
	p3	794	927				
	p5	790	386				
	p14	655	205				

members[5]. The staff names with "*"(asterisks) represent area managers. From Table 4.3, the following tendencies can be seen:

- Three out of seven area manages have the highest $I_p^{<out>}$ value in their branch offices.
- Those who have high $I_p^{<out>}$ values do not always have high $I_p^{<in>}$ values, and vice versa.

To investigate further based on the assumption in Section 4.3, we compared $I_p^{<out>}$ and $I_p^{<in>}$ with the results of Q3 of the questionnaire where staff members were asked to pick up to three staff members to be substantive leaders. The number of votes represents how well the management ability of a staff member was evaluated regardless of their position. The rankings of area managers by votes, $I_p^{<out>}$, and

---

[5] For example, p1 can be seen in every branch office, but it represents distinct staff members.

Table 4.4. Ranking of area managers for each branch office

Branch	Vote	$I_p^{<out>}$	$I_p^{<in>}$
A	1st	1st	3rd
B	2nd	1st	4th
C	N/A	2nd	14th
D	2nd	1st	6th
E	4th	5th	4th
F	N/A	4th	8th
G	1st	2nd	3rd

$I_p^{<in>}$, shown in Table 4.4, support the assumption, that is, area managers with high $I_p^{<out>}$ and $I_p^{<in>}$ values tend to have a number of votes in Branches A, B, D, and G.

By contrast, area managers in Branches C and F did not get any votes. In these cases, we identified the medium rankings of $I_p^{<out>}$ (second in Branch C and fourth in Branch F) and the low rankings of $I_p^{<in>}$ (fourteenth in Branch C and eighth in Branch F). Considering the results with the assumption, the lack of management abilities of area managers in Branches C and F arises from their low level of $I_p^{<in>}$ (i.e., they rarely replied to e-mails sent by other staff members.).

The $I_p^{<out>}$ value of the area manager in Branch E ranked much lower than the one in Branch D, whereas the rankings by $I_p^{<in>}$ were almost uniform. Thinking of the assumption, the low degree of satisfaction with the area manager in Branch E comes from the low level of $I_p^{<out>}$ (i.e., he rarely sent useful e-mails to other staff members.).

In Branch G, the area manager ranked first by votes and ranked high by $I_p^{<out>}$ and $I_p^{<in>}$ as well, although the achievement rate was the worst (only 69%). From this result and the assumption, the degree of satisfaction with the area manager in Branch G comes from the high level of $I_p^{<out>}$ and $I_p^{<in>}$ (i.e., he communicated with other staff members by e-mail.).

In summary, we can conclude that a reliable leader should give supportive messages as well as directive messages to staff members to constract his/her reliable relation with staff members.

4.5.3 Interpretation of Human Influence Networks

The graphical outputs of human influence networks helped us understand the behavior of area managers from a structural point of view. The graphical outputs of the IDM obtained from the archives of e-mails of Branches A, C, F, and G are shown in Figs. 4.4, 4.5, 4.6, and 4.7, respectively. Here, the figures are composed of the top six most influential staff members and influential links between them with values of influence of more than 10. In the figures, area manages are depicted as gray nodes, and the relationships between staff members are shown as directed links with values of influence. From the figures, how area manages behave with other staff members

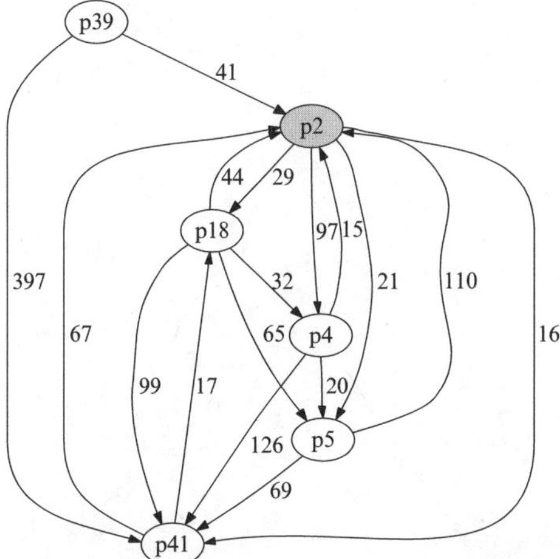

Fig 4.4. Human influence network in Branch A

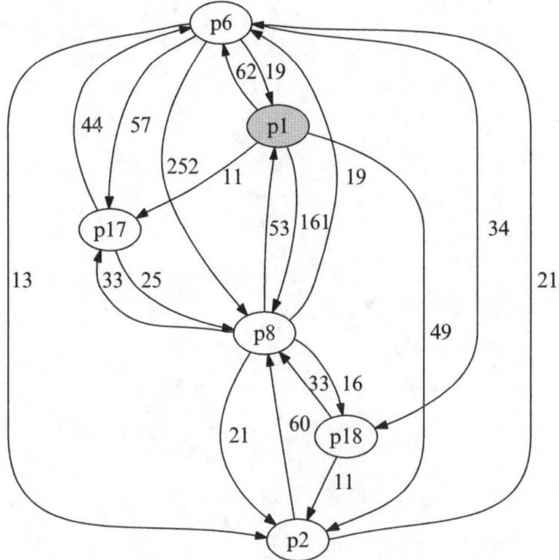

Fig 4.5. Human influence network in Branch C

can be seen. We summarized the leadership behaviors based on human influence networks and the analysis of the questionnaire.

- **Interactive behavior:** In Branch A, B, and D (Figure 4.4 for example), the behaviors of the area manages are "interactive" enough to give and receive influence from other staff members. The questionnaires showed that these area managers are highly trusted.
- **Partially interactive behavior:** In Branch E and G (Figure 4.7 for example), the behaviors of the area managers are "partially interactive", where the area managers receive influence from five staff members while influencing three staff

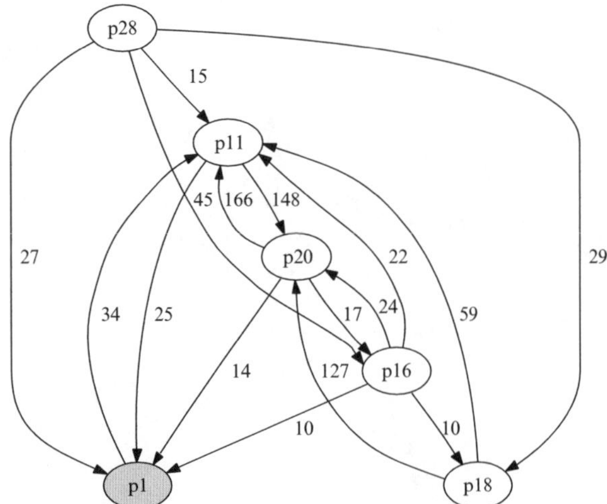

Fig 4.6. Human influence network in Branch F

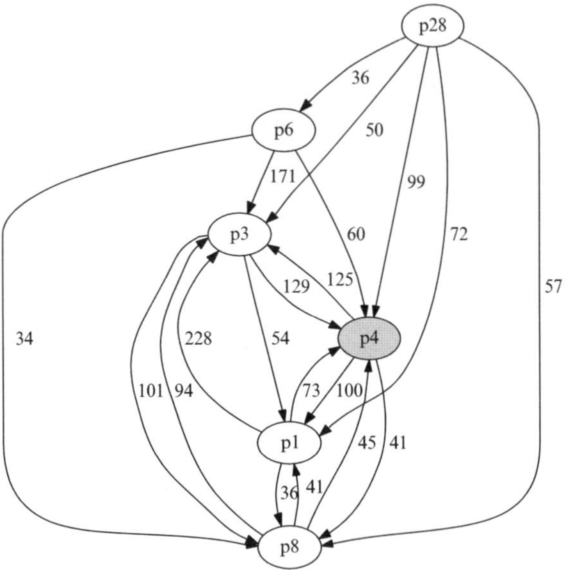

Fig 4.7. Human influence network in Branch G

members. From the results of the questionnaire, the reason that the area manager in Branch G succeeded in managing the Branch came from the high influential and influenced values. By contrast, the area manager in Branch E failed because of the low influential and influenced values as compared to other managers.

- **Preferential behavior:** In Branch C (Figure 4.5), the behavior of the area manager is "preferential", more strict than partially interactive behavior, where the

Table 4.5. Four types for leadership behavior

	High $I_p^{<out>}$	Middle $I_p^{<out>}$	Low $I_p^{<out>}$
High $I_p^{<in>}$	Interactive	Partly interactive	Passive
Middle $I_p^{<in>}$	—	—	—
Low $I_p^{<in>}$	Preferential	—	—

area manager influences four staff members and is influenced by only two staff members. The lack of interaction, specifically from his low values of influence, resulted in the fault of managing the Branch.

- **Passive behavior:** In Branch F (Figure 4.6), the behavior of the area manager was "passive", where the area manager was influenced by four staff members and influenced only one staff member. The apparent lack of communication caused staff members to distrust the area manager in the Branch.

From these results, we concluded that interactive behavior is ideal for leadership. Also, we can classify four types of leadership behavior as in Table 4.5. This classification provides the possibility of other behavioral types, such as $high\ I_p^{<out>}$ $and\ low\ I_p^{<in>}$, but further research is needed.

4.6 Conclusion

The activities of NPOs are based on the voluntarism of staff members, however staff members often lose this motivation and eventually leave organizations as the number of staff members increases. Part of the reason comes from staff members being unsatisfied with the lack of management ability in leaders. The suggestion for reliable leadership behavior obtained in this study would contribute to establishing the way to manage NPOs.

For future work, we will apply the proposed approach to other popular communication tools, such as weblogs or BBSs, to accumulate further knowledge on leadership behavior. In addition, we are planning to establish guidelines regarding leadership behaviors relating to communication to put the results of our research to use.

References

1. D. Cohen and L. Prusak. *In Good Company How Social Capital Makes Organizations Work*. Harvard Business School Press, 2001.
2. P. F. Drucker. *Post-Capitalist Society*. Harper Business, 1993.
3. D. Krackhardt and J. R. Hanson. Informal networks: The company behind the chart. *Harvard Business Review*, pages 104–111, July-August 1993.

4. G. V. Krogh, K. Ichijo, and I. Nonaka. *Enabling Knowledge Creation: How to unlock the Mystery of Tacit Knowledge and Release the Power of Innovation*. Oxford University Press, 2002.

5. N. Matsumura. *Chance Discovery*, chapter Topic Diffusion in a Community, pages 84–97. Springer Verlag, 2003.

6. N. Matsumura. Collaborative communication strategies in online community. In *The Fourth International Workshop on Social Intelligence Design (SID2005)*, 2005.

7. N. Matsumura, X. Llorà, and D. E. Goldberg. Mining directed social network from message board. In *The Fourteenth International World Wide Web Conference (WWW2005)*, pages 1092–1093, 2005.

8. OSSIP. Osipp center for nonprofit research & information:npo 2004 white paper – the japan nonprofit almanac –. Technical report, OSSIP, Osaka University, 2004. http://www.osipp.osaka-u.ac.jp/npocenter/NPO2004.pdf.

9. D. Perkins and D. Wilson. Bridging the idea-action gap. *Knowledge Directions*, 1(2), 1999.

10. G. Salton and M. McGill. *Introduction to modern information retrieval*. McGraw-Hill, New York, 1983.

11. E. Wenger. *Communities of Practice: Learning, Meaning, and Identity*. Cambridge University Press, 1999.

5

Discussion Visualization on a Bulletin Board System

Wataru Sunayama

Hiroshima City University, 3-4-1, Ozuka-Higashi, Asa-Minami-Ku, Hiroshima, 731-3194, Japan

Summary. It is important for a collaborative community to decide its next action. The leader of a collaborative community must choose an action that increases rewards and reduces risks. When a leader cannot make this decision, action will be determined through community member discussion. However, this decision cannot be made in blind discussions, so systematic discussion is necessary to choose effective action in a limited time. In this paper, we propose a bulletin board system framework in which the effective discussion is established through the visualized discussion logs.

Keywords: bulletin board system, discussion visualization, discussion divergence, discussion convergence, communication support

5.1 Introduction

It is important for a collaborative community to decide its next action. The leader of a collaborative community must choose an action that increases rewards and reduces risks. When a leader cannot make this decision, action will be determined through community member discussion. However, this decision cannot be made in blind discussions, so systematic discussion is necessary to choose effective action in a limited time. In addition, the repetition of divergence and convergence in the discussion according to members' common knowledge and background will lead to that effective conclusions.

We propose a bulletin board system (BBS) framework that supports discussion divergence and convergence for deciding a series of actions to be undertaken. This BBS framework provides an environment in which the remote community members can participate in systematic discussions. In this framework, as the scenario must be stream-like, a sub-story model that supposes consecutive appearances of the same words is incorporated.

Rather than interpreting previous discussions, this study aims to smoothly control current and necessary discussions. That is, the system grasps each state of the discussion, indicates the type of opinion, and supplies environments for constructive discussions.

W. Sunayama: *Discussion Visualization on a Bulletin Board System*, Studies in Computational Intelligence (SCI) **123**, 95–109 (2008)
www.springerlink.com

In this paper, backgrounds are described in section 5.2 and a BBS framework in section 5.4. Evaluation criteria for each comment in BBS are determined in section 5.5. Experimental results are shown in section 5.6, and section 5.8 concludes this paper.

5.2 Background

Many relational works have analyzed the co-occurrence of words in each user comment [2] or to visualize the structure of mutual citations [3]. In this study, although co-occurrence is also incorporated, a suggestion system measures discussion streams not only by two comments but also by more than three comments in which consecutive words appear.

Since there is a lexical chain [4] in natural language processing, the system is regarded as the application of a lexical chain without a thesaurus. A topic extraction method [5] extracts relationships in a story with a Text-Tiling method. However, such a system cannot be applied to bulletin board systems that contain various comments by multiple members. In this study, the framework assumes multiple sub-topics, loosely related in comment streams.

To measure the relationship between segments in a document, a Vector Space Model [6] is generally used, where the angle between two vectors is the relationship. However, this is insufficient for estimating a document stream because the angle is defined in only two segments. In addition, although there is a sentence ordering system for summarization [7], our study examines continual document streams.

There is a study [3] that analyzes the discussion structure for finding and comprehending topics. In this study, topics are given, and BBS supplies suggestions to control discussion progress.

5.3 Sub-Story Model

A document contains a main topic, the author's primary subject, and sub-topics related to the main topic. We label the logical stream of these topics main story and sub-story, respectively. A sub-story model is then defined as the words that appear in consecutive segments as the topic of the sub-stories to create a document stream (Fig. 5.1).

Terms related to sub-story models are defined as follows:

- main topic: a topic related to a text mainly described in many segments
- sub-topic: a topic related to a text partly described in more than two segments
- main story: a logical relationship related to a main topic
- sub-story: a logical relationship related to a sub-topic
- main keyword: a word representing a main topic
- sub-keyword: a word representing a sub-topic
- main topic sentence: a sentence strongly related to a main topic
- sub-topic sentence: a sentence strongly related to a sub-topic

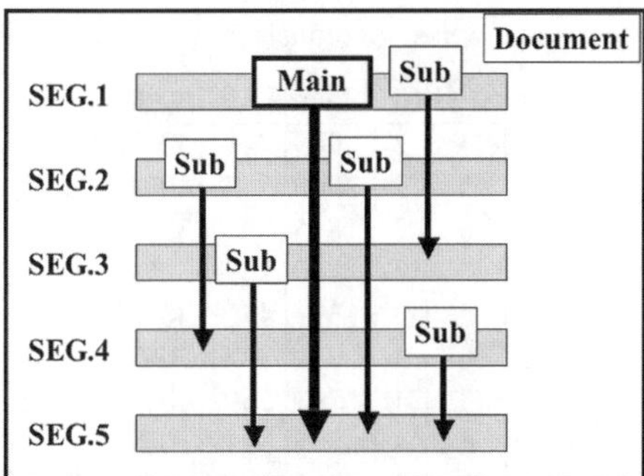

Fig 5.1. Sub-story model

"Main" and "Sub" in Fig. 5.1 denote the main topic with main keywords and sub-topic with sub-keywords, respectively, and arrows indicate the story.

A unit for measuring a document stream is called a "segment." Segments are chapters, sections, or paragraphs that enable the system to observe the consecutive appearance of topic words.

5.3.1 Preliminary Experiments for Sub-Story Model

To verify the sub-story model assumption described in section 5.3, preliminary experiments were executed to investigate how often words appear in texts in consecutive segments.

The texts for these experiments were the 100 papers for the eighteenth annual conference of the Japanese Society of Artificial Intelligence. Each paper consists of four pages with about six segments. The papers were in Japanese, and nouns were extracted in advance by a Japanese morpheme analysys system [8].

Table 5.1 shows the tendency of nouns appearing in consecutive segments. These data are for 90 papers whose segment number (SEG) is from four to six, and the probability values denote the rate of words appearing in consecutive segment frequency (SF) segments over all words that appear in SF segments.

As a result, all real rates (Prob.) are larger than random values (Rnd.)[1] by more than 15%, so a word tends to appear in consecutive segments.

As second preliminary experimental results, Table 5.2 shows the appearing segment range (SR) for all parts of speech: nouns (Nn), verbs (Vb), adjectives (Adj), conjunctions (Con), adverbs only for nouns (Adn), adverbs (Adv), particle (Pc) auxiliary verbs (AV), and Random probabilities (Rn) in 45 texts consist of five segments. The rate of each part of speech for all words are shown in Table 5.3.

As a result, all parts of speech exceed the random values, so all words tend to appear in each text consecutively.

---

[1] In a four segments text, when a word's SF is two, 1-2, 2-3, 3-4 are consecutive appearing patterns, and 1-3, 2-4, 1-4 are not consecutive. If that word appear in random segments, 50% $(= 3/(3 + 3))$ becomes the random value.

Table 5.1. Probability of a noun, whose SF equals N, appearing in N consecutive segments.

SEG	SF	Noun	Prob.	Rnd.
4	2	989	74%	50%
5	2	2940	58%	40%
6	2	1815	49%	33%
4	3	440	73%	50%
5	3	1391	47%	30%
6	3	919	38%	20%
5	4	791	55%	40%
6	4	555	39%	20%
6	5	335	49%	33%

Table 5.2. Cumlative rates of appearing segment range (SR) of a word whose SF equals 2 by a part of speech (%)

SR	Nn	Vb	Adj	Co	Adn	Adv	Pc	AV	Rn
2	58	58	52	50	49	49	52	49	40
3	85	86	84	81	82	84	81	84	70
4	95	96	94	99	97	96	97	100	90
5	100	100	100	100	100	100	100	100	100
word	2940	546	104	128	39	108	286	49	

Table 5.3. Rate of part of speech for all words

part of speech	rate
nouns	69%
verbs	15%
particles	7%
others	9%

According to these preliminary experiments results, similar linguistic expressions are used continuously, and most words in a document tend to appear in a shorter range of segments independent of part of speech. In other words, the words appeared in consecutive segments as often possible.

Especially, nouns and verbs are likely to appear continuously because they are used as subject and predicate in each sentence. On the contrary, the other repetition of parts of speech may be caused unconsciously because of writing habits.

5.3.2 Discussion Control Using a Sub-Story Model

In this study, we apply this sub-story model by regarding a thread in a BBS as a document and a segment as a comment written by a community member. The word for the discussion topic becomes the main topic and main keywords, and the relational significant words of the main topic become sub-topics and sub-keywords. The system supplies implications of each comment status by regarding the consecutive appearing words as the appearance of the document stream.

5.4 Discussion Support Bulletin Board System

In this section, we describe a BBS system framework that supports discussion control by the sub-story model described in the previous section. According to these preliminary experiments results, similar linguistic expressions are used continuously, and most words in a document tend to appear in a shorter range of segments independent of part of speech. In other words, the words appeared in consecutive segments as often possible.

Especially, nouns and verbs are likely to appear continuously because they are used as subject and predicate in each sentence. On the contrary, the other repetition of parts of speech may be caused unconsciously because of writing habits.

In this study, we apply this by regarding a thread in a BBS as a document and a segment as a comment written by a community member. The system supplies implications of each comment status by regarding the consecutive appearing words as the appearance of the document stream.

The person who controls the discussion is not a system but a leader and a member of the community. The system clarifies the placement of each comment in the flow of the topical discussion and implies byway or unrelated comments and urges comments including necessary elements. Terms and environments for the BBS framework are defined as follows:

- BBS: a discussion environment for community members using the Internet; a set of threads.
- Thread: a blackboard system in which members can discuss the topic by writing comments; a set of comments.
- Comment: a member's written opinion.

5.4.1 Discussion Streams

The discussion procedure for problem solving from topic suggestion to a conclusion that has a solution is as follows:

[Discussion Process]

1. Create a thread of the main/sub-topic.
2. Divergence Phase: Discussion to diverge opinions and ideas.
   If a sub-topic is suggested,
   a) For the sub-topic, follow procedures 1 to 5.
   b) Embed its conclusion.
3. Change from divergence phase to convergence phase.
4. Convergence Phase: Discussion to converge the collected opinions.
5. Create conclusions for the main/sub-topic.

This procedure is similar to the relationship among main function and the sub-functions in computer programming, as shown in Fig. 5.2. The tree structure expresses problem occurrence and divisions. A large problem is divided into small problems recursively to solve more concrete problems and to accumulate the solutions.

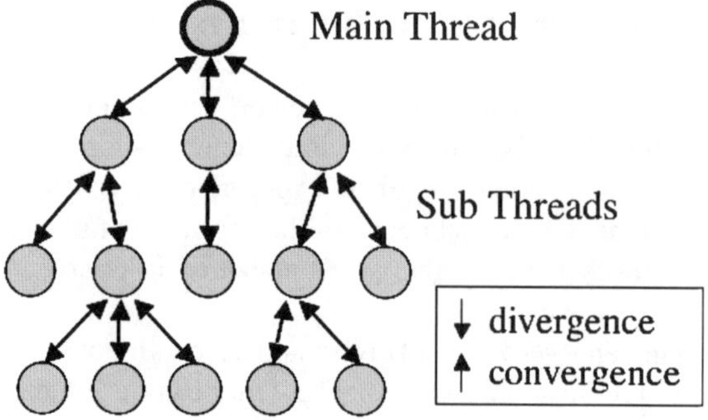

Fig 5.2. Divergence and convergence of threads in BBS

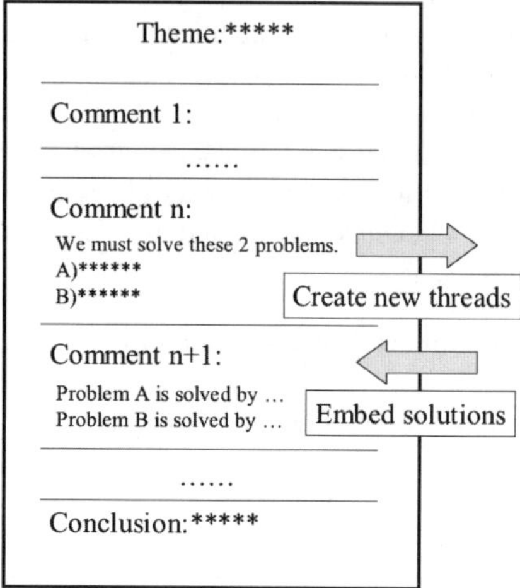

Fig 5.3. Example description of a thread

The system also assumes that problem division in a thread is executed, as Fig. 5.3 When a member find that a problem is divided into more concrete problems, he/she creates a new sub-thread for each. Solutions are embedded after the problems are solved.

In the rest of this section, the procedure of this process is shortly described.

Creation of a Thread

A community leader establishes a main topic and creates a new thread for discussion. A community member creates a new thread by creating a sub-topic in the divergence phase. The following are the types of threads:

- Main: thread about main topic created by the leader of a community.
- Sub: thread about a sub-topic created by a member of the community.

- Byway: thread about a secondary additional topic that does not affect the solution of the main topic.

By creating a thread for each divided distinct topic, members can concentrate on discuss a topic and the system can grasp states of the thread.

Divergence Phase

Discussion is divided mainly into two phases: divergence and convergence. In the divergence phase, members hold discussions to collect opinions. The rate of divergence is simply measured by new words. However, if the topic spreads blindly, discussion will not converge. Therefore, the system supports discussion control in the range of the topic and supports the various opinions of members.

Creation of a Sub-Thread

In the divergence phase, if it is clear that conditions for solving the topic of the current thread exist, and if the members acknowledge that they are valid, a new sub-thread is created by the member who set those conditions.

Transition from Divergence to Convergence Phase

After most opinions have been added to the thread, the discussion phase switches from divergence to convergence.

That is, if there was a discussion that exceeded the threshold, calculated by the number of relational keywords of the topic, the number of sub-threads, and the discussion time from when the thread was created, a transition is suggested to the member who created the thread. If the member agrees, the transition will be executed.

Convergence Phase

In the convergence phase, a new discussion is not needed, but discussion continues to converge opinions and already obtained conclusions. Therefore, the rate of new words is minimized by the suggestions of the system. In addition, to converge the sub-thread conclusions, new comments using multiple topics of sub-threads are recommended; sub-topics supplied by order of occurrence and comments, including adjacent topics, are recommended.

Create Topic Conclusions

When the comments, including the oldest sub-topic, and all topics have appeared, the system urges the member who created the thread to create conclusions. After conclusions are declared, the thread is closed.

5.4.2 Conditions to Comments

Members write their opinions in the threads as comments. However, since it is hard to control discussions written in a free format, conditions are attached to the comments for discussion analysis based on the sub-story model. The system recommends that users write new comments including consecutive words, while the system examines each comment to ensure that it conforms to the topic and the discussion stream. The following conditions are highly recommended for each comment:

- Use the topic word in the thread.
- Use words already used in the same thread.
- Avoid pronouns.
- Use new relational words of the topic in the divergence phase.
- Use multiple words for sub-topics in the convergence phase.

These suggestions help users write comments for systematic discussions, such as retention of logical relationships, divergence and convergence. To write a comment along these conditions, concrete numerical criteria and comments evaluation criteria are calculated and displayed.

5.5 Comment Evaluation Criteria

In this section, eight comment evaluation criteria for discussion control are described. Six criteria are word evaluations used in each comment, and the other two are status evaluations in discussion streams. The criteria are as follows:

1. topic: relevance to the topic
2. flow: continuity related to the topic
3. new: new tips related to the topic
4. inc: incorporation to the topic
5. byway: continuity unrelated to the topic
6. flood: new tips unrelated to the topic

Criteria from 1 to 6, which are calculated by evaluating words in each comment, are mainly a relationship with the topic words. Finally, criteria for divergence and convergence are prepared by a combination of these six criteria.

a. div: divergence for discussion
b. conv: convergence for discussion

In the rest of this section, these criteria are defined precisely, and "words" mean nouns extracted from each comment. In the future, "words" can be replaced by a set of parts of speech or real words depending on BBS type or language.

5.5.1 Distance Between Words

For the definition of criteria, word distance that denotes the relevance between two words is defined. A corpus is not available because words relationships are depend on a topic and contexts.

First, the distance value in comment C between words w_i and w_j is calculated by Eq. (5.1):

$$distance(w_i, w_j, C) = |ia(w_i, C) - ia(w_j, C)|, \qquad (5.1)$$

where $ia(W, C)$ is a function that gives the number of words between the top of comment C to the first appearance of word W. This distance assumes that a word should be defined at its first appearance as a opinion of the comment.

By using this distance in a comment, preliminary distance (pre_dist) between words W_1 and W_2 for n-th comment in a thread is defined as Eq. (5.2):

$$pre\_dist(W_1, W_2, C_n)$$
$$= \min_{i=1...n} \{distance(W_1, W_2, C_i)\}. \qquad (5.2)$$

This preliminary distance assumes that a member writes a comment referring to the former comments and learning relationship among words in the thread.

5.5.2 Word Labeling

Each word in each comment is labeled according to the role of each word. The definition of each label is as follows and Fig. 5.4 shows the relationship among them.

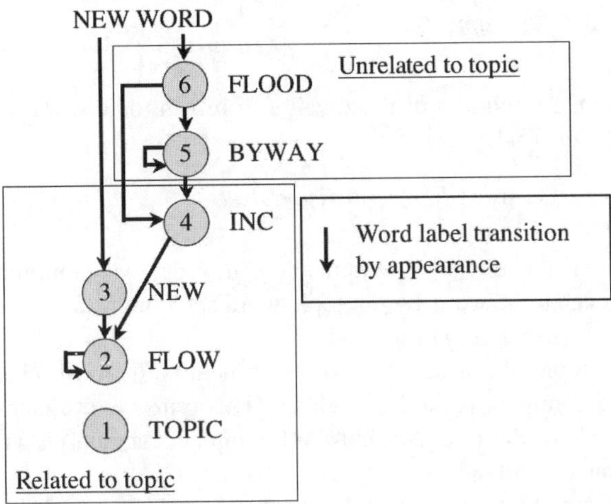

Fig 5.4. Word's labels and its transitions

1. TOPIC: topic words
2. FLOW: already appeared words related to topic
3. NEW: new words related to topic
4. INC: incorporated words from unrelated into related ones
5. BYWAY: already appeared words unrelated to topic
6. FLOOD: new words unrelated to topic

Relevance threshold $dmax$ is defined as 10, which is the maximum distance from one of the topic words. Word W is related to the topic if and only if the distance between W and one of the topic words is smaller than $dmax$. This threshold is defined by preliminary experimental results where a meaningful sentence includes at least seven words [8]. A relational word should be included in three sentences: a sentence containing a topic word, the former, and the latter for the relationship. Therefore, 20 words are relational except a topic word, and the threshold distance is reduced by half.

New words will be labeled either NEW or FLOOD based on the distance from a topic word. For the secondly appeared words, based on the transition in Fig. 5.4, they are labeled FLOW if that word was initially labeled NEW. If a word was initially labeled FLOOD, that word is labeled BYWAY if its distance from a topic word is no less than $dmax$, it otherwise is labeled INC. A word is labeled unrelated to the topic as long as it appears far from topic words.

5.5.3 Criteria for Topic Relevance

Topic relevance of k-th comment C_k in the thread is given by Eq. (5.3) as the sum of the minimum distance to each topic word and its normalization, where n is the number of topic words and $dmax$ is the distance threshold defined in 5.5.2:

$$topic(C_k) = 1 - \frac{1}{n \times dmax} \sum_{i=1}^{n} d_i. \tag{5.3}$$

In addition, the minimum distance value from i-th topic word in k-th comment C_k is defined as Eq. (5.4):

$$d_i = \min\{dmax, \min_j\{pre\_dist(T_i, c_j, C_{k-1})\}\}. \tag{5.4}$$

That is, for each word $\{c_j\}$ in comment C_k, d_i is set to the minimum preliminary distance from each topic word $\{t_i\}$ in topic word set T until $(k-1)$-th comment. If d_i exceeds $dmax$, d_i is set to $dmax$.

As a result, topic relevance is one if a comment is the most related to the topic and is zero if a comment is the least related. This criterion evaluates the relevance to all topic words of the thread because all comments should be written along the established topic of a thread.

Even if a comment does not include any topic words, topic relational words such as labeled FLOW are counted into $topic$. Because of this calculation of secondary topic relationships, a comment should not always include topic words.

5.5.4 Criteria for Each Word

Topic continuity, topic novelty, topical incorporation, unrelated continuity and unrelated novelty of k-th comment C_k in the thread are given by Eq. (5.5) - Eq. (5.9) respectively, where $number(C, LABEL)$ denotes the number of words labeled $LABEL$ in a comment C.

$$flow(C_k) = number(C_k, FLOW). \tag{5.5}$$
$$new(C_k) = number(C_k, NEW). \tag{5.6}$$
$$inc(C_k) = number(C_k, INC). \tag{5.7}$$
$$byway(C_k) = number(C_k, BYWAY). \tag{5.8}$$
$$flood(C_k) = number(C_k, FLOOD). \tag{5.9}$$

Topic continuity $flow$ evaluates the contribution to creating a discussion stream by using words related to the topic of a thread. If a comment is written by many topical words used in former comments, $flow$ becomes large.

Topic novelty new evaluates the contribution to the provision of new tips related to the topic of a thread. If a comment includes new idea or information related to the topic, new becomes large.

Topical incorporation inc evaluates the contribution to the incorporation of unrelated words used in the former comments. If a comment connects unrelated former comments with the topic, inc becomes large.

Unrelated continuity $byway$ evaluates the disturbance to the discussion by the continuous use of unrelated words used in the former comments. If a comment succeeds unrelated former comments without connecting to the topic, $byway$ becomes large.

Unrelated novelty $flood$ evaluates the disturbance to the discussion by providing new tips unrelated to the topic of a thread. If a comment includes new idea or information that does not seem to be related to the topic, $flood$ becomes large.

5.5.5 Criteria for Discussion

Discussion divergence of k-th comment C_k in the thread is given by Eq. (5.10) as the rate of new words above words variety, where $variety$ is the kind of words in comment C_k:

$$div(C_k) = topic(C_k) \times \frac{new(C_k)}{variety(C_k)}. \tag{5.10}$$

This criterion is used for progress and promotion of discussion divergence related to the topic of a thread. If a comment includes new idea or information and is surely related to the topic, div becomes large. For discussion divergence, new relational tips are the most desirable.

Discussion convergence of k-th comment C_k in the thread is given by Eq. (5.11) as the divergence criterion, the rate of FLOW, INC, FLOOD, and BYWAY words above the number of words in the comment C_k as $Allnum(C_k)$.

$$conv(C_k) = topic(C_k)$$
$$\times \{1 - |div(C_k) - (1 - t)|$$
$$- |\frac{flow(C_k) + inc(C_k)}{Allnum(C_k)} - t|$$
$$- \frac{flood(C_k) + byway(C_k)}{Allnum(C_k)}\}. \tag{5.11}$$

That is, discussion convergence consists of t % of FLOW+INC and $(1 - t)$ % of divergence except FLOOD and BYWAY because convergent comments should have well-defined words in the former discussion and some new convergent knowledge. Currently, the rate of t is set to 0.85 by the experimental results in section 5.6. If *convergence* becomes negative, it is set to zero.

This criterion is used for the progress and promotion of discussion convergence related to the topic of a thread. If a comment includes new words representing whole discussion in the thread, *conv* becomes large.

5.5.6 Significance of Comment Criteria

By displaying these described comment criteria, a community leader can eliminate unnecessary comments and encourage members to write more adequate comments because such comment criteria are not visible in an ordinary BBS.

Each member can also do self-contemplation and eagerly to write comments to contribute to discussion progress if such comments are judged as BYWAY or FLOOD, despite their intention to contribute to the discussion.

In addition, all members can grasp how smoothly a discussion evolves because each comment position is obviously revealed. Therefore, members can have a wide vision to the discussion status, such as its stream and progress.

5.6 Experiments for Discussion Divergence and Convergence

We did some experiments on the comment evaluation criteria for discussion divergence and convergence described in the last section.

The texts for these experiments were the 100 papers for the eighteenth annual conference of the Japanese Society of Artificial Intelligence. Each paper consists of four pages with about six segments. The papers were in Japanese, and nouns were extracted in advance by a Japanese morpheme analysys system [8].

Beginning sections in a paper tend to have the large value in the divergence, and end sections tend to have a large value in the convergence value. So, the following texts were prepared for comparison:

- TEXT-O: original papers.
- TEXT-B: papers that exchanged the first and second sections.
- TEXT-C: papers that exchanged the last and the next to last section.

Table 5.4. Averaged divergence values and rates having best value in each paper for first and second segments

text	div-1 (best)	div-2 (best)	Difference
TEXT-O	0.67 (79%)	0.37 (14%)	0.30 (65%)
TEXT-B	0.49 (48%)	0.48 (41%)	0.01 (7%)

Table 5.5. Averaged convergence values and rates having best value in each paper

text	conv-L1 (best)	conv-L2 (best)	Difference
TEXT-O	0.53 (74%)	0.32 (12%)	0.21 (62%)
TEXT-C	0.36 (30%)	0.43 (53%)	-0.07 (-23%)

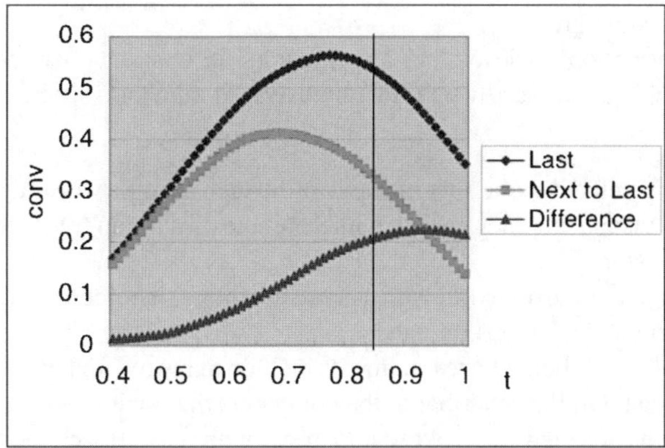

Fig 5.5. Relationship between t-parameter and convergence in TEXT-O

Table 5.4 shows the averaged divergence and the averaged rates of the largest values in each paper for the first and the second segments. In TEXT-O, the rate of the first segment was 79%, which is larger than the second by 65%, though the values were almost the same in TEXT-B. Therefore, the divergence criteria captured that segment "Introduction" was divergent in the relation of each original paper topic.

Table 5.5 shows the averaged convergence and the averaged rates of the largest values in each paper for the last (L1) and the next to last (L2) segments. In TEXT-O, the rate of the last segment was 74%, which is larger than the next to last by 62%, and the next to last was the most in TEXT-C. Therefore, the convergence criteria captured that the segment "conclusions" was the most convergent in the relation of each original paper topic.

Finally, Fig. 5.5 shows the relationship between t-parameter and convergence in TEXT-O. Convergence should become large in the last segment but small in the next to last. Therefore, t-parameter was set to 0.85 for the difference to become large. As a result, a paper consists of 85% already appeared words and 15% new words for concluding summary and future directions.

5.7 Discussion Visualization for Effective Discussion

In disucussion logs, each word is colored according to its label as in Table 5.6. That is, users can see the color of discussion logs and ascertain the state of discussion.

If logs are filled with brown, black, and blue, such discussion is highly concentrated but the conclusion may be normal. If logs are filled with red and purple, such discussion is fragmentary and the contents are not integrated. One of the ideal patterns of discussion is:

All colors are appeared and balanced in the logs.

Besides the brown, black and blue words, red words may lead to new direction of the discussion. Some discussion for red words are needed to decide whether to incorporate or not. This consideration will be appeared by purple words and its incorporation become green.

Other information retrievals are also included in this visualization framework. The discussion logs are narrowed down by users ID numbers, specific keywords or specific colors:

- If comments related to the specific keywords, such as topic words or incorporated words, are visualized, members can interpret the discussion for the keywords and see those history.
- If comments are narrowed down by specific colors, members can read only effective or non-effective comments.
- The logs of a member, whose comments include many red and purple words, may be eliminated. On the other hand, the comments that include green incorporated words can be selected for seeking a member who make the connection between the theme (brown words) and undefined topics (red or purple words).

In addition to above visualization frameworks, a thread is divided into some topical terms. The total number of each label is counted in a thread, and this counted label-balance must vary at the terminal points. Therefore, the system detects such varying points and visualizes the thread structure.

Table 5.6. Label colors and discussion interpretation

label	color	interpretation
TOPIC	brown	theme oriented discussion
FLOW	black	succeeding discussion or repetition
NEW	blue	contribution to discussion divergence
INC	green	relationship definition
BYWAY	purple	barrier to discussion
FLOOD	red	possibility to discussion divergence

5.8 Conclusion

In this paper, a bulletin board system framework for discussion visualization using a sub-story model is proposed. To control the discussion stream of divergence and convergence, consecutive word appearance in a document stream model is applied. Criteria for divergence and convergence are evaluated by research papers as imitative discussion.

In future works, a concrete system will be constructed and discussion support should be realized and evaluated. Visualization methods are also enhanced as a network displaying discussion divergence, convergence, users roles and their relationships. A scenario to the desirable future will not created without cooperation among all members and their consciousness.

References

[1] Y. Ohsawa, H. Soma, Y. Matsuo, N. Matsumura, and M. Usui (2002). Featuring Web Communities based on Word Co-occurrence Structure of Communications. In *Proc. of 11th International World Wide Web Conference (WWW02)*, 2002.

[2] N. Matsumura, Y. Ohsawa and M. Ishizuka (2002). Influence Diffusion Model in Text-based Communication. In *Proc. of 11th International World Wide Web Conference (WWW02)*, 2002.

[3] J. Morris and G. Hirst (1991). Lexical Cohesion Computed by Thesaural Relations as an Indicator of the Structure of Text. *Computational Linguistics*, Vol. 17, No. 1, pp. 21–48, 1991.

[4] M. A. Hearst (1994). Multi-paragraph segmentation of expository text. In *Proc. of the 32nd conference on Association for Computational Linguistics*, pp. 9–16, 1994.

[5] G. Salton, A. Wong, and C. S. Yang (1975). A Vector Space Model for Automatic Indexing. *Communication of the ACM*, Vol. 18, No. 11, pp. 613–620, 1975.

[6] M. Lapata (2003). Probabilistic Text Structuring: Experiments with Sentence Ordering. In *Proc. of the 41st Meeting of the Association of Computational Linguistics*, pp. 545–552, 2003.

[7] W. Sunayama, A. Iyama and M. Yachida (2006). HTML Text Segmentation for Web Page Summarization by Using a Key Sentences Extraction Method. *Systems and Computers in Japan*, John Wiley & Sons, Inc., 2006.

[8] M. Asahara, Y. Matsumoto (2000). Extended Models and Tools for High-performance Part-of-Speech Tagger. In *Proc. of the 18th International Conference on Computational Linguistics*, Vol. 1, pp. 21–27, 2000.

Design of BBS with Visual Representation for Online Data Analysis

Yasufumi Takama and Yuta Seo

Graduate School of System Design, Tokyo Metropolitan University
6-6 Asahigaoka, Hino, Tokyo 191-0065, Japan
ytakama@cc.tmit.ac.jp

Summary. A concept of bulletin board system (BBS) equipped with information visualization techniques is proposed for supporting online data analysis. Although a group discussion is known to be effective for analyzing data from various viewpoints, the number of participants has to be limited in terms of time and space constraints. To solve the problem, this paper proposes to augment BBS, which is one of popular tools on the Web. In order for discussion participants to share the data online, the system provides them with a visual representation of target data, with functions for supporting comment generation as well as retrieving posted comments. In order to show the potential of the concept, a BBS equipped with KeyGraph is also developed for supporting online chance discovery. It has functions for making visual annotations on the KeyGraph, as well as a function for retrieving similar scenarios. The experimental result shows the effectiveness of the BBS in terms of the usefulness of scenario generation support functions as well as that of scenario retrieval engines.

Keywords: Bulletin board system (BBS), online data analysis, information visualization, chance discovery, scenario generation

6.1 Introduction

A concept of bulletin board system (BBS) equipped with information visualization techniques is proposed for supporting online data analysis. Problems to be solved in today's society, such as new product development, environmental problems, and design of large-scale systems, have become large-scale as well as complicated. To solve these problems, communication and discovery from multidisciplinary data is important. That is, vast amount of data relating to a problem should be collected from various fields, and analyzed from a number of experts having different background knowledge.

For example, development of new products will need to collect various data, such as follows.

- Custormers' opinions, such as complaints about existing products, desired facilities, brand image, and rival products.

Y. Takama and Y. Seo: *Design of BBS with Visual Representation for Online Data Analysis*, Studies in Computational Intelligence (SCI) **123**, 111–126 (2008)
www.springerlink.com

- Information about new technologies that could be used for developing new products.
- Estimated cost for development.
- Competitions activities.

Furthermore, the collected data should be analyzed by a number of people having different knowledge and viewpoints, i.e., sales people analyze the data from customers' viewpoints, engineers focus on technical issue, and a manager should evaluate the product from the viewpoint of management.

The above-mentioned process necessarily involves a group discussion for analyzing data from various viewpoints, as well as for collecting various opinions. Actually, having a group discussion is encouraged in activities that involves data analysis, such as cooperative problem solving [5], new ideas/products generation [8, 14], as well as chance discovery [6, 10] including scenario emergence [11]. When a group discussion is to be held, participants have to come to the same place all together. However, the number of participants is limited by the capacity of a meeting room. Furthermore, time constraint for many participants to meet together is also serious. Online discussion can conceptually solve these problems. That is, there is no theoretical limitation on the number of participants, who can virtually meet together via the Internet. Also, the synchronicity of attending discussion is not necessary. As a result, the number of participants can be much more than conventional meeting, which will contribute to the discussion from various viewpoints.

There exist various tools for online meetings, such as video teleconference, chat, instant message and BBS. Among then, this paper employs BBS for online data analysis. The reason why BBS is employed is as follows.

1. It enables asynchronous group discussion.
2. It is widely used on various Web sites, and it is expected that most of discussion participants are familiar with its usage.
3. Its usage is so simple enough for even inexperienced participants to get familiar with its usage.
4. It is originally suitable for obtaining a number of comments from participants.

As for the 4th point, typical BBS consists of threads, each of which is established for a certain topic. A thread contains a series of comments on the corresponding topic, which are posted by many visitors of the BBS. A visitor can read the comments posted by others, based on which s/he writes and posts his/her own comment. Therefore, BBS can collect various comments from a number of discussions participants online.

Although BBS is suitable as a basis for online data analysis, several improvements can be possible for making it specific to online data analysis. In the case of online data analysis, common data should be shared with discussion participants. However, conventional BBS does not have a facility for sharing common data. Furthermore, it is difficult for conventional BBS to compare/integrate comments, which are written as free text information. That is, BBS usually displays a number of comments in a thread in order of arrival. As a comment is often generated as a reply to latest comments, this style is useful to see the comment chain. However, when the number of posted comments becomes large, it becomes difficult to grasp the development of discussion throughout the thread. For example, the relationship between distant comments in a thread tends to be missed by BBS visitors. It is difficult for a visitor to see whether or not a comment relating to his/her opinion exists. In order to solve this problem, functions for retrieving similar comments should be introduced into conventional BBS.

The paper proposes a concept of a BBS integrated with information visualization techniques. A visual representation is generated from a target data for each thread, by which discussion participants can share the data. The system is also equipped with a comment retrieval module, which enables participants to retrieves comments related with theirs.

Based on the proposed architecture, a BBS equipped with KeyGraph is also proposed for supporting chance discovery process. As scenario generation from target data is considered to be one of the important activities in chance discovery process [11], developing the BBS for supporting chance discovery process is suitable for showing the potential of the proposed concept. The discussion participants of the BBS can write scenarios by referring to a KeyGraph generated from target data, and post those to threads. The BBS also has functions for assisting the participants in writing scenarios with visual annotations on the KeyGraph. When a number of scenarios have posted in a thread, it tends to be difficult for readers to find scenarios that are related to their own scenarios. As finding related scenarios are expected to be useful for lively discussion as well as for revising the scenarios, functions for retrieving scenarios are also implemented.

A general architecture of a BBS equipped with information visualization technique is proposed in Section 6.2, then followed by Section 6.3, which describes a BBS equipped with KeyGraph as one of the specific implementations. Experimental results are reported in Section 6.4, which shows the effectiveness of the implemented system in terms of the usefulness of scenario generation support functions as well as that of scenario retrieval engines.

6.2 BBS Equipped with Information Visualization Technique

6.2.1 System Architecture

This section proposes a general architecture of augmented BBS equipped with information visualization techniques for supporting online data analysis. As discussed in previous section, the system should introduce the following facilities into conventional BBS.

- Facility for sharing target data that are to be analyzed.
- Facility for supporting users to compare and integrate multiple comments.
- Facility for retrieving related comments.

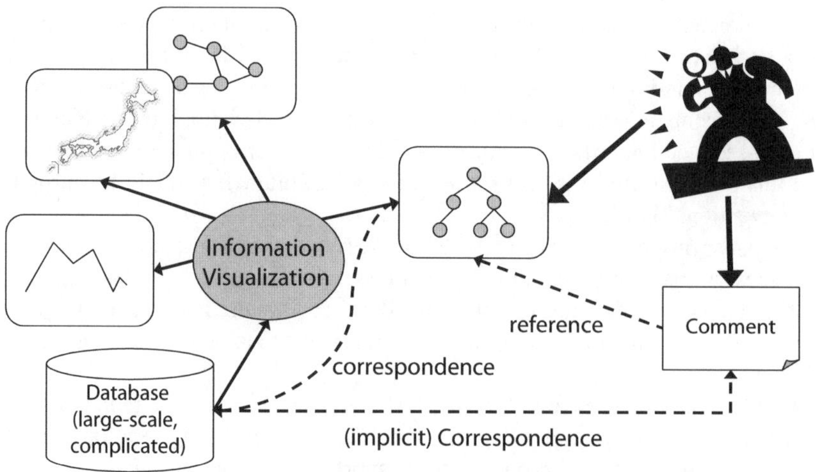

Fig 6.1. Role of information visualization in proposed system

In order to introduce those facilities, the proposed system employs information visualization techniques. Figure 6.1 shows the role of information visualization techniques in proposed system. Providing target data with appropriate visualization style will help discussion participants grasp the data space. Furthermore, as a comment is usually written by referring to some objects in the visual representation, there is a certain correspondence between visual representation and comments. Of course, objects in visual representation have corresponding data in target database. As a consequence, each comment has its corresponding data, based on which comment matching can be possible. Functions for supporting comparison comparison / integration as well as for retrieving comments can be realized based on the comment matching.

Figure 6.2 shows the conceptual architecture of the proposed system, which consists of BBS, database, annotation module, information visualization module, and comment retrieval module. Differently from conventional BBS, the BBS part consists of thread area and visual representation area. A database stores the data to be analyzed, from which a visual representation is generated for each thread by information visualization module. Participants of the group discussion write their own comments based on their interpretations of the visual representation. An annotation module helps this process, by making it easy for the participants to find the relationship between the visual representation and a comment. A comment retrieval module calculates the similarity between comments in a thread, and sorts them in order of similarity.

As a related work, DISCUS [6] employs a visual representation, KeyGraph, in order to visualize a development of online discussion. It successively modifies the displayed graph structure by adding a new comment during a discussion. While it aims to visualize the *subjective data* (comments), the system proposed in this paper employs a visual representation for sharing *target data* during a discussion. Therefore,

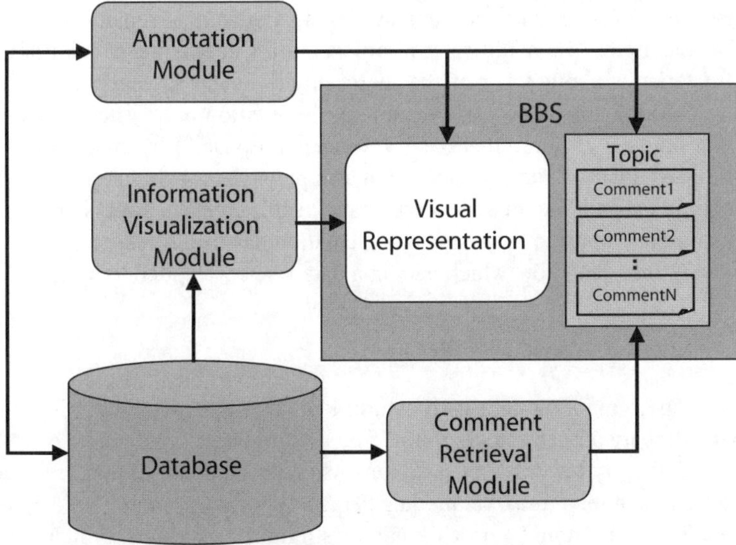

Fig 6.2. System architecture of BBS equipped with information visualization technique

the same visual representation is supposed to be displayed throughout a discussion, so that discussion participants can write their comments referring to the same visual representation. Referring to the same visual representation will make it easy to compare and integrate comments written by different participants [3, 18].

6.2.2 Information Visualization Module

The information visualization module generates a visual representation from a data of interest. The visual representation displayed on BBS can be used for discussion participants to share common data in understandable manner. As there could exist various kinds of data that are to be analyzed, visualization techniques that are used for the module should be selected according to the target data. Various kinds of visual representation [15–17], even a photograph, could be used if the mapping between the representation and target data can be established. The prototype system proposed in Section 6.3 employs KeyGraph [9, 10]. Currently, it is assumed that each visual representation has the corresponding thread. In other words, comments in the same thread refer to the same visual representation.

6.2.3 BBS Module and Annotation Module

The BBS module has the same facilities as the conventional BBS, except that it has a visual representation area and functions for making visual annotation. Visual annotation can be used for referring to a part of visual representation in a comment. When a discussion participant writes a comment from a visual representation, it is expected that s/he mentions the contents of the visual representation in the comment. For example, s/he might refer to objects in a visual representation, or denote the characteristics about a par of the representation such as shape and color. The annotation module enables a participant make annotation in a visual representation, which can be used in a comment as a reference purpose. The annotation does not only help participants write comments, but also plays the role to find relationship between comments and data in a database. As a visual representation is obtained from the data, annotated part in the visual representation has the corresponding data. As a result, the system can know which data in a database correspond to a comment.

6.2.4 Comment Retrieval Module

Although expressing a thought with natural language like a comment is natural for a human, comparison and integration of free text information is not so easy. In particular, the difficulty becomes more serious when vast amount of text information is available. The comment retrieval module helps a user find similar/related comments. The module could employ various kinds of similarity measures, which have been developed in the field of document retrieval, such as vector space model (VSM) [1] with various weighting schemas, phrase-based similarity measure [4, 13], and a measure based on probabilistic model [2, 12]. The similarity calculation based on data annotation as noted in Section 6.2.3 is also available, and it is shown that the similarity calculation can consider data baskets that correspond to comments [3, 18].

6.2.5 Data Sharing based on Information Visualization

The point of the proposed BBS is data sharing based on information visualization techniques. As for the first merit of this approach, the system can find the corresponding data for each comment. The corresponding data cannot be used only for comment retrieval as noted in Section 6.2.4, but also for finding data subset that corresponds to the main topics of the discussion in a thread. By generating new visual representation from the data subset, new thread could be started for more detailed discussion.

Furthermore, finding the correspondence between different threads could be possible by finding corresponding data subset for each thread. In usual BBS, it is difficult to find the relation between different threads. However, the proposed BBS can match the comments that are posted in different threads, if the threads are generated from the same database. This characteristic could be useful for the following purposes.

- Generates new threads dynamically according to the development of discussion.
- Prepares different threads for different groups of participants.

Former purpose is useful for large-scale discussion, which would contain various topics with different level of details. By dividing the discussion into several threads, the development of discussion in each thread is expected to be kept simple. As for the latter purpose, different visual representation could be used for different groups according to their viewpoint and background knowledge, while keeping relationship between threads via database. For example, it would be possible to provide the customers with the image of products and marketers with a KeyGraph that visualizes the customers' opinions posted to the customers' thread.

Although the prototype system in the next section considers neither the generation of new visual representation from data subset nor multiple threads, the characteristics discussed here are supposed to be easily realized based on the proposed BBS.

6.3 KeyGraph-based BBS

6.3.1 Online Data Analysis for Scenario Generation in Chance Discovery Process

One of the important topics in chance discovery is scenario emergence [11], which concerns techniques and a framework for drawing a future scenario from chances that a person or an organization has noticed, and for making decisions based on the scenario. In the context of chance discovery, a scenario means a story about the events or observations that might relate to a chance. The examples of a scenario include a sequence of events that might be caused in the future by a focused event, a hidden relationship between observed events, and interpretation about the reason that a focused event happened. A persona [7], which has become one of the major methods for designing interfaces or services, could also be viewed as a scenario. These scenarios essentially involve uncertainty, and are often generated subjectively.

A data analysis is one of the important activities in chance discovery. When scenarios are to be generated, data relating to the target topic are often collected for analysis purpose. The data concern the related events or observations, which could be obtained in various forms such as questionnaires, POS (Point of Sales) data, and documents. A collection of scenarios can also be considered as a kind of data [6], from which new scenario could be generated. In this case, scenarios are called subjective data.

Although a person can generate his/her own scenario alone, collecting as many scenarios as possible from a number of persons is important mainly in terms of two reasons:

1. Obtaining a final scenario based on the scenarios from all members of an organization will contribute to the consensus on decision making.
2. A number of scenarios bring various viewpoints on the topic, which will lead to deep understanding of the data.

The former merit is applied to decision making in organizations, but the latter is also applied to decision making by a person. That is, a person can make use of scenarios by others as references bringing the viewpoints that s/he could not notice by him/herself.

It is known that group discussion and the introduction of other person's viewpoint is effective for creating new ideas [8, 14] as well as for cooperative problem solving [5]. The process of idea generation and creative problem solving can be divided into divergent thinking process and convergent thinking process. In divergent thinking process, group discussion such as brainstorming is encouraged in order to collect as many ideas/opinions as possible. In convergent thinking process, the comparison and integration of various ideas/opinions are performed to derive new ideas, or optimal/alternative solutions. Comparing the convergent thinking performed by a person alone, group discussion can introduce various viewpoints from participants for evaluating ideas as well as for making decisions. As these advantages of group discussion are expected to be useful for scenario generation as well, chance discovery is one of the most promising field for applying the augmented BBS proposed in Section 6.2.

6.3.2 System Architecture

This section proposes the BBS designed for KeyGraph-based online chance discovery process. The BBS is designed based on the general architecture proposed in Section 6.2, and KeyGraph is employed as visual representation. Figure 6.3 shows the configuration of the BBS. It employs client-server system, and a server is implemented as CGI with Ruby (http://www.ruby-lang.org/). The server stores both logs of BBS (such as thread and scenarios) and the graph data of KeyGraph. The

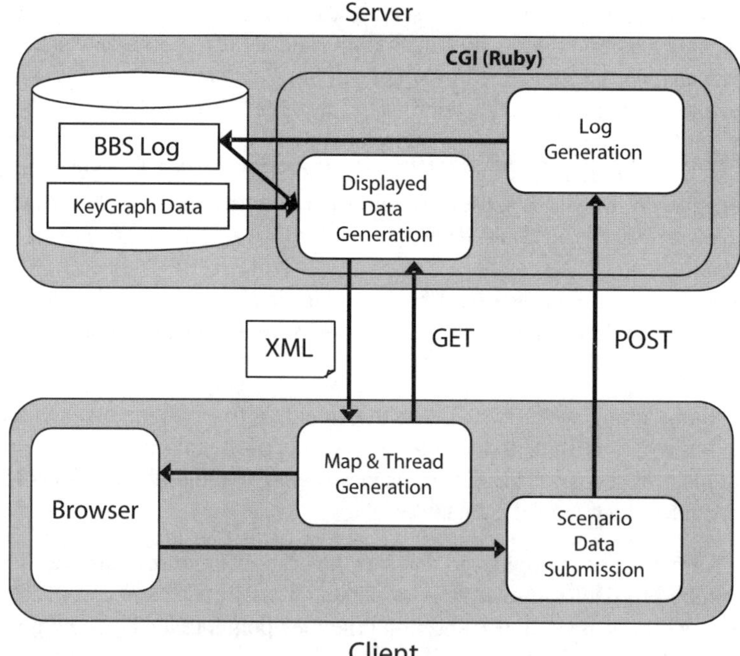

Fig 6.3. System architecture of KeyGraph-based BBS

"Displayed Data Generation" module in the server corresponds to the information visualization module and the comment retrieval module in Fig. 6.2. The displayed data is transmitted from the server to the client with XML format.

A client is implemented with Flash^{TM}, and can be accessed with an ordinary Web browser. The "Map & Thread Generation" module in the client just generates the displayed image including a thread. The "Scenario Data Submission" module plays a role of the annotation module in Fig. 6.2.

6.3.3 Support Functions for Scenario Generation

Figure 6.4 shows a screenshot of the BBS, which is displayed with a Web browser. As the BBS is currently designed for Japanese users, subsequent figures contain Japanese. The screen is divided into 3 areas: a KeyGraph area (upper left), a posting form area (lower left), and a thread area (right). A KeyGraph area displays a Key-Graph to be discussed in a thread. It is displayed as a clickable map, with which users can define islands and bridges that are going to be referred to in their scenarios. Users write a scenario in a posting form, and post it to the server. Posted scenarios are displayed in a thread area in arrival sequence, as displayed in ordinary BBS.

Figure 6.5 shows a description of a scenario, which is displayed in the thread area. The first line of a scenario describes the id, the author name and posted date of a scenario, along with a button to remove the scenario from a thread. From the second line, islands and bridges that are referred to in the scenario are listed, and then followed by the sentences of the scenario. Each island and bridge is highlighted with different color in sentences. Furthermore, when a mouse pointer is being over a scenario in the thread area, the referred islands and bridges are also highlighted

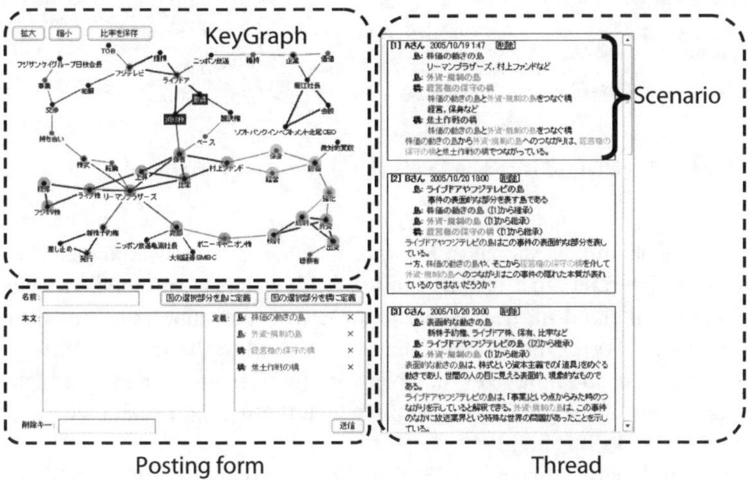

Fig 6.4. Screenshot of KeyGraph-based BBS

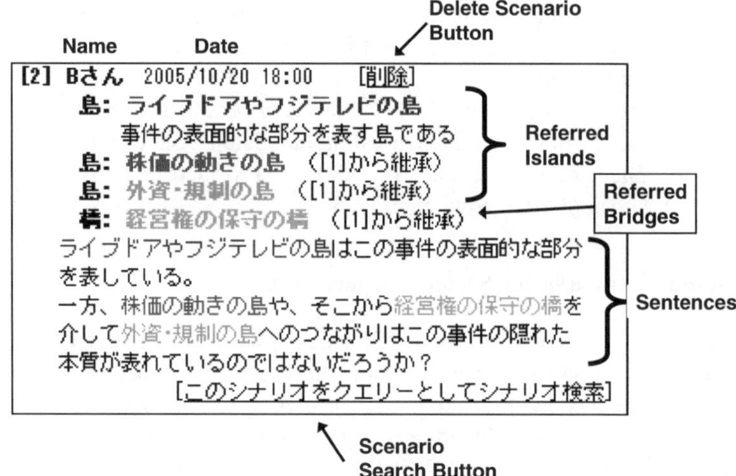

Fig 6.5. Example of scenario description

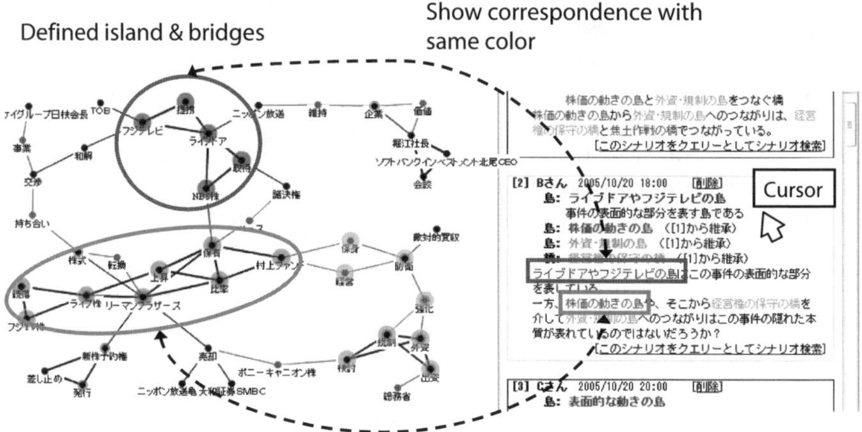

Fig 6.6. Highlighted island on KeyGraph area and the corresponding scenario

on the KeyGraph area as shown in Fig. 6.6. In Fig. 6.6, a mouse pointer is over the scenario 2, which refers to 3 islands and one bridge. The corresponding islands and bridges are highlighted in the KeyGraph area, with the same colors as those used in the sentences. For example, nodes that correspond to the first island ("island of Livedoor Col., Ltd. and Fuji Television Network Inc.") in the scenario are highlighted on the upper part of the graph area, below which the second island ("island of price movement of stocks") is highlighted.

Users cannot only define new islands and bridges, but also inherit those that have been already defined in other scenarios. Using those definition/inheritance facilities make it possible to grasp the topics of a scenario, because readers can confirm the story of a scenario on the KeyGraph with the help of the visual annotation.

The information about defined islands and bridges are considered as metadata of a scenario, which are stored in BBS log at the server, along with a scenario itself.

6.3.4 Scenario Retrieval Engines

The last line in Fig. 6.5 contains a button to retrieve related scenarios from a thread. In order to retrieve related scenarios, this paper employs two retrieval methods; a method based on VSM [1] and a method based on data annotation [3, 18].

The VSM-based method uses keywords that correspond to nodes in a KeyGraph as index terms, based on which a scenario is represented as a vector. The similarity between scenarios is calculated based on cosine value of the corresponding vectors.

The method based on data annotation (called DA-method hereinafter) calculates the similarity between scenarios in terms of overlap of corresponding data in an original data file. Figure 6.7 shows the outline of similarity calculation based on data annotation. As islands and bridges on a displayed KeyGraph are generated based on the co-occurrence of keywords (nodes) in the data file [9, 10], it is possible to find the baskets in the data file that correspond to the islands/bridges referred to in the scenario. In Fig. 6.7, the corresponding data of scenario A (B) is data-set A (B), from which the islands referred to in the scenario are generated.

When a scenario is posted, the corresponding baskets in an original data file are extracted and annotated [3, 18]. The similarity between scenarios is calculated based on Jaccard coefficient, which calculates the overlap of the corresponding baskets between the scenarios. In Fig. 6.7, similarity between scenarios A and B is calculated based on the overlap of data-sets A and B.

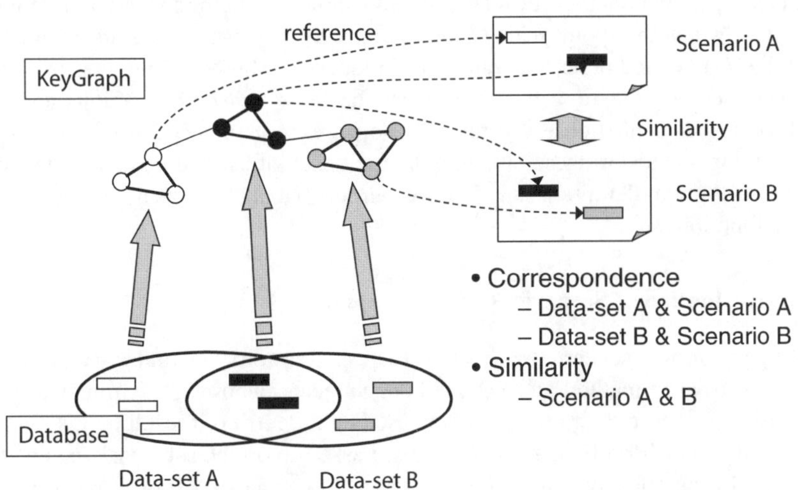

Fig 6.7. Scenario similarity calculation based on data annotation

Compared with VSM-based method that calculates the similarity based on keywords appeared in the sentences, DA-method calculates the similarity based on the factors that are hidden behind scenarios. Therefore, DA-method is expected to retrieve scenarios that are not literally similar to a query scenario, but refer to related topics. As a result, DA-method would provide users with relevant scenarios that have related but different viewpoints, which is important for chance discovery process.

6.4 Experimental Results

6.4.1 Experimental Settings

The implemented BBS is evaluated with test subjects. Thirteen subjects, who were graduate and undergraduate students in Engineering field, used the BBS for discussing economic topic: the M&A issue between Livedoor Col., Ltd. and Fuji Television Network Inc. The headlines relating with the topic were collected from Nikkei News (http://www.nni.nikkei.co.jp/), from 12 Jan. 2005 to 26 Jun. 2005. Total 214 headlines were collected, from which a KeyGraph is generated by considering each headline as a basket. The proposed BBS is evaluated in terms of two viewpoints. Functions for supporting scenario generation as proposed in Section 6.3.3 is evaluated in Section 6.4.2, and scenario retrieval methods as proposed in Section 6.3.4 are evaluated in Section 6.4.3.

6.4.2 Evaluation of Scenario Generation Support Functions

Subjects are asked to write scenarios regarding the above-mentioned topic with using the proposed BBS. After the experiments, they are asked to answer the questionnaires, in which they evaluate the functions for defining and inheriting islands/bridges with 5-point scale (1: poor, 5: good). The evaluations are summarized in Table 6.1. In the table, the value of "Frequency" shows the number of times a subject used the function. It is seen from the table that both functions are given high scores. It is also observed that all subject except subject 2 used the function for defining islands/bridges. Although the frequency of using inheritance function is lower than that of definition function, we can see that all subjects used at least either of the functions.

6.4.3 Evaluation of Scenario Retrieval Engines

After generating a scenario, each subject is asked to retrieve similar scenarios with both VSM-based method and DA-method. Subjects, whose ID is from 1 to 6, are also asked to revise their own scenario based on the retrieved result by DA-method, while subjects whose ID is from 7 to 13 are asked to do the same task using VSM-method. In the questionnaires, they are asked to evaluate each of retrieval method with 5-point scale, as well as to answer the rank of the scenario that is used as a reference to revise the scenario. Table 6.2 summarizes the evaluation, which includes

Table 6.1. Evaluations of scenario generation support functions

Subject ID	Definition Function Score(Frequency)	Inheritance Function Score(Frequency)
1	4(5)	3(0)
2	4(0)	5(1)
3	4(1)	3(2)
4	4(2)	3(0)
5	4(3)	4(0)
6	4(4)	5(1)
7	4(2)	3(0)
8	4(5)	4(2)
9	4(4)	2(0)
10	4(2)	4(0)
11	4(1)	5(2)
12	4(3)	5(2)
13	4(5)	2(0)
1	4.3(2.8)	3.7(0.8)

Table 6.2. Experimental results of retrieval method

Number of reference scenarios	VSM-method	DA-method
1st	3	3
2nd	3	2
3rd	1	0
4th	0	0
5th	1	1
6th	1	1
7th	0	2
Score	3.8	3.7

the number of reference scenarios in each rank of the retrieval result, as well as average evaluation score from questionnaires. Although the score given to both methods are almost similar, the rank of reference scenario is different. When VSM-method is used, subjects tend to refer highly-ranked scenarios, whereas reference scenarios tend to be ranked lower when DA-method is used. For the reason of the difference, there were test subjects who gave comments about VSM-based method that only a few top scenarios were similar to query scenario, but subsequent ones seemed to be less related with query. On the other hand, the scenarios retrieved by DA-method were said to be related with query, even in lower ranks.

Let us consider a case of a subject, who evaluated DA-method better than VSM-based method. He revised a scenario by referring to the 7th-ranked scenario retrieved by DA-method. His initial scenario, reference scenario (generated by other subject), and revised scenario based on the reference scenario, are as follows.

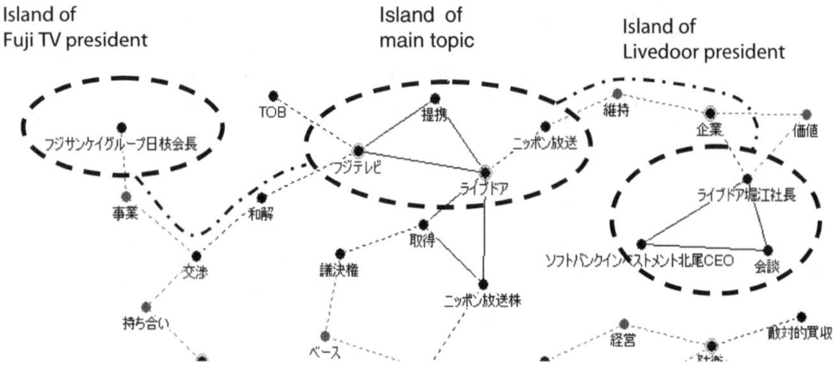

Fig 6.8. Example of scenario expansion

Initial Scenario: *The island of main topic contains keywords that are often appeared in news headlines. These keywords are famous company names, which could
attract readers' attentions.*

Reference Scenario: *The island of main topic, which is a center of the topic, has
connections to the island of Livedoor President Horie, and that of Fuji TV president
Hieda.*

Revised Scenario: *The bridges from the island of Livedoor President to that of Fuji
TV President, via the island of main topic, are supposed to indicate that the confrontation between the presidents would be an essential of this sensational topic
receiving a high degree of media coverage.*

This scenario expansion is illustrated in Fig. 6.8. Before the retrieval, he focused
on only a main topic as shown in Fig. 6.8, i.e. the island of main topic. After he
examined the reference scenario, which mentions the bridges from the presidents
of both companies to the main topic, he noticed that connection and revised the
scenario. It mentions the confrontation of the presidents as the background of the
main topic. This example shows that the scenarios retrieved in low rank with DA-
method do not only contain similar topics but also different viewpoints, which would
lead a subject to notice a new interpretation.

6.5 Conclusions

This paper proposes an augmented BBS for supporting online data analysis. The system integrates information visualization techniques into BBS, so that it can collect
as many comments as possible through online discussion, compare and integrate the
obtained comments with the help of a visual representation generated for sharing target data. Based on the proposed architecture, a BBS equipped with KeyGraph is also
proposed. The discussion participants of the BBS can write scenarios by referring to

the same KeyGraph generated from target data. The system provides the participants with not only functions for assisting them in writing scenarios with visual annotation on the KeyGraph, but also functions for retrieving similar scenarios. The experimental results with test subjects show the usefulness of scenario generation support functions. Furthermore, it is also found the scenario retrieval based on data annotation can let subjects notice a new viewpoint. Future work will include the application of the developed BBS to actual group discussion in business or research projects, as well as the development of other systems based on the proposed architecture.

References

1. Baeza-Yates R, Ribeiro-Neto B (1999) 25. Classic information retrieval. In: Modern Information Retrieval. Addison Wesley 24–33
2. Callan J. P, Croft W. B, Harding S. M (1992) The INQUERY Retrieval System. In: Proc. of 3rd International Conference on Database and Expert Systems Applications (DEXA-92) 78–83
3. Iwase Y, Takama Y (2005) Data annotation based on scenario in chance discovery process. In: Workshop on Rough Sets and Chance Discovery, in 8th Joint Conference on Information Sciences (JCIS2005) 1797–1800
4. Kamel M. S, Hammouda K. M (2004) Efficient Phrase-Based Document Indexing for Web Document Clustering. In: IEEE Transactions on Knowledge and Data Engineering 16(10): 1279–1296
5. Kato N, Kunifuji S (1997) Consensus-making support system for creative problem solving. In: Knowledge-Based Systems 10:59–66
6. Liora X, Goldberg D. E, Ohsawa Y, Ohnishi K, Tamura H, Washida Y, Yoshikawa M (2004) Chance and Marketing: On-line Conversation Analysis for Creative Scenario Discussion. In: 1st European Workshop on Chance Discovery (EWCD'2004) 152–161
7. Merrill C, Feldman D (2004) Rethinking the Path to Usability: How to Design What Users Really Want. In: IT Professional 6(3):51–57
8. Nishimoto K, Sumi Y, Mase K (1996) Toward an outsider agent for supporting a brainstorming session - an information retrieval method from a different viewpoint. In: Knowledge-Based Systems 9:377–384
9. Ohsawa Y, Benson NE, Yachida M (1998) KeyGraph: Automatic indexing by co-occurrence graph based on building construction metaphor. In: Proc. of Advances in Digital Libraries Conference 12–18
10. Ohsawa Y (2003) 18. KeyGraph: Visualized structure among event clusters. In: Ohsawa Y, McBurney P (eds) Chance Discovery. Springer 262–275
11. Ohsawa Y, Fujie H, Saiura A, Okazaki N, Matsumura N (2005) Cooperative Scenario Mining from Blood Test Data of Hepatitis B and C. In: Lecture Notes in Computer Science 3430:312–335
12. Robertson S. E, Walker S, Hancock-Beaulieu M, Gull A, Lau M (1992) Okapi at TREC 3. In: Text REtrieval Conference 21–30
13. Singhal A, Buckley C, Mitra M (1996) Pivoted Document Length Normalization. In: Research and Development in Information Retrieval 21–29
14. Sugimoto M, Hori K, Ohsuga S (1996) A system to visualize different viewpoints for supporting researchers' creativity. In: Knowledge-Based Systems 9:369–376

15. Takama Y, Hirota K (2003) Web information visualization method employing immune network model for finding topic stream. In: New Generation Computing 21(1):49–60
16. Takama Y, Ohsawa Y (2003) 13. Effects of scenic information. In: Ohsawa Y, McBurney P (eds) Chance Discovery. Springer 184–188
17. Takama Y, Kajinami T (2004) Keyword pair extraction for relevance feedback based on interactive keyword map. In: 1st European Workshop on Chance Discovery in ECAI2004 41–50
18. Takama Y, Iwase Y (2005) Scenario to data mapping for chance discovery process. In: Abraham A, Dote Y, Furuhashi T, Koppen M, Ohuchi A, Ohsawa Y (eds) Proc. of 4th IEEE International Workshop on WSTST'05, Soft Computing as Transdisciplinary Science and Technology 470–479

A Study on Web Clustering with Respect to XiangShan Science Conference

Wen Zhang and Xijin Tang

Institute of Systems Science, Academy of Mathematics and Systems Science
Chinese Academy of Sciences, Beijing 100080 P.R. China
{zhangwen,xjtang}@amss.ac.cn

Summary. This paper has presented two clustering results using two different methods to cluster the same Boolean vectors represented the Web documents of XiangShan Science Conference (XSSC). Then, average co-occurrence and average difference are introduced to evaluate the effectiveness of theses two different clustering methods. With these two indicators, the evaluation of experimental results from these two clustering methods is presented. Also, an extended research on Web clustering is presented in this paper, that is, the automatic concepts generation. At last, the reliability of the automatic concept generation is discussed in this paper.

Keywords: Web clustering, average co-occurrence, average difference, automatic concepts generation, Xiangshan Science Cinference

7.1 Introduction

"Clustering is the process of grouping the data into classes or clusters so that objects within a cluster have high similarity in comparison to one another, but are very dissimilar to objects in other clusters" [1]. While Web clustering is the process of clustering Web documents based on its contents or usage records. Generally speaking, five categories of clustering methods have been studied: partitioning methods, hierarchical methods, density-based methods, grid-based methods and model-based methods. Partitioning methods are used in clustering to partition large amounts of objects into small clusters, for example, K-means, K-medoids, etc [2]. A hierarchical method creates a hierarchical decomposition of the given set of data objects [3]. Density-based methods cluster objects according to its density or neighborhood in the spatial database [4]. Grid-based methods quantize the object space into a finite number of cells that form a grid structure and all of the clustering operations are performed on the grid structure [5]. And model-based methods hypothesized a model for each of the clusters and find the best fit of the data to the given model [6].

XiangShan Science Conference (XSSC) is famous for its directive role in Chinese basic scientific research by inviting senior experts in research fields to express their opinions concerning their own researches. After the meeting of experts in XSSC, all

W. Zhang and X. Tang: *A Study on Web Clustering with Respect to XiangShan Science Conference*, Studies in Computational Intelligence (SCI) **123**, 127–136 (2008)
www.springerlink.com

kinds of information related the conference topics are published on the XSSC Website (http://www.xssc.ac.cn) so it is a large valuable repository for us to carry out Web mining on it in order to study the current situation of Chinese basic research. For this reason the augmented information support tool in Group Argumentation Environment [7, 8] (AIS-GAE) is elaborated to explore the significant and interesting knowledge from XSSC website.

In this paper, hierarchical method is employed to cluster the Web documents collected from XSSC website and preprocessed by AIS-GAE. In the process of the hierarchical clustering, two different strategies about preprocessing are proposed to yield the data vectors used to represent the original Web documents and clustered by hierarchical cluster analysis. Thus, different clustering results are obtained by different preprocessing strategies in our test experiments. Next, average co-occurrence and average difference based on the definition of clustering are introduced to evaluate the results. At last, an extended research of Web clustering is presented, that is, automatic concepts generation [9]. Those clustering results of better performance are of out test sample to examine the reliability of the method of our automatic concepts generation. The rest of this paper is organized as follows. A brief description of the working process of AIS-GAE and the preprocessed Web documents collected from XSSC website are provided in Section 2. The two types of Web clustering preprocessing strategies and its respectively experimental results are demonstrated are discussed in Section 3. The average co-occurrence and average difference measure are proposed to evaluate the experiments results derived from the two different Web clustering preprocessing strategies in Section 4. Next, automatic concepts generation is introduced and the reliability of our method on automatic concepts generation is demonstrated in Section 5. At Last, concluding remarks and further research are indicated in Section 6.

7.2 A Brief Description of AIS-GAE and Preprocessed Web Documents of XSSC

AIS-GAE is an intelligent tool which employed the Web text mining technologies to provide an augmented information support for GAE. Figure 7.1 has shown the working process of AIS-GAE which includes four modules: Web crawler, Web indexer, Web summarization and Web clustering. Web crawler is used to collect Web pages from XSSC Website by the traditional spider detecting algorithm [10, 11]. Web page indexer is designed here to filter and reorganize the content of Web pages by information extraction from Web page such as page titles, time stamps, name of person and also the pure texts which are the contents of XSSC topics. Web text summarization is introduced to extract the representative sentences from the pure text of Web page and rearranges the extracted sentences as the abstract of the Web page. Web clustering is employed to cluster the Web pages into classes without supervision by hierarchical methods. For more detailed information about AIS-GAE, please refer to [12].

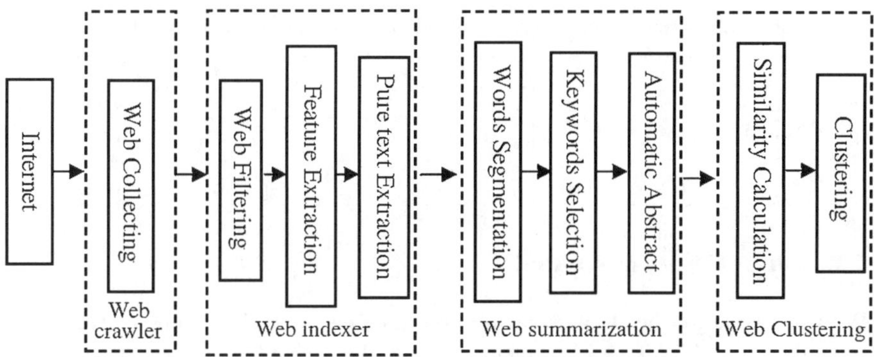

Fig 7.1. The working process of AIS-GAE

动物 影像 生理 临床 科研 手段 射线 脑 优点
光子 图像 检测 挑战 光学 神经 数字 成像 合作

Fig 7.2. Keywords of one Web document

After the process of Web crawler and Web indexer, 192 Web documents were collected from XSSC website. By the process of words segmentation and keywords selection, the keywords which represented the main idea of each Web document are selected as shown in Fig. 7.2. Next, all the keywords of each Web documents were combined into a large words collection, that is,

let k_i^* is the feature keywords set of ith document and the combined keyword collection is $k= \bigcup_{i=1}^{N} k_i^*$ (N=192). So,each of all the Web documents is represented with a Boolean vector using this combined keywords collection, that is,

$$\text{Doc(i)} = (k_{i,1},...,k_{i,j},...,k_{i,m}), \text{ let } k_{ij} = \begin{cases} 1, \text{if keyword j is existing in the } i\text{th document} \\ 0, \text{if keyword j is not existing in the } i \text{ th document} \end{cases}$$

m is the total size of the combined keywords collection and in this application $m = 8392$. Thus, 192 Boolean vectors were obtained to represent the 192 Web documents collected from XSSC Website. And then clustering is undertaken using these vectors.

7.3 Web Clustering on XSSC and its Experimental Results

To carry out clustering on XSSC, two types of preprocessing strategies were examined on the Boolean vectors which are used to represent the XSSC Web documents. One is to cluster the Boolean vectors directly so as to obtain the clusters among the Web documents and another is to transform the Boolean vectors into intermediate vectors using the cosine function firstly and secondly the hierarchical clustering is imposed on the intermediate vectors to cluster the Web documents.

Table 7.1. Distribution of Boolean value in the ith document and the jth document.

		Document j		
		1	0	Total
Document i	1	q	r	q+r
	0	s	t	s+r
	Total	q+s	r+t	q+r+s+t

7.3.1 Direct Clustering on XSSC

With this strategy, the Boolean vectors represented the Web documents were clustered directly with hierarchical method. Here, the Jaccard coefficient was used for computing the similarities between two documents because it is seemed that, by observation, the co-occurrence of one word in two documents is more important than absence of this word in these two documents. Table 7.1 shows how to compute the similarity between two documents using Jaccard coefficient.

In Table 7.1, 1 denotes the co-occurrence of one word in both document i and document j and 0 denotes the absence of one word in both documents. The similarity between the ith document and the jth document is defined as the Jaccard coefficient, that is

$$s(Doc(i), Doc(j)) = \frac{q}{q + r + s} \tag{7.1}$$

With this definition of similarity between documents, a divisive hierarchical clustering is imposed on these 192 Boolean vectors by SPSS. After the hierarchal clustering, a dendrogram is generated and by the observation of XSSC experts, the 192 web documents were separated into 27 clusters.

7.3.2 Hierarchical Clustering using Cosine Transformation on XSSC

In this method, the 192 Boolean vectors are transformed into intermediate vectors employed cosine function firstly and then hierarchical clustering is imposed on these intermediate vectors.

That is, let $\overline{k}_{ij} = \frac{Doc(i) \cdot Doc(j)}{|Doc(i)| \times |Doc(j)|}$, and each of the intermediate vectors is as $\overline{Doc}(i) = (\overline{k}_{i,1}, \overline{k}_{i,2}, ..., \overline{k}_{i,192})$. Then the 192 Web documents is represented by these new 192 intermediate vectors instead of the original 192 Boolean vectors and the hierarchical clustering is imposed on them. Here, the Euclidean distance is employed to compute the similarities between the intermediate vectors, which is defined as

$$s(\overline{Doc}(i), \overline{Doc}(j)) = \left[\left| \overline{k}_{i,1} - \overline{k}_{j,1} \right|^2 + \left| \overline{k}_{i,2} - \overline{k}_{j,2} \right|^2 + \cdots + \left| \overline{k}_{i,192} - \overline{k}_{j,192} \right|^2 \right]^{\frac{1}{2}} \tag{7.2}$$

With this strategy, the 192 Web were clustered employing SPSS and they were divided into 35 clusters.

7.4 Evaluation of Clustering Results

Generally, the clustering result should be evaluated by comparison with a standard clustering result and then a series of indicators are calculated out, such as precision, recall and so on. Unfortunately, none standard clustering result of the Web documents of XSSC is provided. So, it is necessary to find another way to make evaluation.

In order to evaluate the above two hierarchical clustering results, two measures were introduced according to the definition of clustering. And then the above experimental results were evaluated using these two measures in this Section.

7.4.1 Introduction of Average Co-occurrence and Average Difference

From the definition of clustering, it can be drawn that an ideal clustering result is that the objects in a cluster have the most cohesion but the most difference with the objects in other clusters.

To evaluate the cohesion of documents in a cluster, average co-occurrence is introduced which means the average number of representative words existing in all the documents in a cluster. The average co-occurrence in a cluster is defined as:

$$ac = \sum_{w_j \in c_i} f_{w_j}/n_{c_i} \tag{7.3}$$

where c_i is the ith cluster of the clustering results and w_j is the jth representative word of c_i and f_{w_j} is the occurrence frequency of w_j and n_{c_i} is the number of the documents in c_i. In our experiments, the representative words are selected if those words frequency in a cluster is above a threshold that is the half of the number of documents in this cluster. According to the definition of average co-occurrence, it can be deduced that the better is the cluster; the greater is the value of average co-occurrence of this cluster.

Average difference which means the average number of representative words existing in one cluster but not existing in another cluster is introduced to evaluate the difference between two clusters whose definition is as:

$$ad(c_i, c_j) = \frac{\left| S_{c_i} \cup S_{c_j} - S_{c_i} \cap S_{c_j} \right|}{n_{c_i} + n_{c_j}} \tag{7.4}$$

Where S_{c_i} is a set which includes the representative words in c_i and $\left| S_{c_i} \cup S_{c_j} - S_{c_i} \cap S_{c_j} \right|$ is the number of representative words those are existing in one cluster but not existing in another cluster. For the difference between one cluster and the rest of other clusters in the clustering results, the average difference of one cluster is defined as:

$$ad(c_i) = \sum_{j=1}^{m} ad(c_i, c_j)/m \tag{7.5}$$

Here m is the total number of the clusters. It should be noticed that here the frequency of representative words in a cluster is ignored because the number of the

Table 7.2. Average co-occurrence and average difference of direct hierarchical clustering result

Cluster No.	Average Co-occurrence	Average Difference	Cluster No.	Average Co-occurrence	Average Difference
1	6.29	1.26	15	8.00	1.77
2	3.00	0.73	16	3.00	0.91
3	11.75	2.52	17	3.10	0.85
4	11.75	2.56	18	2.00	0.71
5	2.09	0.68	19	2.00	0.74
6	11.25	2.39	20	4.11	0.99
7	8.833	1.78	21	10.29	1.76
8	2.67	0.84	22	9.20	1.92
9	4.60	1.28	23	6.33	1.56
10	5.00	1.57	24	3.29	1.04
11	6.80	1.61	25	6.67	1.62
12	10.00	2.28	26	0.55	0.59
13	7.86	1.45	27	0.00	0.62
14	3.00	1.34			

documents in clusters is different. According to the definition of average difference, it can be deduced that the better performance in clustering, the higher is the value of average difference.

7.4.2 Evaluation of the Test Experiments Results

With the definition of average co-occurrence and average difference, the direct hierarchical clustering result and the result of hierarchical clustering using cosine transformation were evaluated respectively. Table 7.2 shows the average co-occurrence and the average difference of the 27 clusters using hierarchical clustering on XSSC. Also the average co-occurrence and the average difference of the 35 clusters employed hierarchical clustering using cosine transformation is shown in Table 7.3.

From Table 7.2, the mean value of the average co-occurrence of direct hierarchical clustering is 5.68 and the mean value of the average co-occurrence of hierarchical clustering using cosine transformation is 7.33. From Table 7.3, the mean value of the average difference of direct hierarchical clustering is 1.38 and the mean value of the average difference of hierarchical clustering using cosine transformation is 2.18. According to the previous definition of average co-occurrence and average difference, it can be drawn that the clustering result of hierarchical clustering using cosine transformation is better than the result of direct hierarchical clustering about XSSC Web documents.

7.5 Extended Research on Web Clustering

Our extended research on Web clustering is attempting to identify the concepts of the clusters so as to provide a reference of conference classification standard for the XSSC organization committee. It is should be admitted frankly here that the

Table 7.3. Average co-occurrence and average difference of hierarchical clustering using cosine transformation

Cluster No.	Average Co-occurrence	Average Difference	Cluster No.	Average Co-occurrence	Average Difference
1	6.20	1.09	19	9.50	1.94
2	4.67	3.75	20	9.33	1.90
3	12.5	1.82	21	2.57	1.38
4	4.0	2.13	22	4.0	1.93
5	7.67	2.67	23	9.0	2.10
6	8.0	1.44	24	4.33	2.15
7	0.0	2.42	25	3.33	0.96
8	12.0	3.10	26	3.86	1.66
9	6.17	1.24	27	6.4	1.82
10	1.87	1.44	28	4.25	1.61
11	13.67	2.01	29	7.4	2.90
12	3.57	2.11	30	15.0	2.82
13	13.00	2.90	31	8.00	2.90
14	9.67	2.60	32	14.00	4.11
15	6.6	2.99	33	12.75	1.63
16	11.5	4.35	34	2.56	1.44
17	12.75	2.40	35	0.00	1.01
18	6.42	1.50			

accuracy and comprehensibility of the concepts generated automatically by the computer is not comparable with the standard concepts induced by human beings. But the efforts of human beings can be saved with the help of automatic concept extraction.

In Section 4.1, we have pointed out that the representative words for each cluster are selected if those words frequency in a cluster is above the threshold that is the half of the number of documents in this cluster. Here, the representative words can be used for the automatically extracted concepts of each cluster.

To validate our method, experiment is conducted on automatic concepts extraction for each cluster. Firstly, the keywords set given by human for each cluster is established in advance by experts. Then, the representative words of each cluster are evaluated by comparison with the words in the standard set. Last, the precision and recall [13] of each cluster is calculated and these two indicators can tell us in how much extent the automatic concepts extraction of hierarchical clustering using cosine transformation is reliable. Taking cluster No.1 for example, there are 10 Web documents in this cluster so the threshold of co-occurrence for this cluster is 5. By the automatic concepts generation, keywords of this cluster are extracted as is shown in Figure 7.3. And the keywords given by human for this cluster are shown in Figure 7.4. So the precision and the recall of cluster No.1 is 0.2 and 0.25 respectively. Following this method, the automatic concept generation for all the clusters are evaluated and results are demonstrated in Table 7.4.

From Table 7.4, it has been demonstrated that the average precision and recall of our automatic concepts generation method is 0.35 and 0.22 respectively. That is, in

Fig 7.3. The automatic generated concepts for cluster

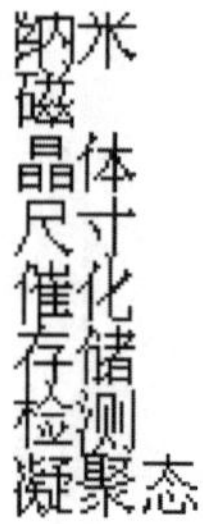

Fig 7.4. The concepts given by human for cluster No.1

Table 7.4. The evaluated results for the automatically generated concepts

Cluster No.	Precision	Recall	Cluster No.	Precision	Recall
1	0.20	0.25	19	0.13	0.25
2	0.33	0.29	20	0.45	0.42
3	0.35	0.88	21	0.25	0.08
4	0.50	0.38	22	0.17	0.07
5	0.27	0.43	23	0.38	0.63
6	0.27	0.36	24	0.17	0.11
7	0.00	0.00	25	0.5	0.33
8	0.20	0.57	26	0.00	0.00
9	0.45	0.50	27	0.40	0.50
10	0.33	0.07	28	0.25	0.40
11	0.10	0.15	29	0.18	0.22
12	0.00	0.00	30	0.20	0.57
13	0.25	0.36	31	0.50	0.55
14	0.31	0.57	32	0.25	0.50
15	0.30	0.27	33	0.25	0.71
16	0.30	0.60	34	0.50	0.18
17	0.30	0.75	35	0.00	0.00
18	0.36	0.18			

a proportion of 0.35 on precision and 0.22 on recall for all the standard concepts set, we can say that our automatic concepts generation method is reliable.

7.6 Conclusion

The motivation of this study is on clustering the Web documents of XSSC in AIS-GAE. Our observation is that if the Boolean vectors represented the Web documents are transformed by cosine function before clustering, the result of clustering would be improved. With this observation, two different clustering experiments are carried out and most importantly, the result of our experiments validates our initial assumption. Furthermore, the average co-occurrence and average difference were proposed based on the definition of clustering to evaluate the clustering results. With both indicators, the clustering results in this paper were evaluated. Also, the extended research of Web clustering on automatic concept generation is presented in this paper and its reliability is demonstrated.

However, although the experimental results on XSSC Web documents have shown that our initial intention looks promising, our current experiments have not fully proved the performance of our clustering strategy.

Further, much work is required for testing our strategy. On one hand, we should apply our strategy on some well known database like WordNet [14], Reuters data set, etc. On the other hand, the average co-occurrence and average difference may also need to be improved with the comparison with traditional evaluation methods to reflect the quality of clustering result more accurately and comprehensively. Another topic is about automatic concepts generation. Although some initial results are demonstrated in this paper, those are far away from the intelligent concepts generation of Web documents.

Acknowledgment

The authors are grateful to the support of the National Natural Science Foundation of China under Grant No. 70571078 and 70221001.

References

1. Han, J.W., Kamber, M.: Data Mining Concepts and Techniques. Mogran Kaufmann Publishers, 2001. 335
2. Macqueen, J.: Some methods for classification and analysis of multivariate observations. In: LeCam, L. M. and Neyman, J. (eds.): Proceedings of the 5th Berkeley Symposium on Mathematics and Statistics, Berkeley: University of California Press, 1967. 281-297.
3. Zhang, T., Ramakrishnan, R., and Livny, M.: BIRCH: An effective data clustering method for very large database. In Proceedings of the 1996 ACM-SIGMOD conference International Conference on Management of Data, Montreal, Canada, 1996. 103-114.

4. Ester, M., Kriegel, H. P. and Sander, J.: Spatial data mining: A database approach. In Proceedings of Symposium on Large Spatial Databases (SSD'97), Berlin, Germany, July 1997. 47-66.

5. Xia, H.X. et al.: Ant-based text clustering using semantic similarity measure: progress report and first stage experiment. In: Gu, J.F. and Chroust, G. (eds.): Proceedings of the First World Congress of the International Federation for Systems Research, Kobe, Japan, 2005. 428

6. Fisher, D.: Improving inference through conceptual clustering. In proceeding of 1987 AAAI Conference, Seattle, Washington, July 1987. 461-465.

7. Liu, Y. J.: Studies on Creativity Support System. Doctoral dissertation, Academy of Mathematics and Systems Science, Chinese Academy of Sciences, 2006. (in Chinese)

8. Tang, X.J., Liu, Y.J., Zhang, W.: Computerized Support for Idea Generation during Knowledge Creating Process. R. In: Khosla, R. J. Howlett, and L. C. Jain (eds.): Knowledge-Based Intelligent Information & Engineering Systems (proceedings of KES'2005, Part IV), Lecture Notes on Artificial Intelligence, Vol.3684, Springer-Verlag, Berlin Heidelberg, 2005. 437-443.

9. Chen, X. and Womersley, R.S.: A parallel inexact Newton method for stochastic programs with recourse. Annals of Operations Research. 64(1996) 113-141. online: http://citeseer.ist.psu.edu/article/chen96parallel.html

10. Lawrence S., Giles C.L.: Searching the World Wide Web. Science, 280 (1998) 98–100.

11. Liu, N.K., Luo, W.D., Chan, M.C.: Design and Implement a Web News Retrieval System. In: R. Khosla, R. J. Howlett and L. C. Jain (eds.): Knowledge-Based Intelligent Information & Engineering Systems (proceedings of KES'2005, Part III), LNAI 3683, Springer, 2005, 149-156.

12. Zhang, W.: Information support tool based on Web Text mining and its application. Master thesis, 2006. Academy of Mathematics and Systems Science, Chinese Academy of Sciences. (in Chinese)

13. Van Rijsbergen, C.J.: Information Retrieval, 2nd Edition, Butterworths, London, UK, 1979, 213, 214.

14. Miller, G.: WordNet: a lexical database for English. Communication of the ACM, 38(11), 1995. 39-41.

Discoveries from Data and Application to Business

A Multilevel Integration Approach for E-Finance Portal Development

Jia Hu and Ning Zhong

Maebashi Institute of Technology 460-1 Kamisadori-Cho, Maebashi 371-0816, Japan
hujia@kis-lab.com and zhong@maebashi-it.ac.jp

Summary. E-finance industry is rapidly transforming and evolving toward more dynamic, flexible and intelligent solutions. The proposed e-finance portal provides an integrated enterprise platform for retail banking services as well as for other financial services. In the meantime, the proposed multilevel solution will keep monitoring and analyzing the huge volume of dynamic information flowing through the portal. In this way, the portal is able to find the useful knowledge for refining business processes and providing personalized services, and detect hidden financial problems as well.

8.1 Introduction

The exponential growth of the Internet over the past few years has essentially altered the landscape of the financial industries with the initiation of continuously available e-trading services and the adoption of e-transactions [6, 12, 16]. The Web has gradually evolved from a service deliver channel to an independent platform for performing financial business.

E-finance encompasses all financial products and services which are available to the consumers through the Internet [6]. The Internet has enabled the expansion of financial services by making it possible to integrate the complex variety of financial data and services, and by providing multiple delivery channels such as mobile, online banking and investment [23], but using it to its full potential is often ignored [3, 32].

The existing infrastructures for financial services are overwhelmingly closed, monolithic, and inward-directed. More dynamic, flexible and intelligent solutions on behalf of consumers and/or business operators are needed. In response to these circumstances, new financial model is beginning to emerge to support the integration of the financial processed across industry and national boundaries.

This paper will discuss how to develop a new type of e-finance portal architecture for integrating data, services and processes in a dynamic way. This solution will be suited not only for supplying effective online financial services for both retail and corporate customers, but also for intelligent risk management and decision making for financial enterprises and partners.

J. Hu and N. Zhong: *A Multilevel Integration Approach for E-Finance Portal Development*, Studies in Computational Intelligence (SCI) **123**, 139–155 (2008)
www.springerlink.com

8.2 Portal Architecture

Figure 8.1 shows the architecture of mining-grid centric e-finance portal (MGCFP) that has been developing by us. The architecture comprises a multi-tired, service-oriented solution that offers a high degree of modularity. The solution is available on the open industry standard platforms J2EE. The portal enables the financial enterprise to have a common infrastructure that encapsulates business rules, back-end connectivity logic and transaction behavior, enabling services to write-once, deploy-everywhere, across channels. The solution ensures a unified view of customer interactions to both the customers and the enterprises.

The Web is exploited as a channel to build, maintain, and develop long-term customer relationships through ready access to a broad and increasing array of products, services and low-cost financial shopping, rapid response to customer inquiries, and personalized product-service innovation. Success in this endeavor would enhance product-service differentiation, which would in turn strengthen customer loyalty, promote cross-selling, increase repeated purchases, and attract new business.

For the individual customers, MGCFP provides them with the facility to check their accounts and do transactions on-line. MGCFP will provide all the bank facilities to its customers when their authentications(user id and password) match, including viewing account information, performing transfers, giving the customer an option of changing address, paying bills online, password retrieval, applying credit card, performing transactions, viewing transactions and locations of the bank and its branches.

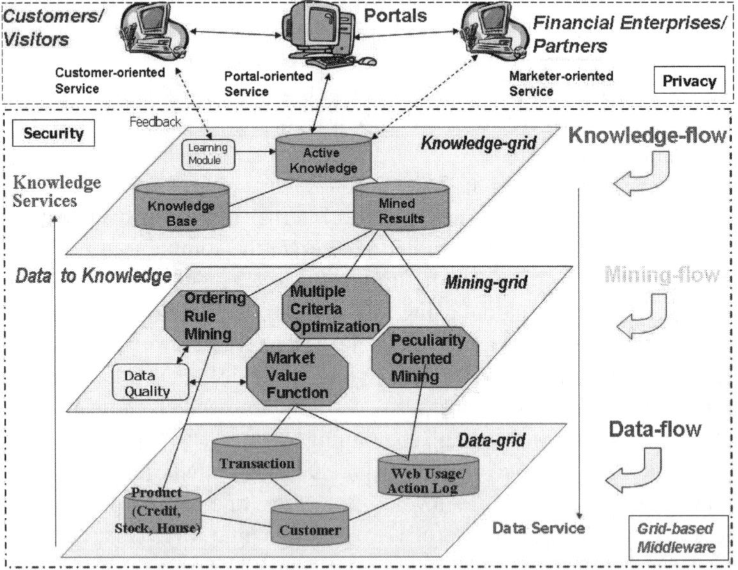

Fig 8.1. The architecture of mining-grid centric e-finance portal

MGCFP could also support an online enrollment facility for credit card customers and should allow customers to view their personnel accounts and to pay bills online from their account.

In MGCFP, there are mainly four kinds of data sources deployed on the data-grid, namely customer, product, transaction, and Web usage dataset. Various data mining methods are employed as agents on the mining-grid for various service-oriented, multi-aspect data analysis [37]. Furthermore, the rules and hypotheses mined from multiple data sources are stored on the knowledge-grid, so that they will be refined into active knowledge by reasoning and inferring with the existing knowledge. The active knowledge is employed to provide personalized financial services for potential customers, portals, and financial enterprises through the three-layer Grid.

The status-based financial services in the MGCFP are dynamically organized by using the workflow management system. The workflows are divided into three levels, namely data-flow, mining-flow, and knowledge-flow, corresponding to the three-layer Grid, respectively. They are generated dynamically, based on the conditions (situations), data quality analysis, mining process, and available knowledge sources. In this model, lower level applications and services provide middleware support for higher level applications and services, thereby opening the door to developing more complex, flexible, and effective systems.

The architecture is deployed on the top of Grid middleware and services, i.e., it uses basic Grid services to build specific knowledge services. Following the integrated Grid architecture approach, these services can be developed in different ways using the available Grid toolkits and services. The current implementation is based on the Globus toolkit, so that the Grid-enabled applications can be accessed by any end users via a standard Web browser.

8.2.1 Data-Grid vs. Data Warehouse

Banks worldwide are subject to a major trend towards multi-channel access, mainly by means of the Internet, mobile phones and other wireless devices using WAP (Wireless Application Protocol) and SMS (Short Messaging Service), and call centers. Customers are also demanding that the multi-channel framework should provide them with a consistent user experience. In addition, the banks themselves need the tools to obtain a unified view of customer interactions through all channels and for all products. As banks expand their service channels to include Internet service, call centers, and mobile banking through wireless devices, they can easily be drawn into piecemeal solutions which impose a penalty in quality of service and in cost of deployment.

There is a strong competitive advantage in acquiring solutions that integrate the channels used and enable a unified management view of customer interactions for all bank services. Traditional e-finance information system is shown in Fig. 8.2. Banking has traditionally been an integrated business, Where financial institutions exclusively distributed self-developed products via proprietary channels and fulfilled all transaction and support services in-house.

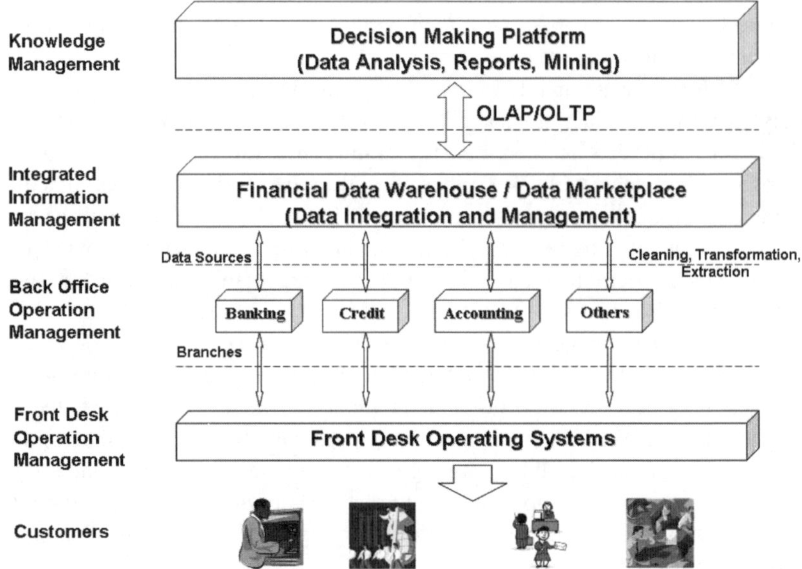

Fig 8.2. A traditional e-banking architecture

Most e-finance information systems are based on data warehouse. Using the popular definition of a data warehouse as a collection of subject-oriented, integrated, time variant, non-volatile data in support of management decisions. Along with the strengths of the data warehouse architecture are many limitations. The first is that it can be very inflexible. The Second limitation is the amount of development and maintenance effort needed to manage a star-oriented data warehouse. And once the data warehouse is built, the maintenance needed to add tables, columns, aggregates and indexes is an ongoing, time-consuming and extremely expensive process.

Therefore, we need a new distributed, flexible infrastructure to develop e-finance portal. As the huge and multiple data sources are coupled with and geographic distribution of data, users, systems, resources, and services in the typical types of enterprises, Grid platform is an ideal middleware or platform for e-finance portals development.

Data storing and retrieving are deployed on the Grid platform, like Globus, as a standard Grid service. OGSA-DAI is used to build database access applications [40]. The architecture is both scalable and transparent. MGCFP is scalable because its design allows for the distribution of different services (represented by objects) among different grid nodes. This is not only means that more nodes can be added to the portal as the number of users increases, but also means that services themselves can be distributed.

8.2.2 Mining-Grid vs. OLAP

E-finance portal is not simply about cheap transaction platform, although that will become more and more important. There is a whole range of all types of risks - credit, liquidity, interest rate risk and market risk - that need to be taken into account. In some ways, the Internet may heighten these risks.

For example, the ability to transfer funds between different bank accounts may increase deposit volatility and could, in extreme situations, lead to "virtual bank runs". Banks will need to build this possibility into their liquidity management policies. Similarly, it is possible that credit risks could increase in the future if the relationship with customers becomes more distant and more transitory, and if the banks relax credit standards because of competitive pressures. On the other hand, banks, will be better placed to obtain and organize information about their customers in an electronic banking environment, and this could help to improve credit evaluation techniques as well as to assist marketing. It all depends how much information each bank has about a given customer.

The mining-flow is a collection of processes related to planning, organizing, controlling, and managing a data mining process dynamically for different mining tasks on the mining-grid. From the top-down perspective, different data mining methods are deployed on the mining-grid as agents for mining services. On the mining-grid, different mining methods work just like agents, that is to say, they are working in an autonomic, distributed-cooperative mode. There are following three main reasons for developing multiple data mining agents on the mining-grid:

1. Businesses rely on data-driven analysis for decision making. Data mining is increasingly recognized as a key to analyzing, digesting, and understanding the large and growing data collected by financial applications.
2. We cannot expect to develop a single data mining method to solve all problems since complexity of the real world applications. Hence, various data mining agents need to be cooperatively used in the multi-step data mining process for performing multi-aspect analysis as well as multi-level conceptual abstraction and learning.
3. When performing multi-aspect analysis for complex problems, a data mining task needs to be decomposed into sub-tasks. Thus these sub-tasks can be solved by using one or more data mining agents that are distributed over different computers and multi-data repositories on the Grid. Thus the decomposition problem leads us to the problem of distributed cooperative system design.

8.2.3 Knowledge-Grid for Knowledge Management

Knowledge-grid allows an integrated management of risks such as credit and market risks. From the top-down perspective, the knowledge level, also the application-oriented level, is supported by both the mining level and data level for serving the customers, portals, and enterprise marketers. From the bottom-up perspective, the data level supplies data services for the mining level, the mining level produces new rules and hypotheses for the knowledge level to generate active knowledge.

In general, several kinds of rules and hypotheses can be mined from different data sources by multi-aspect mining. The results cannot be utilized for knowledge services until they are combined and refined into more general ones to form *active knowledge*, by meta-learning and reasoning. Distributed Web inference engines on the knowledge-grid will employ such active knowledge with various related knowledge sources together to implement knowledge services and business intelligence activities [31, 34].

8.3 Security Concerns

Operational risk, including security risk, is of course one of the more frequently mentioned risks in connection with electronic banking. Security is not a new risk. We are all familiar with the various security issues that banks are facing on a day-today basis, e.g. robberies, thefts of ATM machines, frauds. However, banking transactions over the Internet do pose new issues.

Given the open nature of the Internet, transaction security is likely to emerge as the biggest concern among the e-finance's customers (actual and potential). Since transaction risk would create a significant barrier to market acceptance, its management and control are crucial for business reputation and the promotion of consumer confidence as well as operational efficiency [21]. The customers must be assured that the confidentiality in their transactions must be maintained.

In the empirically most important areas of transaction security and accuracy, encryption protocols such as the Secure Sockets Layer (SSL) have been widely adopted.

8.3.1 Security Objective

Accordingly, the fundamental objectives that e-finance portal security arrangements should try to achieve are to:

- restrict access to the system to those users who are authorized;
- authenticate the identity and authority of the parties concerned to ensure the enforceability of transactions conducted through the Internet;
- maintain the secrecy of information while it is in passage over the communications network;
- ensure that the data has not been modified either accidentally or fraudulently while in passage over the network; and
- prevent unauthorized access to the central computer system and database (intrusion).

8.3.2 Three-Level Security

Security should be integrated into an organization in such a way as to enhance and safeguard each facet in the least intrusive yet most effective way possible at a given time [18]. Therefore, the e-finance portal have to incorporate three-level security measures to protect the transactions from being abused or hacked.

- *Transactional Level Security*. Transactional level security refers to the ability of two entities on the Internet to conduct a transaction; privately and with authentication. SSL (Secure Socket Layer) provides encryption on all data transmitted between Internet, which helps ensure privacy of the data and authentication of the session while preserving the integrity of the message. Moreover, Grid middleware has supplied GSI (Grid Security Infrastructure) layer for security solution in the Grid environment.
- *System Level Security*. System level security helps to protect against corruption of service, and control user access to portal resources. Firewall and password are usually employed in this security level.
- *Application Level Security*. Besides the transaction level and system level security concerns, more concerns are coming from the upper-level applications (such as session management in Web forms) which is highly dependent on the portal architecture and design.

The dynamics at play seem to confirm the aphorism that "security is a process, not a technology." Instead of trying to arrive at a set of universally applicable, absolutely bulletproof security practices, organizations should apply flexible security policies that best serve their goals using whatever information technologies are available to them at the time [18].

8.4 A Case Study: Credit Card Risk Management

Credit card transactions continue to grow in number, taking an ever-larger share of the e-finance system and leading to a higher rate of stolen account numbers and subsequent losses by banks [3]. Large-scale data mining techniques can improve on the sate of the art in commercial practice. Scalable techniques to analyze massive amounts of transaction data that efficiently compute fraud detectors in a timely manner is an important problem, especially for e-finance.

In this section, we present a case study on credit card analysis for demonstrating how to use the model proposed above in an e-finance portal. Data mining for credit card portfolio management decisions is to classify the different cardholder behaviors in terms of their payment to the credit card companies, such as banks and mortgage loan firms. In realty, the common categories of the credit card variables are balance, purchase, payment and cash advance. Some credit card company may consider residence state category and job security as special variables. In the case of FDC (First Data Corporation), there are 38 original variables from the common variables over the past seven months. Then, a set of 65-80 derived variables is internally generated from the 38 variables to perform the precise data mining [27–29].

Our objective is to find:

- which individuals are likely to go bankrupt by looking at the way they use their credit cards,

- who will be interested in buying certain products,
- how valuable a particular customer is,
- who is a good risk for an auto loan,
- what actions should be taken to get someone to pay their bill,
- what tax returns are likely to be fraudulent,
- the probability that a particular credit card stolen.

We looked at the behavior of the cardholder - how much they spent each month and on what, how they paid their bills, how often they revolved (did not pay the full amount each month), and other behavioral information. We then used a set of mining methods developed in our group to predict how much money could be expected to be made or lost from any one account [17, 19].

Figure 8.3 shows a framework of the behavior-based credit card portfolio management, corresponding to the three-layer Grid architecture shown in Fig. 8.1, for fast and effective online customer segmentation, and performing multi-level targeted marketing strategies. From this figure, we can see that the mining process can be divided into the following three phases:

1. Two kinds of datasets, *profile* and *purchasing*, are deployed on the data-grid, and their relationship is connected by the data-flow. The profile dataset is generated by carefully cleaning the customer dataset. On the other hand, the purchasing dataset is generated from transaction dataset, which indicates the purchase number of a customer to a product in a time period.

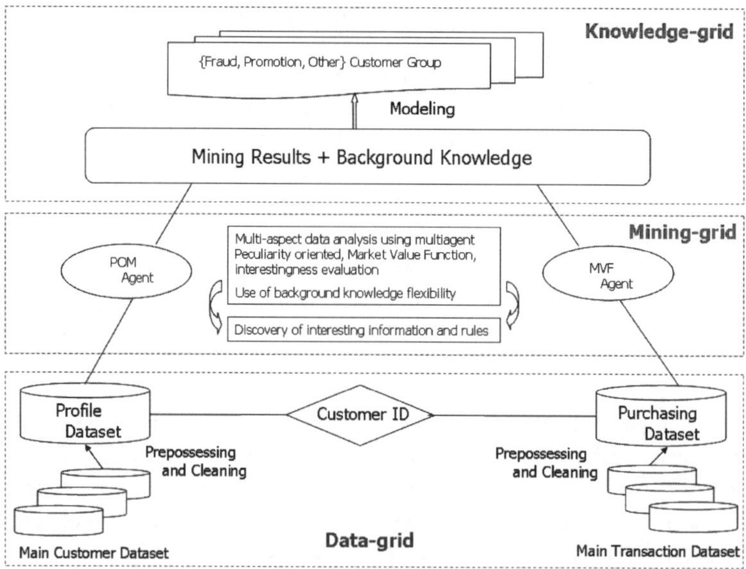

Fig 8.3. A framework of the credit card risk management, corresponding to MGCFP shown in Fig. 8.1

2. Two kinds of mining agents, called POM (peculiarity oriented mining) and MVF
   (targeted market value function), deployed on the mining-grid, are employed to
   mine in the profile and purchasing datasets, respectively, under the mining-flow
   management;
3. The mined results by the POM and MVF agents are combined and refined into
   more general ones to form *active knowledge*, by meta-learning and reasoning
   with various related knowledge sources on the knowledge-grid.

8.4.1 POM Agent

The basis process is as follows, in the profile dataset, let $A_1, A_2, \ldots, A_m$ represents
the different products shown as columns. Let x_{ij} represents the attributes of customer
i to product j, and n is the number of tuples. The peculiarity of x_{ij} can be evaluated
by the *Peculiarity Factor*, $PF(x_{ij})$,

$$PF(x_{ij}) = \sum_{k=1}^{n} N(x_{ij}, x_{kj})^{\alpha} \tag{8.1}$$

where N denotes the conceptual distance, α is a parameter which can be adjusted by
a user, and $\alpha = 0.5$ is used as default.

The peculiarity factor is calculated by the conceptual distances, $N(x_{ij}, x_{kj})$,
with the following equation,

$$N(x_{ij}, x_{kj}) = |x_{ij} - x_{kj}| \tag{8.2}$$

The major method used for testing if the peculiar data exist or not (it is called
selection of peculiar data) after the evaluation for the peculiarity factors is based on
a threshold value in Eq. (8.3),

$$\begin{aligned} threshold \;=\; & mean\ of\ PF(x_{ij}) + \beta \times \\ & standard\ deviation\ of PF(x_{ij}) \end{aligned} \tag{8.3}$$

where β can be adjusted by users, and $\beta = 1$ as default. The threshold indicates that
a data is a peculiar one if its PF value is much larger than the mean of the PF set.
In other words, if $PF(x_{ij})$ is over the threshold value, x_{ij} is a peculiar data. The
details about peculiarity oriented mining refer to [39].

8.4.2 MVF Agent

Targeted marketing involves the identification of customers having potential market
value by studying the customers' characteristics and needs, and selects certain customers to promote. Underlying assumption is similar type of customers tend to make
similar decisions and to choose similar services or products.

Formally, an information table is a quadruple:

$$S = (U, At, \{V_a | a \in At\}, \{I_a | a \in At\})$$

where U is a finite nonempty set of objects, At is a finite nonempty set of attributes, V_a is a nonempty set of values for $a \in At$, $I_a : U \to V_a$ is an information function for $a \in At$. Each information function I_a is a total function that maps an object of U of exactly one value in V_a. An information table represents all available information and knowledge. Objects are only perceived, observed, or measured by using a finite number of properties.

A market value function is a real-valued function from the universe to the set of real numbers, $r : U \to \Re$. For the targeted marketing problem, a market value function ranks objects according to their potential market values.

A linear market value function is of the form:

$$r(x) = \sum_{a \in At} w_a u_a (I_a(x)) \tag{8.4}$$

where w_a is the weight of attribute a, and $u_a : V_a \to \Re$ is a utility function defined on V_a for an attribute $a \in At$. x is the one of the elements in U.

Consider an attribute taking its value from V_a. For $v \in V_a$,

$$u_a(v) = \frac{\Pr(v|P)}{\Pr(v)} \tag{8.5}$$

where $\Pr(v|P)$ denotes the probability distribution of attribute value v in P. $\Pr(v)$ denotes the probability distribution of attribute value v in U.

Consider an attribute taking its value from V_a. For $v \in V_a$,

$$\omega_a = \sum_v \Pr(v|P) \log \frac{\Pr(v|P)}{\Pr(v)} \tag{8.6}$$

where $\Pr(v|P)$ denotes the probability distribution of attribute value v in P. $\Pr(v)$ denotes the probability distribution of attribute value v in U.

For each customer, the purchasing history can be recorded in the transaction dataset and then is transformed to generate the purchasing dataset. Using Eqs. (8.4) to (8.6), we can calculate each customer's market value based on each specific product and category. As stated in [38], the MVF agent is effective to sort customers based on some attributes, like, possibility to buy some product. The MVF agent can use not only demographic information, but also the past purchase information of the customers.

8.4.3 Learning Active Knowledge on the Knowledge-grid

The mined results from two different mining agents are stored on the knowledge-grid, respectively, and they are combined and refined into more general ones to form *active knowledge* by meta-learning and reasoning. Once we obtain behavioral credit cardholder segmentation, any existing recommendation algorithm for cross-sell and up-sell can be employed for the targeted groups. Our segmentation is based on the propensity to consume. Hence, the products that the customer already owned should be filtered out to avoid seemingly trivial recommendations. In this case, association rules could be employed to find other related products.

8.4.4 Experimental Results

The real-life credit card dataset used in this paper is provided by First Data Corporation (FDC), the largest credit card service industry in the world. The raw data came originally from a major US bank which is one of FDC's clients. It contains 5,000 records and 102 variables (38 original variables and 64 derived variables) describing cardholders' behaviors. The data were collected from June 1995 to December 1995 (seven months) and the cardholders were from twenty-eight states in USA. This data has been used as a classic working dataset in FDC for various data analyses to support the bank's business intelligence.

Each record has a class label to indicate its' credit status: either Good or Bad. Bad indicates a bankrupt credit card account and Good indicates a good status account. The 38 original variables can be divided into four categories: balance, purchase, payment, cash advance, in addition to related variables. The category variables represent raw data of previous six or seven consecutive months. The related variables include interest charges, data of last payment, times of cash advance, and account open date. The detailed description of these variables is given in Table 8.1.

The 64 derived variables are created from the original 38 variables to reinforce the comprehension of cardholder's behaviors, such as times overlimit in the last two years, calculated interest rate, cash as percentage of balance, purchase as percentage to balance, payment as percentage to balance, and purchase as percentage to payment. Tables 8.2 gives a brief description of these attributes (variables).

Table 8.1. Original attributes of credit card dataset.

Variables		Description	Variables		Description
1	Balance	Jun 95	20	Cash advance	Jul 95
2		Jul 95	21		Aug 95
3		Aug 95	22		Sep 95
4		Sep 95	23		Oct 95
5		Oct 95	24		Nov 95
6		Nov 95	25		Dec 95
7		Dec 95	26	Interest Charge: Mechanize	Dec 95
8	Purchase	Jul 95	27	Interest Charge: Cash	Dec 95
9		Aug 95	28	Number of purchase	Dec 95
10		Sep 95	29	Number of cash advance	Dec 95
11		Oct 95	30	Cash balance	Dec 95
12		Nov 95	31	Cash balance	Nov 95
13		Dec 95	32	Number of over limit in last 2 years	
14	Payment	Jul 95	33	Credit line	
15		Aug 95	34	Account open date	
16		Sep 95	35	Highest balance in last 2 years	
17		Oct 95	36	Date of last payment	
18		Nov 95	37	Activity index	Nov 95
19		Dec 95	38	Activity index	Dec 95

Table 8.2. Derived attributes of credit card dataset.

Var	Description
1	Balance of Nov 95
2	Max balance of Jul 95 to Dec 95
3	Balance difference between Jul 95 to Sep 95 and Oct 95 to Dec 95
4	Balance of Dec 95 as percent of max balance of Jul 95 to Dec 95
5	Average payment of Oct 95 to Dec 95
6	Average payment of Jul 95 to Dec 95
7	Payment of Dec 95 ($\leq\$20$)
8	Payment of Dec 95 ($>\$20$)
9	Payment of Nov 95 ($\leq\$20$)
10	Payment of Nov 95 ($>\$20$)
11	Payment of Oct 95 to Dec 95 minus Payment of Jul 95 to Sep 95 (Min)
12	Payment of Oct 95 to Dec 95 minus Payment of Jul 95 to Sep 95 (Max)
13	Purchase of Dec 95
14	Purchase of Nov 95
15	Purchase of Dec 95 as percent of max purchase of Jul 95 to Dec 95
16	Revolve balance between Jul 95 to Sep 95 and Oct 95 to Dec 95
17	Max minus Min revolve balance between Jul 95 to Dec 95
18	Cash advance of Jul 95 to Dec 95
19	Cash as percent of balance of Jul 95 to Dec 95 (Max)
20	Cash as percent of balance of Jul 95 to Dec 95 (Min)
21	Cash as percent of balance of Jul 95 to Dec 95 (Indicator)
22	Cash advance of Dec 95
23	Cash as percent of balance of Jul 95 to Dec 95
24	Cash as percent of balance of Oct 95 to Dec 95
25	Cash as percent of payment of Jul 95 to Dec 95 (Min)
26	Cash as percent of payment of Jul 95 to Dec 95 (Log)
27	Revolve balance to payment ratio Dec 95
28	Revolve balance to payment ratio Nov 95
29	Revolve balance to payment ratio Oct 95 to Dec 95
30	Revolve balance to payment ratio Jul 95 to Dec 95
31	Revolve balance to payment ratio Jul 95 to Sep 95 minus Dec 95 to Dec 95 (Max)
32	Revolve balance to payment ratio Jul 95 to Sep 95 minus Dec 95 to Dec 95 (Min)
33	Revolve balance to payment ratio Jul 95 to Dec 95, Max minus Min (>35)
34	Revolve balance to payment ratio Jul 95 to Dec 95, Max minus Min (≤ 35)
35	Purchase as percent of balance Dec 95
36	Purchase as percent of balance Oct 95 to Dec 95
37	Purchase as percent of balance Jul 95 to Dec 95
38	Purchase as percent of balance Jul 95 to Sep 95 minus Oct 95 to Dec 95
39	Purchase as percent of balance, Max minus Min, Jul 95 to Dec 95
40	Purchase as percent of payment, Jul 95 to Dec 95
41	Purchase as percent of payment, Nov 95
42	Purchase as percent of payment, Jul 95 to Sep 95 minus Oct 95 to Dec 95 (Max)
43	Purchase as percent of payment, Jul 95 to Sep 95 minus Oct 95 to Dec 95 (Min)
44	Purchase as percent of payment, Dec 95 as percent of Jul 95 to Dec 95
45	Purchase as percent of payment, Max minus Min, Jul 95 to Dec 95

Var	Description
46	Interest charge Dec 95
47	Interest charge Dec 95 as percent of credit line
48	Calculated interest rate ($\leq 5\%$)
49	Calculated interest rate ($>5\%$)
50	Number of months since last payment
51	Number of months since last payment squared
52	Number of purchases, De 95
53	Number of cash advances, Dec 95
54	Credit line
55	Number of cash advances, Dec 95
56	Over limit indicator of Dec 95
57	Open to buy, Dec 95
58	Utilization, Dec 95
59	Number of times delinquency in last two years
60	Residence state category
61	Transactor indicator
62	Average payment of revolving accounts
63	Last balance to payment ratio
64	Average OBT revolving accounts

For the purpose of credit card classification, the 64 derived variables were chosen to compute the model since they provide more precise information about credit card account's behaviors. The dataset is randomly divided into one training dataset and one verifying dataset. The training dataset has class label for each record and is used to calculate the optimal solution. The verifying dataset, on the other hand, has no class labels and is used to validate the predicting accuracy. The predicting accuracy of a classification method is not judged by the accuracy of training dataset, but the accuracy of verifying dataset. The goal of classification is to apply solutions obtained from the training phase to predict future unknown data objects.

The objective is to produce a "black list" of the credit cardholders. This means we seek a classifier that can identify as many *Bad* records as possible. This strategy is a basic one in credit card business intelligence., Theoretically speaking, we shall first construct a number of classifiers and then choose one more *Bad* records.

The past experience on the experiment showed that the training results of a data set with balanced records (number of *Good* equals number of *Bad*) may be different from that of an unbalanced data set. Given the unbalanced 1,000 accounts with 860 as *Good* and 140 as *Bad* accounts for the training process, the absolute accuracy rate of MVF agent is 41.53% and the absolute accuracy rate of POM agent is 37.69%.

A balanced data set was formed by taking 140 *Good* accounts from 860 of the 1,000 accounts used before and combing with the 140 accounts. The absolute accuracy rate of MVF agent is 48.53% and the absolute accuracy rate of POM agent is 33.69%. Generally, in both balanced and unbalanced data sets, MVF agent is better that POM agent.

Table 8.3. Comparisons on prediction of 5,000 records.

	T_{good}	T_{bad}	Total
Decision Tree			
Good	**2180**	2005	4185
Bad	141	**674**	815
Total	2321	2679	**5000**
Neural network			
Good	**2814**	1371	4185
Bad	176	**639**	815
Total	2990	2010	**5000**
MVF			
Good	**2280**	1905	4185
Bad	514	**301**	815
Total	2794	2206	**5000**
POM			
Good	**1906**	2279	4185
Bad	550	**265**	815
Total	2456	2544	**5000**

The other two known classification techniques, decision tree and neural network have been used to test (or predict) the 5,000 records. The results were shown in Table 8.3.

As we see, the best training comparison on Good (non-bankruptcy) accounts is the Neural network with 2,814 out of 4,184 (67.24%) while the best for Band (bankruptcy) accounts is the Decision Tree with 674 out of 815 (82.7%).

8.5 Related Work

Financial enterprises face significant challenges on both the supply side and demand side, associated in particular with competition, product-service quality and differentiation, transaction security, cost efficiency, and demographic change [16,20,21]. Traditional integrated financial enterprises, which exclusively distribute self-developed products via proprietary channels and fulfill all transaction and support services in-house, are no longer adequate for the changing demands of the environment [24].

This has resulted in banks moving to deliver their products and services through low-cost multiple channels that automate the process and remove the expensive human element [33]. The Web-based e-banking systems are well accepted and developed in the real world [9,11,22].

The challenge Web-based financial intermediaries face is how to supply appropriate services to right customers on demand by organizing and analyzing multiple data sources through different channels in a dynamic way. These two questions are highly coupled with each other.

In one way, customer-related data must be captured across every channel touched by a customer (brick-and-mortar branch bank office, call center, Web site, mobile device, ATM) [33]. For portals with personalized interactive content, the process must take the highly dynamic nature of the content into account [25].

In another way, knowing the customers can help the business to target its products and services to the best effect of both the customer and the business and helping customers satisfy their goals effectively helps to develop customer loyalty [33].

Therefore, properly engineered portals are needed to capture and process information describing customer behavior, desires, and needs and then use that information to present it back in formats that draw a customer in to return and buy more services from the financial institution [33].

A new approach to system architecture is needed that reduces the complexity and costs of coupling information systems as well as increases flexibility to accommodate change. As the financial services and applications is requested to be bundled or integrated, the corresponding technologies have to be integrated to support this.

In [10], authors described a theoretical architecture of a multiagent system that offers an integrated access to the data and knowledge that is available in a bank system. In [7], authors proposed a agent-based data mining system for e-trading through World Wide Web. The real-time data mining technology is integrated into the system for tapping information flows in the e-marketplace and delivering timely information at the right granularity. In [26], the author introduced Service-Oriented Architecture for building e-banking platform. Therefore, a multi-layered model was proposed to converge the related technologies such as portal, process orchestration, Web service, Grid computing, business rules, etc.

8.6 Conclusions

The real world needs request the financial services and applications to be bundled or integrated more than ever before. Therefore, it is clear that the value of e-finance portals cannot be fully realized unless we integrate all the interrelated processes and services effectively through the whole e-marketplace. Moreover, in the end, it is the performance of the entire process that will determine the effectiveness and success of the portal.

In this paper, we proposed a financial portal model with dynamic multi-level workflows corresponding to a multi-layer Grid architecture, for multi-aspect analysis in distributed, multiple data sources, and for dynamically organizing status-based financial services. It is clear that the value of e-finance technology cannot be fully realized unless we streamline all the interrelated processes and services in the marketplace effectively [12].

References

1. G. Alonso, F. Casati, H. Kuno and V. Machiraju, Enterprise Application Integration, *Web Services - Concepts, Architectures and Applications*. Springer, 67-92, 2004.

2. J. Blythe, E. Deelman and Y. Gil, Automatically Composed Workflows for Grid Environments, IEEE Intelligent Systems, 19(4), 16-23, 2004.

3. P. K. Chan, W. Fan, A. L. Prodromidis and S. J. Stolfo, Distributed Data Mining in Credit Card Fraud Detection, IEEE Intelligent Systems, 14(6), 67-74, 1999.

4. A. Congiusta, A. Pugliese, D. Talia and P. Trunfio, Designing Grid Services for Distributed Knowledge Discovery, *Web Intelligence and Agent Systems: An International Jounal*, 1, 91-104, 2003.

5. V. Curcin, M. Ghanem, Y. Guo, M. Köhler, A. Rowe, J. Syed and P. Wendel, Discovery Net: Towards a Grid of Knowledge Discovery, *Proc. KDD'02*, 658-663, 2002.

6. K. Dandapani, Success and Failure in Web-based Financial Services, *Communications of the ACM*, 47(5), 31-33, 2004.

7. J. Debenham and S. Simoff, An e-Market Framework for Informed Trading, Proc. 15th International Conference on World Wide Web (WWW'06), 651–658, 2006.

8. E. Deelman, J. Blythe, Y. Gil and C. Kesselman, Workflow Management in GriPhyN, in J. Nabrzyski et al. (eds.), *Grid Resource Management*. Kluwer, 99-116, 2003.

9. R. Dewan and A. Seidmann, Current Issues in E-Banking, *Communications of the ACM*, 44(6), 31-32, 2001.

10. R. Dewan and A. Seidmann, Information Management at a Bank Using Agents: Theory and Practice, *Applied Artificial Intelligence*, 14(7), 677–696, 2000.

11. M. Fan, J. Stallaert and A. B. Whinston, A Web-Based Financial Trading System, IEEE Computer, 32(4), 64-70, 1999.

12. M. Fan, J. Stallaert and A. B. Whinston, The Internet and the Future of Financial Markets, *Communications of the ACM*, 43(11), 82-88, 2000.

13. I. Foster and C. Kesselman, Concepts and Architecture, in I. Foster and C. Kesselman (eds.), *The Grid 2: Blueprint for a New Computing Infrastructure*. Morgan Kaufmann, 37-64, 2003.

14. Y. Gil, E. Deelman, J. Blythe, C. Kesselman and H. Tangmunarunkit, Artificial Intelligence and Grids: Workflow Planning and Beyond, *IEEE Intelligent Systems*, 19(1), 26-33, 2004.

15. J. Gordijn and H. Akkermans, Designing and Evaluating E-Business Models, IEEE Intelligent Systems, 16(4), 11-17, 2001.

16. S. H. Ha and S. C. Park, Matching Buyers and Suppliers: An Intelligent Dynamic-Exchange Model, IEEE Intelligent Systems, 16(4), 28-40, 2001.

17. J. He, X. Liu, Y. Shi, W. Xu and N. Yan, Classifications of Credit Cardholder Behavior by using Fuzzy Linear Programming, *International Journal of Information Technology and Decision Making*, Vol. 3, 633-650, 2004.

18. C. L. Huntley, A Developmental View of System Security, IEEE Computer, 39(1), 113-115, 2006.

19. G. Kou, X. Liu, Y. Peng, Y. Shi, M. Wise and W. Xu, Multiple Criteria Linear Programming to Data Mining: Models, Algorithm Designs and Software Developments, *Optimization Methods and Software*, Vol. 18, 453-473, 2003.

20. K. Kumar and J. Hillegersberg, New Architectures for Financial Services, *Communications of the ACM*, 47(5), 27-30, 2004.

21. Z. Liao and M. Cheung, Challenges to Internet e-banking, Communications of the ACM, 46(12), 248-250, 2003.

22. W. G. Li, V. Guilherme and F. Branco, A Web Information System for Determining the Controllers of Financial Entities in Central Bank of Brazil, Web Intelligence and Agent Systems, 4(1), 99–116, 2006.
23. N. M. Mallat, M. Rossi and V. K. Tuunainen, Mobile Banking Services, *Communications of the ACM*, 47(5), 42-46, 2004.
24. A. Pan and A. Vina, An Alternative Architecture for Financial Data Integration, *Communications of the ACM*, 47(5), 37-40, 2004.
25. E. Schonberg etc., Measuring Success, *Communications of the ACM*, 43(8), 53–57, 2000.
26. T. Ch. Shan, Building a Service-Oriented eBanking Platform, Proceedings of the 2004 IEEE International Conference on Services Computing (SCC'04), 237–244, 2004.
27. Y. Shi, Y. Peng, G. Kou and Z. Chen, Classifying Credit Card Accounts for Business Intelligence and Decision Making: A Multiple-Criteria Quadratic Programming Approach, *International Journal of Information Technology and Decision Making*, Vol. 4, 581-600, 2005.
28. Y. Shi, Y. Peng, W. Xu and X. Tang, Data Mining via Multiple Criteria Linear Programming: Applications in Credit Card Portfolio Management, *International Journal of Information Technology and Decision Making*, Vol. 1, 131-151, 2002.
29. Y. Shi, M. Wise, M. Luo and Y. Lin, Data Mining in Credit Card Portfolio Management: A Multiple Criteria Decision Making Approach, in M. Koksalan and S. Zionts (eds.), *Advance in Multiple Criteria Decision Making in the New Millennium*, Springer, Berlin, 427-436, 2001.
30. T. Slewe and M. Hoogenboom, Who Will Rob You on the Digital Highway?, *Communications of the ACM*, 47(5), 56-60, 2004.
31. K. Tomita, N. Zhong and H. Yamauchi, Coupling Global Semantic Web with Local Information Sources for Problem Solving, *Proc. SWMR'04*, 66-74, 2004.
32. R. T. Watson, G. M. Zinkhan and L. F. Pitt, Integrated Internet Marketing, *Communications of the ACM*, 43(6), 97-102, 2000.
33. N. Wells and J. Wolfers, Finance with a Personalized Touch, *Communications of the ACM*, 43(8), 31-34, 2000.
34. N. Zhong, Developing Intelligent Portals by Using WI Technologies, in J. P. Li et al. (eds.), *Wavelet Analysis and Its Applications, and Active Media Technology*. World Scientific, 2, 555-567, 2004.
35. N. Zhong, J. Liu, and Y. Y. Yao, Envisioning Intelligent Information Technologies (iIT) from the Stand-Point of Web Intelligence (WI), *Communications of the ACM*, 50(3), 89-94, 2007.
36. N. Zhong, J. Liu, and Y. Y. Yao, In Search of the Wisdom Web, *IEEE Computer*, 35(11), 27-31, 2002.
37. N. Zhong, J. L. Wu and Ch. N. Liu, Building a Data Mining Grid for Multiple Human Brain Data Analysis, *Proc. KGGI'03*, 36-45, 2003.
38. N. Zhong, Y. Y. Yao, Ch. N. Liu, Ch. X. Ou and J. J. Huang, Data Mining for Targeted Marketing, in N. Zhong and J. M. Liu (eds.), *Intelligent Technologies for Information Analysis*. Springer Monograph, 109-131, 2004.
39. N. Zhong, Y. Y. Yao and M. Ohshima, Peculiarity Oriented Multi-Database Mining, *IEEE TKDE*, 15(4), 952-960, 2003.
40. The OGSA-DAI project: *http://www.ogsadai.org.uk/*.

9

Integrated Design Framework for Embedded GUI System

Noriyuki Kushiro and Yukio Ohsawa

School of Engineering, The University of Tokyo, 7-3-1 Hongo, Bunkyo, Tokyo 113-8656, Japan
`kushiro@mx1.ttcn.ne.jp` and `ohsawa@sys.t.u-tokyo.ac.jp`

Summary. There is a growing interest to introduce a Graphical User Interface (GUI) into embedded systems for improving their usability. However, both development and implementation cost hinder GUI from penetration to real products. In this paper, we propose an integrated GUI design framework for cooperation among specialists related to the GUI development and a high performance GUI software and hardware platform for embedded systems to solve these issues. We confirm their availability and impact for GUI design process by applying the framework and the platform to four kinds of GUI products.

9.1 Introduction

There is a growing interest to introduce a Graphical User Interface (GUI) into embedded systems for improving their usability. However, the following two issues hider GUI from penetrating into embedded systems.

1. Development cost: A GUI development requires collaboration with different kinds of specialists. However, the process and the methods for collaboration have not been established. The GUI development is obliged to carry out by trial and error.
2. Implementation cost: A GUI requires large memory and high performance CPU (Central Processing Unit) to handle image and sound data for implementation. They push the GUI hardware cost expensively.

We propose an integrated GUI design framework for supporting cooperative activities among specialists related to the GUI devel opment and a high performance GUI software and hardware platform for embedded systems to solve these issues.

N. Kushiro and Y. Ohsawa: *Integrated Design Framework for Embedded GUI System*, Studies in Computational Intelligence (SCI) **123**, 157–170 (2008)

9.2 Issues for Realizing GUI on Embedded Systems

Issues of the GUI development on embedded systems are described as the followings:

1. Each design activity in the GUI development requires a different kind of specialist. Communication errors between specialists are often occurred. The gap in communication between specialists causes back track of the GUI development.
2. A big modality conversion from a system design specifications described in literature to a screen design expressed in pic tures. The gap in modality conversion also causes back track of the GUI development.
3. A system engineer does not understand clearly what should be defined in system specifications for a screen design. Omission of the system specifications is often detected in the succeeding steps of the GUI development. The omission worsens the GUI development cost.
4. A screen designer is obliged to design a screen by trial and error, because the method for designing a screen has not been established.
5. A usability engineer takes a lot of time to identify issues detected in a usability test, because the gap in modality conversion makes the identification of the issues difficult.
6. A programming engineer takes a lot of trouble for GUI implementation on embedded systems with many restrictions.

9.3 Solutions

The following two solutions are proposed to solve these issues.

1. An integrated GUI design framework (A methodology and tools) for supporting cooperation among specialists related to the GUI development
2. A high performance GUI software and hardware platform for embedded systems

9.3.1 GUI Design Framework

There are two major approaches for improving efficiency of GUI development. One is a quick prototyping approach [1] and the other is a design reuse approach [2]. The quick prototyping is an approach to shorten GUI development cycle. However, it is difficult to apply the embedded system, because it requires rich hardware resource with large memory and high performance CPU. The design reuse is an approach to reduce work by reusing property accumulated in the past development. Pattern [3] and Application framework [4] are typical researches of the design reuse. In the last few years, several researches about the pattern and the application framework have been devoted. However, many approaches have ended in failure for the following reasons:

1. Difficulties for understanding: It is difficult for engineers to understand concepts of the pattern and the framework. Because, it requires deep knowledge about the domain targeted by the pattern and the framework themselves.

2. Difficulties for applying: It is difficult for engineers to apply the pattern and the framework to their own applications, because the descriptions of the pattern and the framework are very polymorphic.

An integrated design framework for embedded GUI systems is tried to construct to solve the above mentioned issues. The integrated design frame work is consists of the following mechanisms.

1. A common design field, where engineers share each specification mutually
2. Design support tools, which supports each engineerfs design activities

In this paper, the common design field is defined as a common structure of knowledge required for GUI design and knowledge arranged in the structure. Engineers with different kinds of expertise communicate and share mutual knowledge on the common design field.

9.3.2 Common Design Field and Design Support Tools

A common design field is a place for transferring the design knowledge between engineers. Nishida et al. modeled engineers knowledge as three layers shown in Fig. 9.1 [5]. The common design field is designed on the frame.

The communication between engineers is indispensable for cooperation in GUI design. The necessary condition for realizing communication between engineers is that a common protocol exists between each knowledge layer shown in Fig. 9.1. The common design field plays as a gateway for communication between engineers with different expertise (Fig. 9.2).

A GUI design team is typically composed of the following engineers: system engineers, screen designers, programmers and usability test engineers. The last target

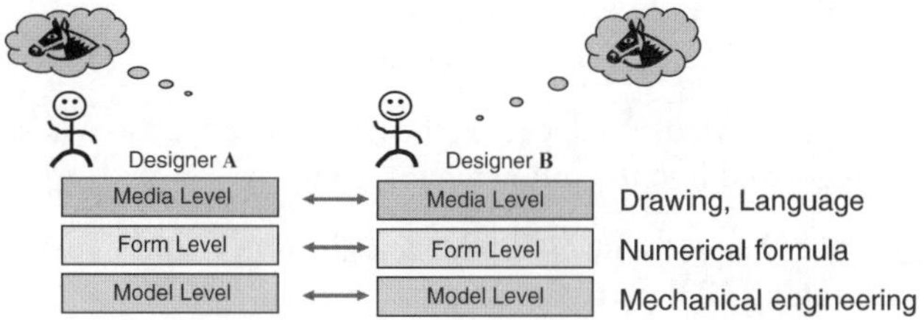

Fig 9.1. Communication model between engineers (Nishidafs model)

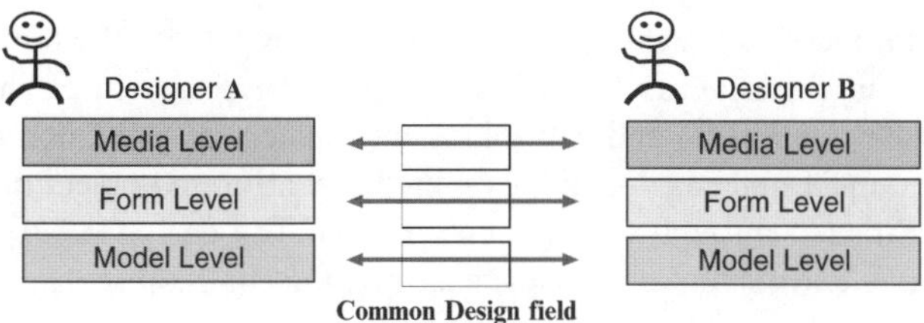

Fig 9.2. Common design field

of these engineers' activities is screens expressed in pictures. Therefore, it is appropriate to apply knowledge required for drawing a picture as the common design filed between engineers.

Generally, drawing a picture is seemed not to be so difficult besides performance of the work. However, it is still enough creative not to explain how we can draw a picture [6]. Only few attempts have so far been made at the point. In this study, we apply the knowledge structure of linguistics as a substitution of the structure of drawing, because writing a text is regards as similar creative activity of drawing a picture and the research for linguistics is progressing in comparison with drawing a picture.

In linguistics, a text has a hierarchical structure composed of the following elements: text, sentence, word, morpheme and phoneme. And a text has rules, i.e. grammar and syntax, which specify mutual relationships between each element.

Each element of GUI is mapped on the element of linguistics.

> *Phoneme* = Character, pictogram and sound etc.
> *Morpheme* = Simple GUI parts like a push button and a slide switch etc.
> *Word* = Compound GUI parts combined with simple GUI parts
> *Sentence* = A screen composed of both simple GUI parts and compound GUI parts.
> *Text* = A group of screens to realize a certain scenario

Furthermore, the rules are mapped as follows.

> *Phoneme* : *Morpheme* = Rules for simple GUI parts operation, i.e. rules for alternation of pictogram when a button pushed
> *Morpheme* : *Word* = Rules for compound GUI parts operation, i.e. rules for reflection of the value when a slide switch operated.
> *Word* : *Sentence* = Rules of changes in a screen, i.e. rules for a pop-up screen at the time of button operation.
> *Sentence* : *Text* = Rule of screen changes

Concrete design knowledge mapped on the frame is defined as a design rule. The design rule is categorized into the following three kinds:

1. Correspondences between the GUI parts and the object in the real world
2. Rules for designing GUI parts and screens
3. Correspondences between the GUI parts and the GUI class library in the framework

The domain modeling technique [7, 8], which is a pattern elicitation technique in the software engineering, is applied to construct the design rule. The embedded GUI systems for monitoring and controlling appliances equipped in a building and a home are set up as analysis domain. Common scenarios and object models for the domain are extracted through the domain analysis. The design rules are extracted trough designing and evaluating screens based on the common scenarios and object models.

Table 9.1. Design rules arranged in common design field

Linguistics		Phoneme	Morpheme	Word	Sentence	Text
GUI		Interface phoneme	Simple GUI parts	Compound GUI parts	Screen	Group of Screen
			Rules for operation of Simple GUI parts	Rules for operation of Compound GUI parts	Rules for operation in a Screen	Rules for transition of Screens	
Design Rules	Correspondences between the GUI and the object	Correspondence between GUI parts and the object in the real world Example: – On/Off uses "Push button" in GUI parts – Tempareture setting uses "Analague volume"				
	Rules for designing GUI parts and screens	Designing rules for GUI screen and parts based on the ergonomics and cognitive engineering. It consists of a design guideline and concrete design rules(Design catalogue).				
	Correspondences between the GUI and the GUI class	Correspondence between GUI parts and the class library for Example: – "Push button" is implemented by "Push button" in class library				

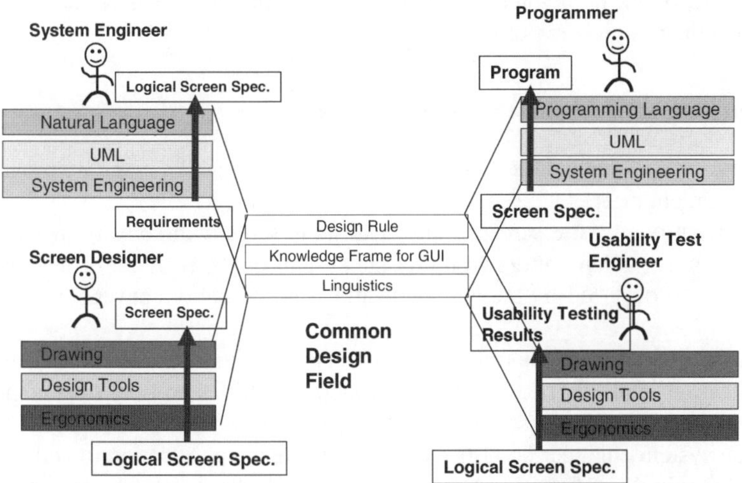

Fig 9.3. Integrated framework for GUI design and evaluation

The extracted design rules are arranged in the frame shown in Table 9.1. System engineers, screen designers, programmers and usability test engineers utilize the design rules as common knowledge for communication and cooperation.

9.3.3 Design Support Tools

Design support tools are group of tools to support each engineerfs activity based on the common design field (Fig. 9.3). For example, the system designers are provided a process and tools for designing the logical panel design specification from the requirements. The programmers are also provides GUI class library for implementation based on the logical panel design specifications.

Table 9.2. GUI design and evaluation supporting process and tools

Target Engineer	Design Support Tools
System Engineers	Design Process and description methods for the logical screen design on the system specifications
Screen Designer	Design Process and design rules for the screen design on the logical screen specifications
Programmer	GUI hardware and software platform for implementation on the logical screen specifications
Usability Test Engineers	Process and design rule for evaluating the implemented screen

Engineers realize the embedded GUI system by utilizing the design support tools (Table 9.2) on the common design field. Thus, engineers can easily understand why and how the design is realized.

Design Support Tools for the System Engineer

The process and method for describing logical screen specifications are provided for the system engineers.

Conventionally, the screen designer designs screens directly from the system specifications, which are drawn up by the system designer. A lot of defeats of the specification occurred by the communication and modality conversion gap in this step.

In this research, an intermediate specification document, namely logical screen specification is placed between the system specifications and the screen specifications for making the gaps small.

The system engineer describes the logical screen specification to add the information required for designing a screen after drawing the system specification based on requirements of stakeholders.

An object-oriented-analysis tool is enhanced for describing the logical screen specifications(Table 9.3). Recently, UML (Unified Modeling language) [10] is widely uses as an object oriented analysis tool. However it is difficult for screen designers or usability test engineers to understand.

Coadfs object analysis tool [9] is selected for intuitively understanding. The describing methods are enhanced for describing the logical screen specifications. The example of enhancements is shown in Fig. 9.4.

Design Support Tool for Screen Designer

A screen designer design a screen from the logical screen specifications by using the design rule specified at the common design field. The following two kinds of design rule are prepared for the screen designer.

Table 9.3. GUI specification description methods

Description method		GUI specification
Scenario		Operation mode
Class Chart	Relationship between objects	Screen transition, Operating procedure
	Object	Object for operation and monitoring
	Property of Object	Properties for operation and monitoring in the object
	Relationship between objects	Cluster of properties shown in a screen
	Priority in Properties	Priority for showing properties
	Method	Menu for operation and monitoring
Event Trace Chart		Interaction between man and GUI
Data Dictionary		Definition of value of property

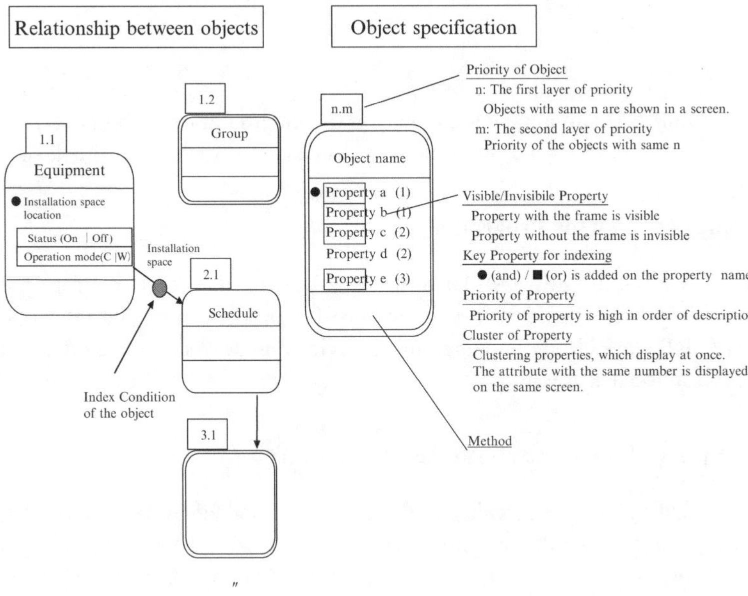

Fig 9.4. Example of GUI specification description

1. GUI parts catalog: the catalog of GUI parts, which have already designed and evaluated as parts for reusing.
2. GUI design guideline: meta design rules for designing new GUI parts and screen.

These design rules are arranged in the frame based on linguistics (Table 9.1). The screen designer designs a screen to select GUI parts from the catalogue. When a new GUI part is necessary to design, the engineer designs a part with the GUI design

Table 9.4. Example of design guideline

Category	Contents of Design Guideline
Interface Phoneme	
@ Icon and Button	Basic Size of Icon and Button
	ENormal Size: From 5 to 8 grids square
	ESize for Navigation button: Flexible on the caption length
	i2jBasic Color of Icon and Button
	EButton: Gray
	EIcon: Gray
	ENavigation Button: Sky Blue
	EHelp, Guide: Green
	EEEEEE
Rules of operation for simple GUI parts	
Do/Undo	EDo: Navigation Button with OK icon
	EUndo: Navigation Button with U turn icon
	EEEEE
	EEEEE

guideline(Table 9.4) and appends the part on the catalog after usability testing. The integrated framework provides continuously process for brushing up the design rule.

Design Support Tool for Programmer

A GUI software and hardware platform (hereafter it called GUI platform) shown in Fig. 9.4, is provided as design support tool for programmer. The GUI platform consists of GUI class library and its running environment. Detail of the GUI platform is described in the next session.

Design Support Tool for Usability Testing Engineer

A usability testing engineer evaluates the screen embodied by the screen designer and the programmer. All the design activities are performed on the common design field, to trace a series of design activities, i.e. what kind of requirements are requested for the screen, which design rules and which GUI class library are selected to satisfy the requirements, becomes possible. That is, the usability testing engineer easily identifies in which steps problems have been occurred and minimizes back track for reflecting the results of usability test.

9.3.4 GUI Platform for Embedded Systems

GUI requires large memory and high CPU performance CPU to handle a log of image and sound data in comparison with the character based interface. In this research, we develop a GUI platform [11] for embedded systems. The GUI platform is composed of the following three key technologies:

1. Memory reduction GUI-OS
2. GUI class library
3. High performance hardware platform

Memory Reduction GUI-OS

According to our previous prototyping, memory cost came up to about 60% of a GUI hardware cost. Memory reduction is important to decrease total hardware cost. Especially, reducing RAM is critical, because RAM is four times expensive than ROM per the same memory space. The following object implementation technologies are developed for reducing memory:

ROM Based Object Implementation Method

An object is designed as working mainly on ROM for reducing RAM. Most parts of an object (non-variable properties and method) are mapped on ROM. Only a few variable properties are mapped on RAM. The allocation of ROM and RAM are declared in the analysis step and is fixed in execution step (Fig. 9.5).

Hierarchical Object Structure

Resource management is required to use limited memory efficiently. All resources are defined as objects for reusing. GUI parts (such as button and icon) tend to access same image and sound data. These data are separated from GUI parts and assigned them to the independent object interface phoneme object which commonly used by GUI parts. Similarly, same operation procedures (such as time setting procedure or equipment control procedure) tend to be utilized repeatedly on a screen. For reusing these stereotyped operation procedures, operation procedures are defined as operation procedure objects, i.e. Slide control and Scenario Control. The linkage between GUI parts object and database object decides the semantic content of the operation procedure. To make the operation procedure object independent from the particular semantic content, the linkage has been separately defined as link object (Fig. 9.6).

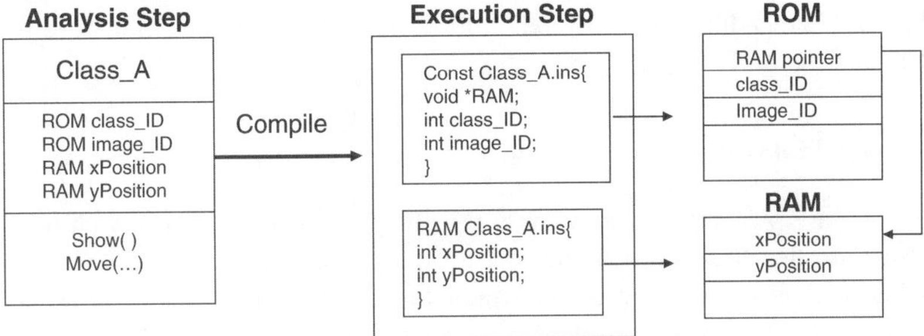

Fig 9.5. ROM based objects implementation

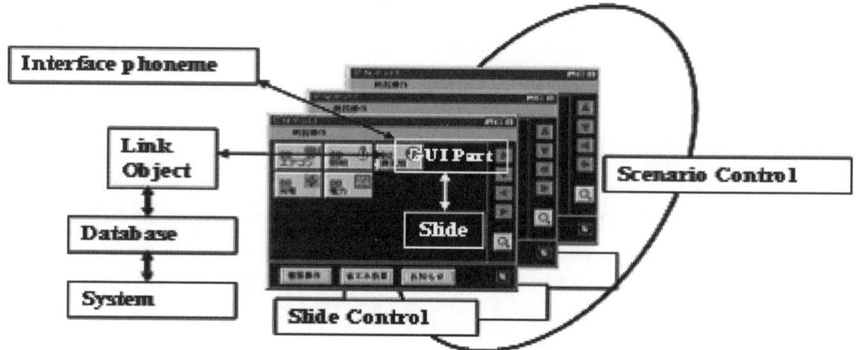

Fig 9.6. ROM based objects implementation

As for a work station or a personal computer, most of GUI software is implemented with an object oriented programming language (OOPL) like Java and C++. But for the embedded systems, GUI software must be implemented on a small hardware to minimize cost. The OOPL is difficult to apply the embedded systems as the following reasons: OOPL requires large RAM space and the class library is too redundant to apply to the embedded GUI systems [14].

C-language is selected to minimize memory. However, it does not support mechanisms for object oriented programming. To realize object oriented programming environment on the non-OOPL, only two indispensable mechanisms for object oriented programming are developed: Object management (class and instance management) and Message management mechanism.

Consequently, the GUI platform is realized about 1/100 memory size as small as the case realized with the object oriented language (C++).

GUI Class Library

The class library is developed based on the structure of the common design field. The class library classified into 9 categories and consists of 100 kinds of objects. The class library is understood intuitively with analogy of writing a text. Furthermore, the class library, which inherits the flexible structure of language, minimizes the influence of amendment/improvement of the class library and maximizes reusing of the library.

Hardware Platform

The hardware platform is very simple (Fig. 9.8). It consists of only 4 chips, 32 bit micro controller, ROM, RAM and LSI for the peripheral interfaces. Sound control and tablet control are achieved by using timer, pulse width modulation and A/D converter circuits equipped in the micro controller to reduce cost. Small size multi task operating system is installed on the hardware platform.

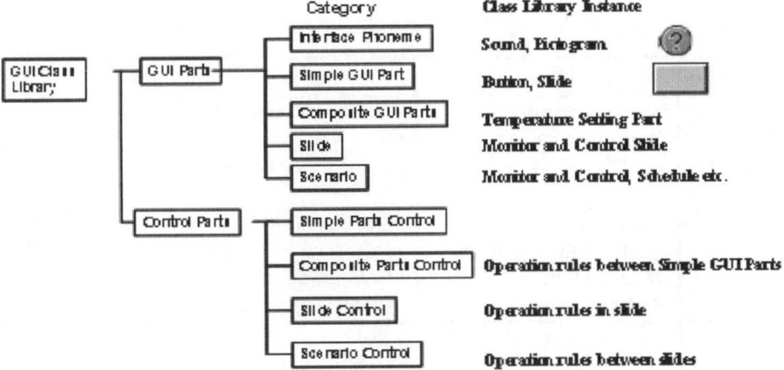

Fig 9.7. GUI class library

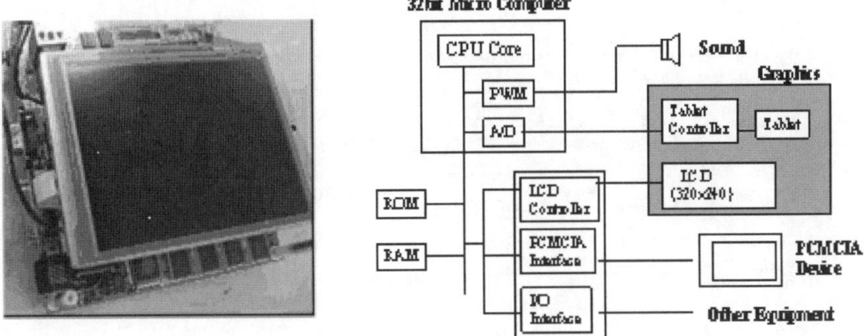

Fig 9.8. GUI hardware platform

9.4 Applications for Integrated Framework

The integrated framework is applied to four kinds of embedded GUI systems shown in Table 9.5: one system is for building facility management, two systems for residential control [12] and one system for remote controller [13].

9.5 Evaluation

By applying the integrated design frame work to 4 kinds of products, we confirm that the integrated design framework impacts design, implementation and evaluation of the embedded GUI systems.

9.5.1 Impact for Design and Evaluation

A usual development process for the embedded GUI systems is obliged to perform by trial and error approach and the efficiency of the development is bad (Fig. 9.9).

The development process is improved from trial and error to water fall model by introducing the integrated framework (Fig. 9.10).

Table 9.5. Application of the integrated framework

Product	Target	Functions
Building Facility Management System	Central Controller for Building Facilities: Lighting and Air conditioner	• GUI operation panel • Real time controller for equipments • Remote monitoring and controlling via Internet
Residential Control System	Controller for Home Appliances: Water-heater, Air-conditioner etc.	• GUI operation panel • Real Time controller for equipment • Gateway between Internet and home-network • Program download via Internet
Remote Controller for Home Appliance	Remote Controller for Elderly Persons	• GUI operational panel • Multi model response • Real Time Controller for home appliances

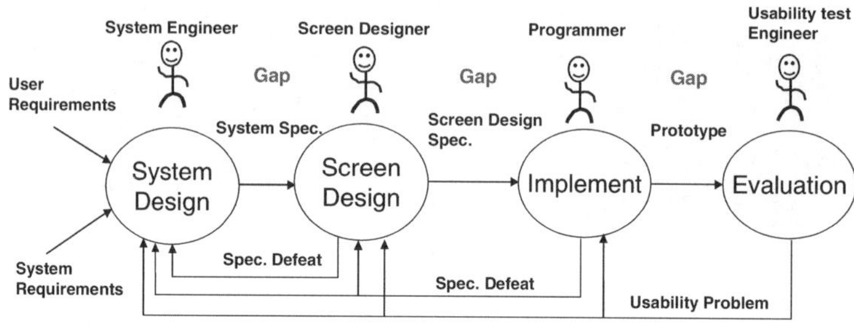

Fig 9.9. Usual design and evaluation process

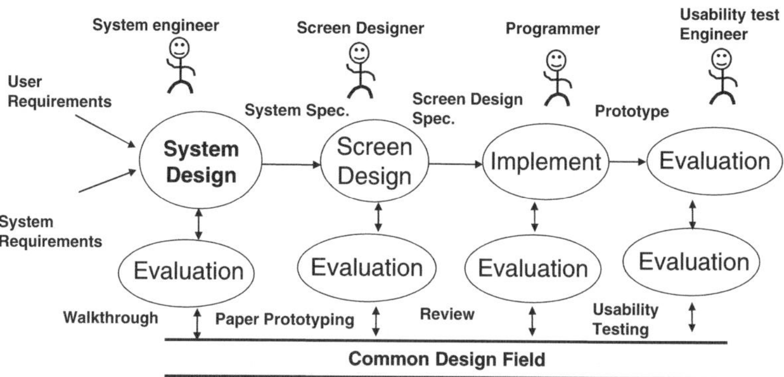

Fig 9.10. Improved design and evaluation process

1. All the results of engineers activities are shared on the common design field. It decreases the gaps both in the communication and the modality conversion between engineers.

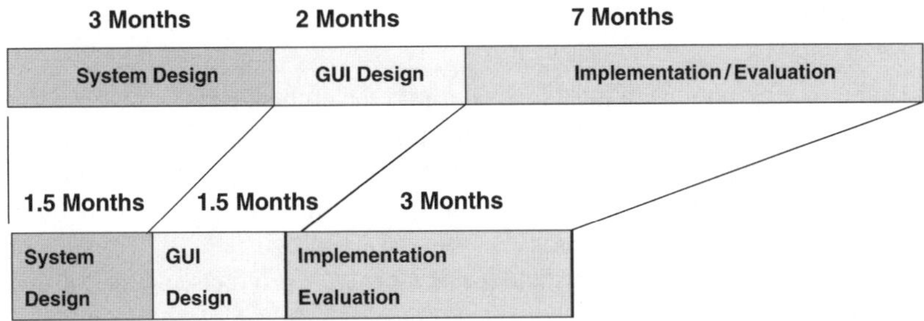

Fig 9.11. Results of estimation for the impact of GUI design

2. Each result is evaluated step by step in every design process. Hierarchical evaluation process improves the GUI development process drastically by shortening feedback loop of the evaluation.

In order to estimate the impact of the design process quantitatively, a feasibility study are performed to compare man-month in cases that the conventional method and the proposed method applied to almost equivalent embedded GUI systems. The results of the feasibility study are shown in Fig. 9.11.

1. System design: man-month of system design is shortened by reusing requirements and system specifications accumulated by the domain modeling.
2. GUI design: reuse of design rule improves man-month 30%. Designing new icons and their evaluation based on the design guideline restrain the improvement of man-month.
3. Implementation and evaluation: the platform improves the efficiency of the implementation 40%. The reuse of well evaluated GUI parts decreases fatal problems of usability, which detected in the evaluation process.

9.5.2 Impact for Implementation

The memory reduction GUI-OS and the class library minimize memory consumption and increase reuse rate of library.

1. ROM based software implementation method: Most of the software is mapped on ROM. The ratio ROM to RAM in total memory consumption is 80 to 20.
2. Hierarchical object structure: the interface phoneme object reduces about 100KB memory and the operation procedure objects reduce about 10MB memory consumption.
3. Performance: the class library provides sufficient performance. GUI parts with sound and animation performs briskly. The average rate of CPU load for GUI library is under 10%.

9.6 Conclusion

There is a growing interest to introduce GUI into the embedded systems for improving their usability. However, the development and implementation cost hider GUI from penetrating into the embedded systems. We have developed the integrated GUI design framework to support cooperative activities among specialist and the GUI platform for embedded systems.

We have applied the framework and the platform to 4 products and the followings have been confirmed:

1. The framework has availability enough to apply several embedded GUI products.
2. The framework impacts the design and evaluation process. The process changes from cut-and-try to water-fall model. The changes drastically improve the efficiency of GUI development.
3. The platform minimizes memory in use, especially RAM. The hierarchical GUI class library increases the reuse rate of the library.

We believe that the framework clears the way to introduce GUI to the embedded systems.

References

1. R.Acosta, C.Burns, W.Rzepka and J.Sidoran, *Applying Rapid Prototyping Techniques in the Requirements Engineering Environment*, IEEE, 1994.
2. C.Kruger, Software reuse, *ACM Computing Surveys*, 1992, 24(2), pp.131-183.
3. E.Gamma, R.Helm, R.Johnson and J.Vlissides, *Design Patterns: Elements of Reusable Object-Oriented Software*, Addison-Wesley, 1995.
4. T.Reenskaug, *Working with Objects: The OORam Software Engineering Method*, Manning Pubns Co, 1996.
5. T.Nishida,T.Tomiyama, T.Kiriyama and H.Takeda, *Management on the Engineering knowledge*, Asakura Shoten(in Japanese),1998.
6. R.A.Finke, B.Thomas and S.M.Smith, *Creative Cognition*, MIT press, 1992.
7. R.Prieto-Diaz, Domain Analysis: An Introduction, *ACM SIGSoft Software Engineering Notes*, 1990, 15(2), pp.47-54.
8. W.Tracz and J.Poukin, *A Domain Specific Software Architecture Tutorial*, 3rd ICSR, 1994.
9. P.Coad, *Object Oriented Analysis*, Yourdon press, 1990.
10. UML,*http://www.uml.org/*.
11. N.Kushiro et.al, *Agent Based Graphical User Interface Architecture for Consumer Products*, IEEE Transaction on Consumer Electronics, 1996.
12. N.Kushiro et.al, *Integrated Residential Gateway Controller for Home Energy Management System*, IEEE Transaction on Consumer Electronics, 2003.
13. N.Kushiro et.al, *Design and Evaluation of Universal Controller with Multi-modal Response on Andragogy Theory*, Journal of Human Interface Society, 2006.
14. B.J.Cox, *Object-Oriented Programming, An evolutionary Approach*, Addison-Wesley,1998.

10

A Unified Probabilistic Inference Model for Targeted Marketing

Jiajin Huang[1] and Ning Zhong[1,2]

[1] The International WIC Institute, Beijing University of Technology
100022, Beijing, China
hjj@emails.bjut.edu.cn

[2] Department of Life Science and Informatics, Maebashi Institute of Technology
Maebashi-City 371-0816, Japan
zhong@maebashi-it.ac.jp

Summary. Targeted marketing is a low cost, low risk, and profit-driven strategy. It aims to recommend right products to right customers or select right customers for right products. It is an important area of application for data mining and Web intelligence. It typically involves two components, customers and products. This paper employs a probabilistic inference model to model the relationships between customers and products. The relationships are interpreted by precision-oriented and recall-oriented interpretations in this model. According to available different information, naive bayes and the latent semantic model could be used to measure the relationships. As an illustrative example, this paper present inductive methods to show how to employs the probabilistic inference model to measure customer preference. Experiments on real world data show that the proposed methods are effective.

10.1 Introduction

Targeted marketing typically involves two components, customers and products, which lead to two kinds of problems. On the one hand, there may exist a large number of customers to whom marketers want to promote a particular product. The problem is how to select customers who want to buy the product. On the other hand, there may also be a large number of products which marketers want to recommend to a particular customer. The problem is how to select products that will interest the customer. The former problem involves the identification of a relative small group of potential customers from a huge pool of customers to sell a certain product and the latter problem involves the identification of a relative small set of products from a huge pool of products for a certain customer. In general, the former is called product-oriented targeted marketing problem [9] or direct marketing problem [3]; the latter is called customer-oriented targeted marketing problem [9] or product recommendation problem [1].

In recent years, the advances in computer technology offer new challenges and opportunities for targeted marketing. Targeted marketing is an important area of

J. Huang and N. Zhong: *A Unified Probabilistic Inference Model for Targeted Marketing*, Studies in Computational Intelligence (SCI) **123**, 171–186 (2008)
www.springerlink.com

application for data mining and a crucial area of web intelligence for e-business portals [11, 22, 26–28]. Many methods have been proposed and studied to solve the targeted marketing problems. They employ either customer profiles, product profiles or transaction data to develop targeted marketing systems.

In product-oriented marketing (or direct marketing) systems, customer profiles are used to select customers who have potential market values. Information of a product may be sent to non-buyers who share the same features of buyers or similar to buyers. Many data mining techniques have been used to solve the direct marketing problem with customer profiles as input [25], such as neural network [14], decision tree [13], market value function [23], and so on.

There are two main methods in product-oriented marketing (or recommender) systems, namely collaborative filtering and content-based filtering. Collaborative filtering [6] uses transaction data to find the preference of a customer for a given product. Under the assumption that similar type of customers tend to make similar decisions and to choose similar products, collaborative filtering aims to recommend products to a given customer based on the purchase history of both the customer and like-minded customers. There are two classes of collaborative filtering techniques: memory-based method and model-based method [15]. The memory-based method uses transaction data to find customers with similar preference. The model-based method uses transaction data to learn models, including association rule [2], Bayesian network, *etc.*, and employs these models for prediction. Content-based filtering uses product profiles to identify products which a certain customer likes. Content-based filtering tends to recommend products similar to what customers have bought in the past [1]. Many methods have been proposed to enhance the performance of a recommender system. An common solution is to combine collaborative filtering and context-based filtering techniques by using customer profiles, product profiles and transaction data [4, 5, 17].

For a more complete and realistic model of targeted marketing, one needs to consider at least three types of information. They are the customer features represented by customer profiles, the product characteristics represented by product profiles, and the transaction data that link customers and products. One of the important issues in targeted marketing is to build the relationships between customers and products. Inspired by the probabilistic inference model for information retrieval [16, 18, 19], this paper employs the probabilistic inference model to build the relationships by using customer profiles, product profiles and transaction data. This model interprets the relationships between customers and products by the notions of precision and recall. And at the same time, this model combines customer-oriented targeted marketing and product-oriented targeted marketing. As an illustrative example, this paper uses the probabilistic inference model to measure customer preference and then employs an inductive method to solve targeted marketing problem.

The rest of the paper is organized as follows. Section 2 briefly reviews the probabilistic inference model. Section 3 discusses the formulation of the probabilistic inference mode for targeted marketing according to available different kinds of information. Section 4 uses inductive methods to illustrate the application of the model shown in Section 3. By using real world examples, Section 5 evaluates the result of the method proposed in Section 4. At last, Section 6 gives conclusions.

10.2 A Probabilistic Inference Model

This section briefly reviews the probabilistic inference model by using the notion from [18, 19]. In the probabilistic inference model, it is assumed that there exists an ideal concepts space U. A proposition is a subset of U.

A probability function Pr is defined on the concept space U. Given two propositions E and H in the knowledge space U, the single implication, written $E \to H$, indicates that E supports H. The double implication, written $E \leftrightarrow H$, means that E and H support each other.

If we regards E to be the evidence, we have

$$\psi(E \to H) =_{def} Pr(H|E) = \frac{Pr(H \cap E)}{Pr(E)}. \tag{10.1}$$

$\psi(E \to H)$ provides a plausible measure of the degree of belief or confirmation of proposition H given evidence E [18].

The function $\psi(E \to H)$ is an asymmetric measure. That is, the degree of E supporting H may not necessarily be the same as H supported E. Thus, a symmetric measure for mutual support is defined as

$$\psi(E \leftrightarrow H) =_{def} \frac{Pr(H \cap E)}{Pr(H \cup E)}. \tag{10.2}$$

The double implication $H \leftrightarrow E$ can be described as follows in terms of the two single implications, $E \to H$ and $H \to E$ [19],

$$\psi(E \leftrightarrow H) = \frac{\psi(H \to E)\psi(E \to H)}{\psi(H \to E) + \psi(E \to H) - \psi(H \to E)\psi(E \to H)}. \tag{10.3}$$

The subsequent sections borrow the above discussions and results from [18, 19] to analyze the targeted marketing problem.

10.3 The Formulation of the Probabilistic Inference Model for Targeted Marketing

10.3.1 Information Sources

There are at least three types of information and knowledge for targeted marketing. They are the customer features represented by customer profiles, the product characteristics represented by product profiles, and the transaction data. The customer profile is mainly about demographic information which is given explicitly by customers. The product profile is generally related to inherent properties of a product.

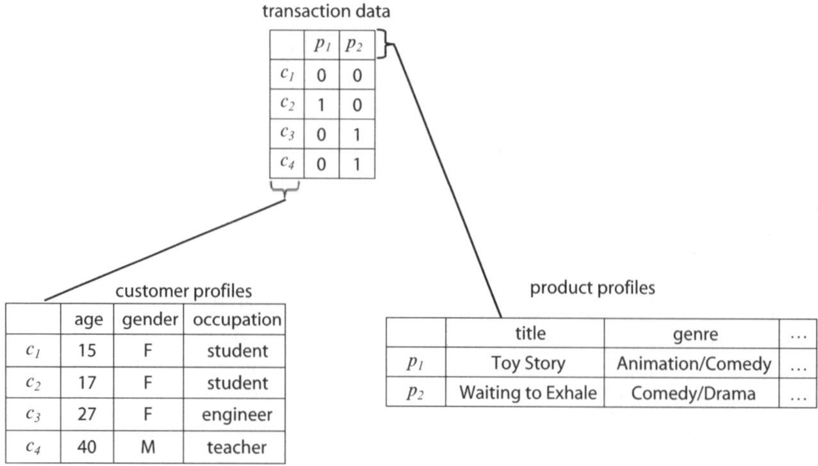

Fig 10.1. An example of information sources for targeted marketing

For transaction data, it represents customers' judgements about products. In general, the judgements could be represented by customers' options and opinions about products. Options represent whether a customer has bought a product, and opinions represent whether a customer likes a product [8].

Formally, the three types of information can be represented as information tables. Take a movie database as an example [29]. Customer profile table shows information about age, gender, education, etc. of customers. The rows of the table corresponds to customers, the columns correspond to their attributes, and each cell is the value of a customer with respect to an attribute. Product profile table shows properties of movies, such as title, genre, and so on. Transaction table shows customers' options about movies. In the transaction table, each row corresponds to a customer, each column to a movie and each cell to a customer's judgement about a movie. To simplify the questions, we represent judgements using options as "0" or "1", where "0" denotes that the customer has not bought the movie while "1" denotes that the customer has bought the movie. Let c_1, c_2, c_3, c_4 denote four customers and p_1, p_2 denote two movies. The relationship among the three tables is shown in Fig. 10.1. From Fig. 10.1, we can see that customer profile table and product profile table represent information about customers and movies in transaction table, respectively, and on the other hand, transaction table links customer profile table and product profile table. Together, they provide useful information for targeted marketing.

10.3.2 Probabilistic Inference Models for Targeted Marketing

In the context of targeted marketing, a product p and a customer c can be viewed as propositions in a probability event space U. Proposition p represents the knowledge contained in the product, while proposition c represents the information contained in the customer. Proposition $p \cap c$ represents the portion of knowledge in both p

and c. The probability $Pr(c)$ is interpreted as the degree to which U is covered by the knowledge contained in c. The probability $Pr(c \cap p)$ represents the degree to which U is covered by the knowledge contained in both c and p.

On the one hand, if product p is regarded as evidence, the degree of support for customer c provided by c is given by

$$\psi(c \to p) = Pr(p|c) = \frac{Pr(p \cap c)}{Pr(c)}. \tag{10.4}$$

On the other hand, if customer c is regarded as evidence, the degree of support for product p provided by c is given by

$$\psi(p \to c) = Pr(c|p) = \frac{Pr(p \cap c)}{Pr(p)}. \tag{10.5}$$

A set of customers on one side and a set of products on the other are the subjects of the studies in a marketing problem. By focusing on each of them, we can identify two marketing strategies called the product-oriented marketing strategy and customer-oriented marketing strategy respectively. On the one hand, product-oriented marketing attempts to find groups of customers with respect to a product. On the other hand, customer-oriented marketing attempts to find targeted groups of products with respect to a customer. Precision and recall are basic measures of targeted marketing [7]. In product-oriented targeted marketing, precision is defined as the probability that the selected customer is interested in the given product and recall is defined as the probability that a customer who is interested in the given product could be selected. In customer-oriented targeted marketing, precision is defined as the probability that a recommended product interests the given customer and recall is defined as the probability that a product which the given customer is interested in could be selected. Equation (10.4) provides a precision-oriented interpretation of the relationships between customers and products for product-oriented targeted marketing and a recall-oriented interpretation for customer-oriented targeted marketing. Equation (10.5) provides a recall-oriented interpretation of the relationships between customers and products for product-oriented targeted marketing and a precision-oriented interpretation for customer-oriented targeted marketing. Table 10.1 summarizes the relationship between the measures and the probabilistic inference model in targeted marketing.

Based on results from [18, 19], the relationship between the above precision-oriented and recall-oriented is given by

$$\psi(c \to p) = \frac{Pr(p)}{Pr(c)} \psi(p \to c). \tag{10.6}$$

Table 10.1. Precision-oriented and recall-oriented interpretations for customer-oriented and product-oriented targeted marketing

	precision-oriented	recall-oriented
product-oriented	$\psi(c \to p)$	$\psi(p \to c)$
customer-oriented	$\psi(p \to c)$	$\psi(c \to p)$

The mutual support between p and c is given by

$$\psi(p \leftrightarrow c) = \frac{Pr(p \cap c)}{Pr(p \cup c)}. \tag{10.7}$$

According to Eq. (3), we have

$$\psi(p \leftrightarrow c) = \frac{\psi(c \rightarrow p)\psi(p \rightarrow c)}{\psi(c \rightarrow p) + \psi(p \rightarrow c) - \psi(c \rightarrow p)\psi(p \rightarrow c)}. \tag{10.8}$$

We obtain the probabilistic inference model for targeted marketing by transplanting and modifying results from models in information retrieval proposed by [16, 18, 19]. In this model, we can obtain the precision-oriented and recall-oriented interpretations for the customer-oriented and product-oriented targeted marketing strategies. By using Eqs. (10.6) and (10.8), the precision-oriented and recall-oriented interpretations are unified in the probabilistic model. As mentioned before, customer profiles, product profiles and the transaction data are involved in targeted marketing. The subsequent sections will discuss how to combine the probabilistic inference model and the available information for targeted marketing.

Customer Information Available for Targeted Marketing

In general, customer information includes age, sex, etc. Let $a_1, a_2, \cdots, a_m$ denote customer attributes and $v_{a_1}^c, v_{a_2}^c, \cdots, v_{a_m}^c$ denote values on customer attributes. So customer c could be represented as $c = v_{a_1}^c \cap v_{a_2}^c, \cdots, \cap v_{a_m}^c$. According to Eq. (10.5), we have

$$\begin{aligned}
\psi(c \rightarrow p) &= Pr(p|c) \\
&= \frac{Pr(c|p)Pr(p)}{Pr(c)} \\
&= \frac{Pr(v_{a_1}^c \cap v_{a_2}^c \cdots \cap v_{a_m}^c |p)Pr(p)}{Pr(c)}.
\end{aligned} \tag{10.9}$$

Under the assumption that each attribute is independent on each other, based on naive bayes method, we have

$$\psi(c \rightarrow p) = \frac{Pr(p)}{Pr(c)} Pr(v_{a_1}^c |p)Pr(v_{a_2}^c |p) \cdots Pr(v_{a_m}^c |p). \tag{10.10}$$

From Eq. (10.10), we can see that naive bayes method could be interpreted as a precision oriented method in probabilistic inference model for product-oriented targeted marketing when customer zer profiles are available.

At the same time, according to Eq. (10.1), we have

$$\psi(p \rightarrow v_{a_t}^c) = Pr(v_{a_t}^c |p), \tag{10.11}$$

then

$$\psi(c \to p) = \frac{Pr(p)}{Pr(c)}\psi(p \to v^c_{a_1})\psi(p \to v^c_{a_2}) \cdots \psi(p \to v^c_{a_m}). \qquad (10.12)$$

The above analysis demonstrates that naive bayes method can be unified into the probabilistic inference model for targeted marketing.

Product Information Available for Targeted Marketing

Let $a^p_1, a^p_2, \cdots, a^p_n$ denote product attributes and $v^p_{a_1}, v^p_{a_2}, \cdots, v^p_{a_m}$ denote values on product attributes. By transplanting the results from section 10.3.2, we have

$$\psi(p \to c) = \frac{Pr(c)}{Pr(p)}\psi(c \to v^p_{a_1})\psi(p \to v^p_{a_2}) \cdots \psi(c \to v^p_{a_n}). \qquad (10.13)$$

The above analysis demonstrates that naive bayes method could be also unified into the probabilistic inference model for targeted marketing when product information is available.

Transaction Information Available for Targeted Marketing

When only transaction are available, there exists a latent class variable set Z to connect customers and products in the latent semantic model [8]. In the latent semantic model, we have

$$Pr(c|p) = \sum_{z \in Z} Pr(z|p)Pr(c|z). \qquad (10.14)$$

By submitting Eq. (10.14) to Eq. (10.5), we have

$$\psi(p \to c) = Pr(c|p) = \sum_{z \in Z} Pr(z|p)Pr(c|z). \qquad (10.15)$$

According to the definition in Eq. (10.1), we have $\psi(p \to z) = Pr(z|p)$ and $\psi(z \to c) = Pr(c|z)$ and then

$$\psi(p \to c) = Pr(c|p) = \sum_{z \in Z} \psi(p \to z)\psi(z \to c). \qquad (10.16)$$

Let $Z = \{z_1, \cdots, z_s\}$, $\mathbf{p} = (\psi(p \to z_1), \cdots, \psi(p \to z_s))$, and $\mathbf{c} = (\psi(z_1 \to c), \cdots, \psi(z_s \to c))$, we have

$$\psi(p \to c) = \mathbf{p} \cdot \mathbf{c}; \qquad (10.17)$$

$$\psi(c \to p) = \frac{P(p)}{P(c)}\mathbf{p} \cdot \mathbf{c}; \qquad (10.18)$$

$$\psi(c \leftrightarrow p) = \frac{P(c)\mathbf{p} \cdot \mathbf{c}}{P(c) + P(p) - P(c)\mathbf{p} \cdot \mathbf{c}}. \qquad (10.19)$$

Equations (10.17), (10.18), and (10.19) unify the latent semantic model into the probabilistic inference model by using the product representation scheme $\mathbf{p} = (\psi(p \to z_1), \cdots, \psi(p \to z_s))$ and the customer representation scheme $\mathbf{c} = (\psi(z_1 \to c), \cdots, \psi(z_s \to c))$. Hofmann [8] measured $\psi(p \to z)$ and $\psi(z \to c)$ $(z \in Z)$ by EM method.

Three Kinds of Information Available for Targeted Marketing

The above customer-oriented and product-oriented marketing strategies focus on a particular customer and a specific product, respectively. In real applications, it may be difficult to obtain enough data to construct a reliable model for a single customer or a single product. Collaborative filtering and recommendation methods are potential solutions to this difficulty. The central idea is to divide customers and products into similar groups. Such a granulation of customer and product spaces enables us to consider we could group customers or products firstly. If we divide customers or products into g groups $\{\xi_1, \xi_2, \cdots, \xi_g\}$, based on the complete probability formula, we have

$$
\begin{aligned}
\psi(c \to p) &= Pr(p|c) \\
&= \frac{Pr(p \cap c)}{Pr(c)} \\
&= \frac{\sum_i Pr(p \cap c|\xi_i)}{Pr(c)}.
\end{aligned}
\tag{10.20}
$$

As product p is independent on customer c in the concept space $\{\xi_1, \xi_2, \cdots, \xi_g\}$, we have $Pr(c \cap p|\xi_i) = Pr(p|\xi_i)Pr(c|\xi_i)$. So we have

$$
\begin{aligned}
\psi(c \to p) &= \frac{\sum_i Pr(p \cap c|\xi_i)}{Pr(c)} \\
&= \sum_i \frac{Pr(p|\xi_i)Pr(c|\xi_i)Pr(\xi_i)}{Pr(c)} \\
&= \sum_i Pr(\xi_i|c)Pr(p|\xi_i).
\end{aligned}
\tag{10.21}
$$

Let $\psi(c \to \xi_i) = Pr(\xi_i|c)$ and $\psi(\xi_i \to p) = Pr(p|\xi_i)$, we have

$$
\psi(c \to p) = \sum_{\xi_i} \psi(c \to \xi_i)\psi(\xi_i \to p).
\tag{10.22}
$$

Let customer c be vector $\mathbf{c} = (\psi(c \to \xi_1), \cdots, \psi(c \to \xi_g))$ and product p be vector $\mathbf{p} = (\psi(\xi_1 \to p), \cdots, \psi(\xi_g) \to p)$, we have

$$
\psi(c \to p) = \mathbf{c} \cdot \mathbf{p}.
\tag{10.23}
$$

According to Eq. (10.6), we have

$$\psi(p \to c) = \frac{Pr(c)}{Pr(p)} \sum_{\xi_i} \psi(c \to \xi_i)\psi(\xi_i \to p)$$

$$= \frac{Pr(c)}{Pr(p)} \mathbf{c} \cdot \mathbf{p}. \tag{10.24}$$

As mentioned in the probabilistic inference model in information retrieval [18], the quantities $\psi(c \to \xi_i)$ and $\psi(\xi_i \to p)$ should be provided by customers based on their preference. Certain conditions must be satisfied to make them consistent and meaningful. For example, if customer c belongs to ξ_i, $\psi(c \to \xi_i)$ should be assigned to 1, i.e. $\psi(c \to \xi_i) = 1$. Given ξ_i and ξ_j, there should exist $\psi(c \to \xi_i) > \psi(c \to \xi_j)$ if customer c is more similar to customers in ξ_i than customers in ξ_j. In a similar way, $\psi(\xi_i \to p) = 1$ should be assigned to represent that all customers in ξ_i likes product p; $\psi(\xi_i \to p) > \psi(\xi_j \to p)$ should be assigned if customers in ξ_i like product p more than customers in ξ_j. In practice, it may be difficult to obtain these axiom relationship directly [18]. Zeng et al. [24] employed text classifying methods to classify product and then obtain $\psi(\xi_i \to p)$; they also employed customer information and visiting information to obtain $\psi(c \to \xi_i)$.

The above sections mainly demonstrate how to apply the probabilistic inference model to targeted marketing. These methods don't consider the customer preference. The following sections will apply the probabilistic inference model to measuring the customer preference. Inductive methods are employed to solve the targeted marketing problem. In the following section, we show the product oriented targeted marketing for available customer information as an illustrative example.

10.4 Adaptive Linear Market Value Functions for Targeted Marketing

We first briefly review the notion of a customer preference for the product-oriented targeted marketing. Let C denote a finite set of customers and P denote a finite set of products. In product-oriented targeted marketing, given a product $p \in P$, the customer preference can be formally described by a binary relation $\succ_p$ on C as follows, where $c', c'' \in C$.

$$c' \succ_p c'' \Leftrightarrow \text{customer } c' \text{ likes product } p \text{ more than } c''. \tag{10.25}$$

It is possible that both customers like the product. So a binary relation $\sim_p$ on C is given by

$$c' \sim_p c'' \Leftrightarrow \neg(c' \succ_p c'') \text{ and } \neg(c'' \succ_p c'). \tag{10.26}$$

Intuitively, it is impossible that c' likes the product p more than c'' and at the same time c'' likes the product p more than c'. Moreover, we can assume that it is not more that p is preferred by c' than c''' if it is not more that p is preferred by c'

than c'' and at the same time it is not more that p is preferred by c'' than c'''. So it seems reasonable that the preference relation $\succ_p$ satisfies the following two axioms. The preference relation $\succ_p$ satisfying these two following axioms is called a weak order.

(1) $\forall c', c'' \in C$, if $c' \succ_p c''$, then $\neg(c'' \succ_p c')$;
(2) $\forall c', c'' \in C$, if $\neg(c' \succ_p c'')$ and $\neg(c'' \succ_p c''')$, then $\neg(c' \succ_p c''')$.

After regarding $\succ_p$ as the weak order, we have the following theorem [20].

Theorem 1. If C is a countable customer set and $\succ_p$ a relation on C. There exists a real-valued function $f_p : C \to \Re$ satisfying

$$c' \succ_p c'' \Leftrightarrow f_p(c') > f_p(c''), \tag{10.27}$$

if and only if $\succ$ is a weak order. Moreover, f_p is defined up to a strictly monotonic increasing transformation.

From theorem 1, the notions of customer ranking are linked with the customer preference $\succ_p$. In detail, if $\succ_p$ is a weak order, there is a function to measure the customer preference. We call the function as the market value function. After the market value function ranks customers in descendent order, the sequence of customers reflects the order of customers under the preference $\succ_p$. It is worth mentioning that the market value function f_c satisfied Eq. (10.27) provides an perfect ranking. For a real application, we sometimes need an acceptable ranking which means the customers having more market value can not be ranked after the the customers having less market value. The weaker condition could be defined by

$$c' \succ_p c'' \Rightarrow f_p(c') > f_p(c''). \tag{10.28}$$

If we assume that the market value function f_p is measured by $\psi(c \to p)$ or $\psi(p \to c)$ in product-oriented targeted marketing, then for the precision-oriented measure, we have
$$c' \succ_p c'' \Leftrightarrow \psi(c' \to p) > \psi(c'' \to p), \tag{10.29}$$
and for the recall-oriented measure, we have

$$c' \succ_p c'' \Leftrightarrow \psi(p \to c') > \psi(p \to c''). \tag{10.30}$$

As the difference between the precision-oriented and recall-oriented measures could be eliminated by normalization [19], the following description only focuses on the precision-oriented measure.

Now let us show an inductive method to solve the product-oriented targeted marketing. In previous sections, we can assume that $\psi(c \to p)$ could be measured by the dot product between the customer vector $\mathbf{c}$ and the product vector $\mathbf{p}$, that is $\psi(c \to p) = \mathbf{c} \cdot \mathbf{p}$. We can easily describe customer c as a vector $\mathbf{c}$ by using the customer attribute value. After $\mathbf{c}$ is available, the focus is how to find the product vector $\mathbf{p}$. According to the acceptable ranking shown in Eq. (10.28), we have

$$c' \succ_p c'' \Rightarrow \mathbf{c'} \cdot \mathbf{p} > \mathbf{c''} \cdot \mathbf{p}. \tag{10.31}$$

In general, the preference $\succ_p$ satisfying Eq. (10.31) is called a weakly linear preference relation [20].

Let

$$\mathbf{B} = \{\mathbf{b} = \mathbf{c'} - \mathbf{c''} | c', c'' \in C, c' \succ_p c''\}. \tag{10.32}$$

According to Eq. (10.31), we have

$$\mathbf{b} \cdot \mathbf{p} > 0, \quad for \; \forall \mathbf{b} \in \mathbf{B}. \tag{10.33}$$

We can see that if Eq. (10.33) holds, $\mathbf{p}$ is correct, and if $\exists \mathbf{b}' \in \mathbf{B}, \mathbf{b}' \cdot \mathbf{p} < 0$, an error occurs. In this case, the value $-\mathbf{b}' \cdot \mathbf{p}$ is a measure of the error. Let $\mathbf{B}' = \{\mathbf{b}' | \mathbf{p} \cdot \mathbf{b}' < 0, \mathbf{b}' \in \mathbf{B}\}$. We aim to minimize the total errors $\sum_{\mathbf{b} \in \mathbf{B}'}(-\mathbf{p} \cdot \mathbf{b})$. Based on the above analysis, we can get the algorithm AMV (Adaptive Market Value) to search the weight vector $\mathbf{p}$ by using gradient descent to minimize the total errors.

In AMV, the set of customers C are divided into two disjoint subsets C^+ and C^-. C^+ denotes the set of customers who have bought the product p and C^- denotes the set of customers who have not bought the product p. Based on C^+ and C^-, AMV defines the customer preference as

$$c' \succ_p c'' \Leftrightarrow c' \in C^+, c'' \in C^- \tag{10.34}$$
$$\Leftrightarrow c' \text{ has bought the product } p, \text{ while } c'' \text{ has not.}$$

Algorithm 1: AMV (C^+, C^-)

Data: a positive set C^+, a negative set C^-;
Result: a product vector $\mathbf{p}$.
    (1) Choose an initial vector $\mathbf{p}$
    (2) do$\{$
            for each customer c' in C^+
                for each customer c'' in C^-
                    if $\mathbf{c'} \cdot \mathbf{p} < \mathbf{c''} \cdot \mathbf{p}$
                    then vector $\mathbf{b} = \mathbf{b} + (\mathbf{c'} - \mathbf{c''})$
          $\mathbf{p} = \mathbf{p} + \mathbf{b}$;
      $\}$while Eq. (10.33) holds

If the gradient descent is to consider the total error defined for each instance in P, we can get the algorithm SAMV (Stochastic Adaptive Market Value) as follows:

Compared with the algorithm AMV, the algorithm SAMV updates attribute weights upon examining each positive instance.

In real world, we want to update the current weight when a new instance occurs and Eq. (10.33) does not also always hold during the loop. The first problem can be

---

Algorithm 2: SAMV (C^+, C^-)

Data: a positive set C^+, a negative set C^-;
Result: a product vector **p**.
    (1) Choose an initial vector **p**
    (2) do{
           for each customer c' in C^+{
              for each customer c'' in C^-
                  if $\mathbf{c'}\cdot\mathbf{p} < \mathbf{c''}\cdot\mathbf{p}$
                  then vector **b=b+(c'-c'')**
             p=p+b;
        }
    }while Eq. (10.33) holds

---

solved by only repeating one time in the algorithm SAMV (called SAMV1). In general, there are too few positive instances and too many negative instances in C [13]. Under the assumption that the current negative instances are enough, the attribute weights can be updated by the algorithm SAMV1 when a new positive instance occurs. The second problem can be solved by setting the iteration numbers.

10.5 Experiments

Two datasets on potential customers have been used for our approaches. These datasets are divided into the training set and the testing set, respectively. The training set is used to learn the model and the learned functions are used to rank the testing examples. The first dataset is about members in a club. Each member is described by 96 attributes, such as sex, age, hobby, income and so on. We selected randomly 18964 (894 positive instances) as the training set and 24134 (647 positive instances) as the testing set, respectively. The second dataset is about a RV insurance policy [12]. Each customer has 85 attributes. The dataset is divided into a training set with 5822 examples and a testing set with 4000 examples, respectively.

The lift index [13] is used as the evaluation criterion. After all testing examples are ranked using the proposed algorithms, we divided the ranked list into 100 equal deciles. Let S_i denote how many positive examples are in the ith deciles. The lift index is defined as

$$S_{lift} = (1 \times S_1 + 0.99 \times S_2 + \ldots + 0.01 \times S_{100})/\sum S_i. \tag{10.35}$$

Thus, we can show the distribution of the positive examples in the ranked list for the intuitive evaluation.

Table 10.2 shows the lift index results of the market value model (MV) [23], the algorithms AMV, SAMV1 and simple Naive Bayes (SNB) on the two datasets. The SNB means that $Pr(c)$ is a constant. In this table, the lift index of MV is a little better than AMV and is similar to SAMV1 on dataset 1. The AMV is better than other methods on dataset 2.

Table 10.2. Lift value results on two datasets

Datasets	SAMV1	AMV	MV	SNB
Dataset 1	63.1%	62.8%	63.2%	50.3%
Dataset 2	71.7%	73.0%	71.7%	58.8%

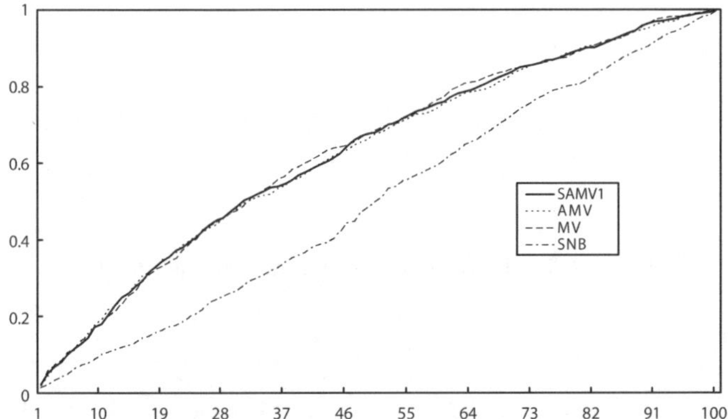

Fig 10.2. The lift curve of four methods on dataset 1

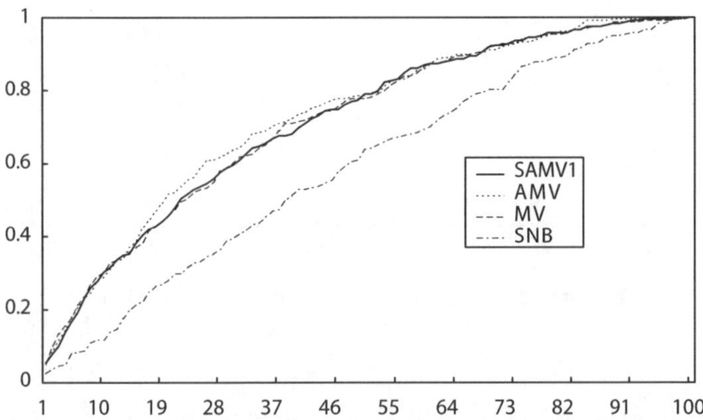

Fig 10.3. The lift curve for four methods on dataset 2

Figures 10.2 and 10.3 show the lift curves of the above four methods on the two datasets, respectively. The horizontal axis shows the top proportion of customers in the ranked list and the vertical axis shows the proportion of responses based on the total positive instances in the testing set. In Fig. 10.2, the SAMV1 and AMV are better than MV over the targeting range between 11% and 24% on dataset 1. In Fig. 10.3, the results of AMV are better than SAMV1 and MV between about 15% and 48% of targeted customers.

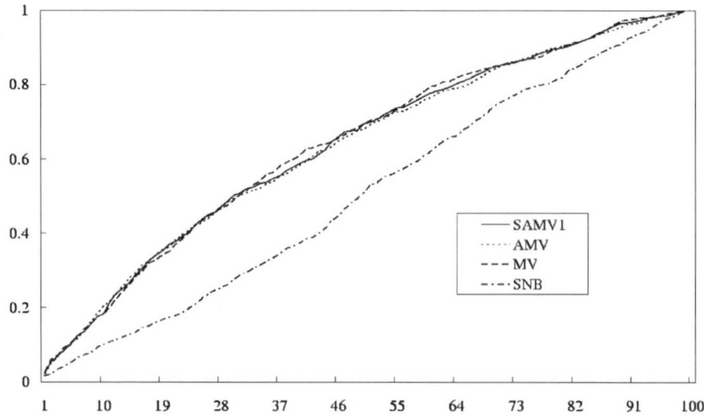

Fig 10.4. The ROC curves for four methods on dataset 1

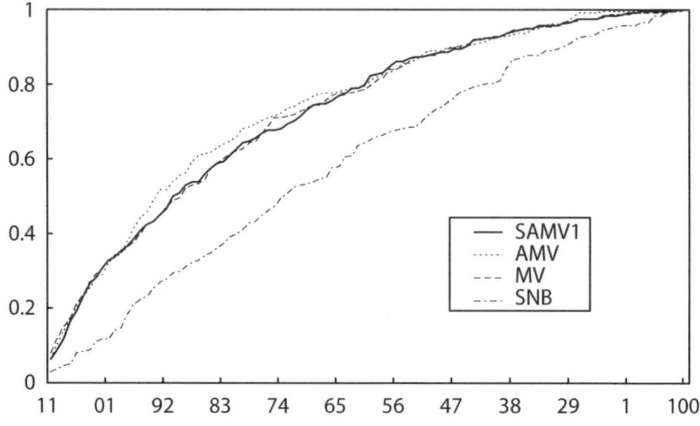

Fig 10.5. The ROC curves for four methods on dataset 2

Figures 10.4 and 10.5 show the ROC curves of the above four methods on the two datasets, respectively. The horizontal axis shows a percentage of the total number of negatives and the vertical axis shows a percentage of the total number of positives. In Fig. 10.4, the results of AMV, SMV1 and MV are similar on dataset 1. In Fig. 10.5, the results of AMV are better than SAMV1 and MV if marketers aim to cover about between 36% and 77% of positives because the AMV gives less false positive rate than others.

In total, we can get the approximate results by using SAMV1, AMV and MV, respectively. These methods outperform simple Naive Bayes. The SAMV1 and AMV provide a new method to estimate market value function. Under the assumption that the current negative instances are enough, the SAMV1 can be regarded as an incremental algorithm.

10.6 Conclusion

Targeted marketing typically involves two components, customers and products. There are at least three kinds of information available for targeted marketing, namely customer profiles, product profiles, and transaction data. Inspired by the probabilistic inference model for information retrieval, this paper uses the probabilistic inference model to analyze the relationships between customers and products in targeted marketing. This model interprets the relationships between customers and products by the notions of precision and recall in customer-oriented and product-oriented targeted marketing strategies. The further theoretical analysis could be included in future work.

As an illustrative example, this paper studied the adaptive market value functions in targeted marketing. Algorithms AMV, SAMV and SAMV1 are presented. These methods are evaluated using real world examples. Experiments show that these methods are effective in targeted marketing. However, the proposed methods in this paper are based on two classes (buy or not buy). In fact, there are multilevel preference (i.e. multi-class) in sets of customers such as buy, not buy, likely buy, and so on. In the future, we will research the multilevel preference in the market value function model.

References

1. G. Adomavicius and A. Tuzhilin, "Toward the Next Generation of Recommender Systems: A Survey of the State-of-the Art and Possible Extensions", *IEEE Transactions on Knowledge and Engineering*, 17(6): 734-749, 2005.
2. R. Agrawal, T. Imielinski, and A. Swami, *Proc. the ACM SIGMOD International Conference on the Management of Data*, 207-216, 1997.
3. D.S. Associates, *The New Direct Marketing*, McGraw-Hill, 1999.
4. M. Balabanovic and Y. Shoham, "Fab: Content-based, Collaborative Recommendation", *Communications of the ACM*, 40: 66-72, 1997.
5. J. Basilico and T. Hofmann, "Unifying Collaborative and Content-Based Filtering", *Proc. the International Conferece on Machine Learning*, 2004.
6. J. Breese, D. Hecherman, and C. Kadie, "Empirical Analysis of Predictive Algorithms for Collaborative Filtering", *Proc. the Conference on Uncertainty in Artificial Intelligence*, 43-52, 1998.
7. J. Herlocker, J.A. Konstan, L.G. Terveen, and J.T. Riedl, "Evaluating Collaborative Filtering Recommender Systems", *ACM Transactions on Information Systems*, 22(1): 5-53, 2004.
8. T. Hofmann, "Latent Semantic Models for Collaborative Filtering", *ACM Transactions on Information Systems*, 22(1): 89-115, 2004.
9. J. Huang, N. Zhong, Y.Y. Yao, and C. Liu, "A General Framework of Targeted Marketing", *Proc. the Atlantic Web Intelligence Conference*, 197-203, 2005.
10. J. Huang, N. Zhong, C. Liu, and Y.Y Yao, "Adaptive Linear Market Value Functions for Targeted Marketing", *Proc. the International Conference on Rough Sets and Current Trends in Computing*, 743-751, 2004.

11. J. Hu and N. Zhong, "Organizing Multiple Data Sources for Developing Intelligent e-Business Portals", *Data Mining and Knowledge Discovery, an International Journal*, 12 (2-3), 127-150, 2006.
12. Y.S. Kim and W.N. Street, "Coil challenge 2000: Choosing and Explaining Likely Caravan Insurance Customers", Technical Report 2000-09. Sentient Machine Research and Leiden Institute of Advanced Computer Science, June 2000. http://www.wi.leidenuniv.nl/putten/library/cc20000/.
13. C.X. Ling and C. Li, "Data Mining for Direct Marketing: Problem and Solutions", *Proc. The International Conference on Knowledge Discovery and Data Mining*, 73-79, 1998.
14. R. Potharst, U. Kaymak, and W. Pijls, "Neural Networks for Targeted Selection in Direct Marketing", *Networks in Business: Techniques and Applications*, Idea Group Publishing, 2001.
15. B. Sarwar, G. Karypis, J. Konstan, and J. Riedl, "Item-Based Collaborative Filtering Recommendation Algorithms", *Proc. the International Conference on World Wide Web*, 285-295, 2001.
16. C.J. Van Rijsbergen, "A Non-classical Logic for Information Retrieval", *The Computer Journal*, 29: 481-485.
17. M. Vozalis and G. Margaritis, "On the Enhancement of Collaborative Filtering by Demographic Data", *Web Intelligence and Agent Systems, An International Journal*, 4(2): 117-138, 2006.
18. S.K.M. Wong and Y.Y. Yao, "A Probabilistic Inference Model for Information Retrieval", *Information Systems*, 16(3): 301-321 , 1991.
19. S.K.M. Wong and Y.Y. Yao, "On Modeling Information Retrieval with Probabilistic Inference", *ACM Transactions on Information Systems*, 13(1): 38-68, 1995.
20. S.K.M. Wong and Y.Y. Yao, "Query Formulation in Linear Retrieval Models", *Journal of the American Society for Information Science*, 41(5): 334-341, 1990.
21. S.K.M. Wong and Y.Y. Yao, "Evaluation of an Adaptive Linear Model", *Journal of the American Society for Information Science*, 42(10): 723-730, 1991.
22. N. Zhong, J. Liu, and Y.Y. Yao, (eds.) *Web Intelligence*, Springer, 2003.
23. Y.Y. Yao, N. Zhong, J. Huang, C. Ou, and C. Liu, "Using Market Value Functions for Targeted Marketing Data Mining", *International Journal of Pattern Recognition and Artificial Intelligence*, 16(8): 1117-1131, 2002.
24. C. Zeng, C.X. Xing, and L.Z. Zhou, "Similarity Measure and Instance Selection for Collaborative Filtering", *Proc. the International Conference on World Wide Web*, 652-658, 2003.
25. N. Zhong, Y.Y. Yao, C.N. Liu, J.J. Huang, and C.X. Ou, "Data Mining for Targeted Marketing", *Intelligent Technologies for Information Analysis*, Springer, 109-131, 2004.
26. N. Zhong, J.M. Liu, and Y.Y. Yao, "Envisioning Intelligent Information Technologies Through the Prism of Web Intelligence ", *Communications of the ACM*, 50(3): 89-94, 2007.
27. N. Zhong, J. Liu, and Y.Y. Yao, "In Search of the Wisdom Web" Special issue on Web Intelligence (WI), *IEEE Computer*, 35(11): 27-31, 2002.
28. J. Liu, N. Zhong, Y.Y. Yao, and Z.W. Ras, "The *Wisdom Web*: New Challenges for Web Intelligence (WI)", Special issue on Web Intelligence (WI), *Journal of Intelligent Information Systems*, 20(1): 5-9, 2003.
29. http://www.movielens.org/

Computational Methods for Discoveries from Integrated Data - Human-Interactive Annealing for Multilateral Observation

Yoshiharu Maeno[1], Kenichi Horie[2], and Yukio Ohsawa[2]

[1] Graduate School of Systems Management, Tsukuba University,
3-29-11 Otsuka, Bunkyo-ku, Tokyo, 112-0012 Japan
maeno.yoshiharu@nifty.com

[2] School of Engineering, the University of Tokyo,
7-3-1 Hongo, Bunkyo-ku, Tokyo, 113-8563 Japan

Summary. Unobserved events play an important role in the dynamics of observed events. They are either something hidden intentionally, or something new not yet recognized. Such invisible events are named dark events. A new method named human-interactive annealing is presented to understand the threat arising from the dark events and to invent a scenario to turn the threat to opportunity. The method is extended for integrated data from multilateral observation. A scenario invention experiment is demonstrated using patent documents in the field of knowledge acquisition to design corporate R&D strategies. Multilateral observation provides more clues when the engineers and strategists obtain an idea on emerging technologies.

11.1 Problem - Observation and Discovery

Unobserved events often play an important role in the dynamics of observed events. The unobserved events are either something hidden intentionally, or something new not yet recognized. Unobservability or invisibility is more difficult nature to understand than exception [Suzuki 2005] or rareness [Weiss 1998]. The followings are typical examples. In terrorism, information protection is important to terrorism and criminal organizations. A commander tries to conceal himself from leaving any traces in communication and meeting logs, which are the basic intelligence to the police. Unobservability is critical. Otherwise, exposure and arrest of a relevant pilot would have been a fatal damage to the terrorist organization in the 9/11 attack [Klerks 2002], [Krebs 2002]. In corporate R&D [Batallas 2006], patents provide companies with technological means to solve an engineering design problem [Chakrabarti 1993], [Chakrabarti 1991]. Unknown but significant technological means element is potential threat to corporate R&D. It is a technology hidden by a rival company like a submarine patent, an emerging technology from other field of expertise, a technology owned by a niche company or a small technician community, or a completely new technology etc.

Y. Maeno et al.: *Computational Methods for Discoveries from Integrated Data - Human-Interactive Annealing for Multilateral Observation*, Studies in Computational Intelligence (SCI) **123**, 187–203 (2008)
www.springerlink.com

It is important to discover a clue on the unobserved events, which do not appear in the observed records, but is significant in affecting the behavior of the observed activities of individual, companies, organization etc. We call the latent structure governed by the unobserved events dark events after dark matter in cosmology. The dark matter refers to hypothetical particles which do not emit or reflect radiation to be detected directly. But its presence can be inferred from gravitational effects on visible matter such as stars and galaxies. The dark events are one of the origin of potential threat in engineering, social, or economic problems. By becoming aware of them, we may invent a scenario to turn the threat to opportunity. We aim at inferring the dark events from the observation and to invent a scenario which indicates the significance of the dark events. The scenario is a sequence of events which can be achieved by our present decision-making rather than a sequence of events predicting the future. Multilateral observation or the integrated data provides us with additional clues on the dark events. We introduce human-interactive annealing process [Maeno 2007a] which visualizes observation in a graph (nodes, links, and clusters) named scenario map, and indicates the candidates of dark events within the empty space among the clusters on the graph. Visualization depends on the degree of the human's prior understanding. It means that the process is adaptive to the individual human and aims at providing clues on something surprising beyond the individual human's understanding. The process includes an algorithm [Ohsawa 2005] to analyze observation and to draw a graph as well as human's interpretation.

The problem we address is similar to the node discovery problem for a network. The node discovery problem for a scale-free network [Maeno 2006a] and for a homogeneous network [Maeno 2007b] was studied. The process can also be applied to help humans understand their own preference or the tendency in behavior (in the field of marketing and cognitive science) [Maeno 2007c]. In this contribution, we focus on a problem to analyze existing patent documents to get an idea on a new technology (corporate R&D) [Maeno 2006b]. The significance of the individual technology depend on various background context and circumstantial knowledge. Human's prior understanding is more important in the problem than in the node discovery problem. We utilize the multilateral observation consisting of identifier, field, document text etc. of the patents. Experiments are demonstrated; on inventing scenarios on a new technology in the field of knowledge acquisition.

11.2 Method - Human Interactive Annealing

11.2.1 Environment

We describe the details of the human-interactive annealing process. A scenario map is used as a tool to help actors communicate with each other in the process. The scenario map visualizes the relationship among the observed events as a graph (nodes and links) structure. Clues on the dark events are indicated as a potential structure on our scenario map. The presence of the links between nodes represents co-occurrence

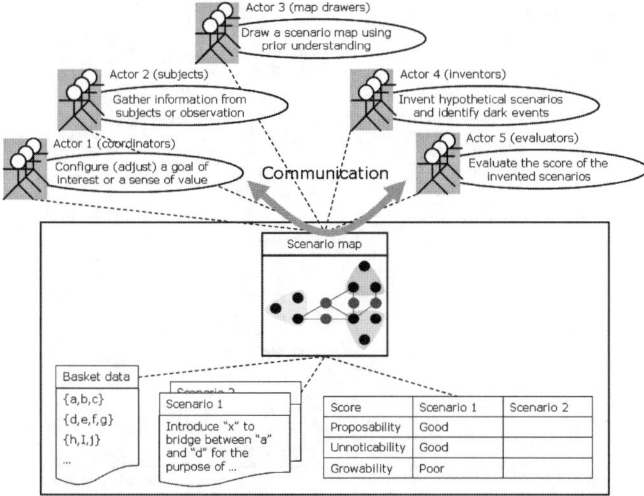

Score	Scenario 1	Scenario 2
Proposability	Good	
Unnoticability	Good	
Growability	Poor	

Fig 11.1. Five actors appearing in the human-interactive annealing process and communication with the aid of a scenario map (observation, invented scenarios, and evaluation scores).

of the events. The process includes an algorithm to draw a scenario map from the observation records as well as human's interpretation (understanding of the scenario map and invention of scenarios).

Five actors appear in the human-interactive annealing process. Communication among the five actors is fostered by the scenario map. They are illustrated in Fig. 11.1. Map drawers (actor 3) and inventor (actor 4) are usually the same. This is because the human-interactive annealing process requires inventor's prior understanding to draw a scenario map. Coordinators (actor 1) and evaluators (actor 5) are sometimes the same. This is because evaluation is according to the goal or sense of value configured by the coordinators. In meta-cognitive procedure to understand the user's own cognition, all of the five actors are the same.

11.2.2 Process

The annealing process is a combination of two complementary elements; human's interpretation and algorithm. The two elements are illustrated in Fig. 11.2 with five scenario map examples. The dark events are made visible, owing to the algorithm. The horizontal axis is the number of iteration. The vertical axis is like temperature ruling the structural nature of the topological relationship among events. In other words, the temperature is a parameter controlling the number of clusters, or the granularity of the clusters in the scenario map. The granularity of the clusters which is easy to understand is closely related to the system of concepts acquired by a user in advance. The temperature is, therefore, associated to the degree of human's prior understanding of the problem domain.

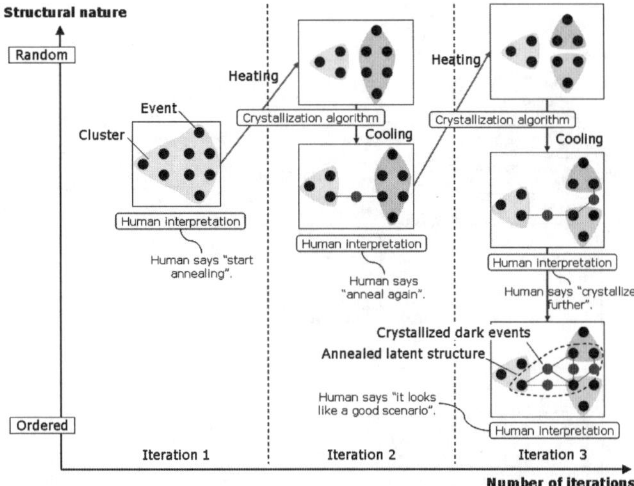

Fig 11.2. Iteration of scenario map drawing drawn in the human-interactive annealing process. The scenario map grows from coarse-grained one to fine-grained one as the process proceeds. The vertical axis represents temperature, or randomness. (Copyright, IEEE 2007)

The process includes a heating step followed by cooling step. In the heating step, up to the specified peak temperature, the number of clusters and edges between visible events decrease. Weak associations are destroyed. The crystallized dark events disappear. Then, a cooling step comes, where event structures are solidified as temperature goes down. With the algorithm used in the cooling step, the computer analyzes co-occurrence between events. It is the basis to draw a scenario map. The number of crystallized dark events between clusters of visible events increases on an scenario map. The clusters are connected to each other to form a single large structure. The crystallization is followed by human's interpretation.

After human's interpretation, when the understanding of the problem is believed to get richer, the temperature shall be set higher. More complex higher-order hidden structures shall be revealed. This will lead to the discovery of unique and unexpected scenario. When the understanding is believed to get poorer, the temperature shall be set lower. The user should try to understand the basic lower-order structures from the scenario map. The iteration in the annealing process is continued until human converges into complete posterior understanding. It should be noted that the annealing process is adaptive to the degree of human's prior understanding by letting the user adjust the temperature.

Steps for algorithm and human interpretation in the human-interactive annealing process is shown in Fig. 11.3.

11.2.3 Algorithm

The input to the algorithm is the records observed for events. The records are in the form of baskets in eq. (11.1).

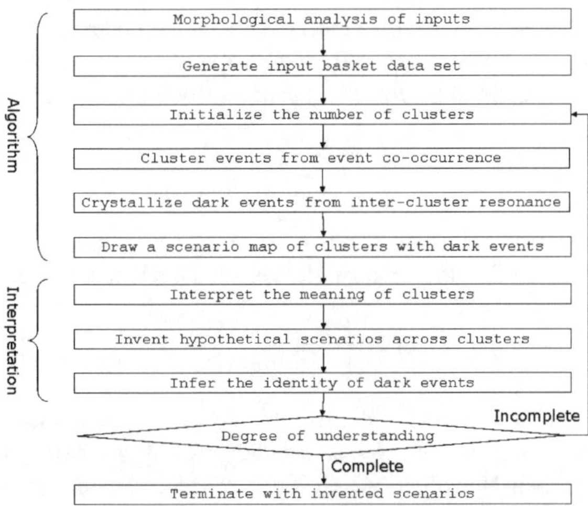

Fig 11.3. Steps for the algorithm and human interpretation in the human-interactive annealing process. Morphological analysis is necessary when the input is text document such as patents, newspaper articles, sentences on the web pages etc. (Copyright, IEEE 2006)

$$b_j = \{e_i\}. \tag{11.1}$$

The content of the basket b_j are a set of events e_i which are observed simultaneously or grouped under a specific subject. The output from the algorithm is a clue on dark events. More specifically, our aim is to identify the basket b_i which is related to the dark events the most likely. The core of our algorithm is, therefore, to design a predictor function $W(b_i)$ to evaluate the likeliness of the individual baskets b_i. The basket b_i evaluated as the most likely should have included the dark events ($\mathrm{DE}_i = e_x$). There are links r_{xj} between the events e_x and e_j. The event e_j is the gateway event $e_{\mathrm{gtw}(j)}$ ($e_{\mathrm{gtw}(j)} \in b_i$ and $e_{\mathrm{gtw}(j)} \in c_j$), where c_j denotes a cluster found within the observation.

In a scenario map, a group of dark events can be visualized as a structure representing a bridge (e_x and r_{xj}) between the gateway events $e_{\mathrm{gtw}(j)}$ in the clusters c_j. Before processing the baskets, the number of clusters $|c|$ is calculated using the specified temperature, which corresponds to the human's prior understanding. The temperature is initialized to be a small number and increased. As shown in Fig. 11.3, the algorithm consists of six steps. When the input data consists of texts, morphological analysis is applied to generate basket data. Morphology is the study branch in linguistics which analyzes the internal structure of words. Morphological analysis investigates the relationship between the words within phrases and sentences, and decomposes them into a sequence of words which are the smallest units of syntax with the aid of computers automatically. For example, a phrase: *computational methods for discoveries from integrated data* becomes $b_0 = \{$compute, method, for, discover, from, integrate, datum$\}$.

At first, events are identified. The all events in the baskets b_j are picked up. An individual event is denoted by e_i. Then, the events are clustered into vertex groups based on the distance. Distance (or closeness) between nodes are defined according to the occurrence and co-occurrence frequency of the events. Occurrence frequency of an event $F(e_i)$ is defined by eq. (11.2) using a Boolean function $B(s)$.

$$F(e_i) \equiv \sum_j B(e_i \in b_j). \tag{11.2}$$

It is the number of the baskets where e_i appears. The frequency is increased by 1 when e_i appears multiple times in a single basket. This is formulated in eq. (11.3).

$$B(s) = \begin{cases} 1 & \text{if } s \text{ is TRUE} \\ 0 & \text{otherwise} \end{cases}. \tag{11.3}$$

We use Jaccard's coefficient defined by eq. (11.4) as a measure of the co-occurrence. Jaccard's coefficient is used widely in link discovery, web mining, or text processing [Liben-Nowell 2004]. Co-occurrence or dependence coefficient may also be used instead of Jaccard's coefficient.

$$J(e_i, e_j) \equiv \frac{F(e_i \cap e_j)}{F(e_i \cup e_j)} = \frac{\sum_k B((e_i \in b_k) \wedge (e_j \in b_k))}{\sum_k B((e_i \in b_k) \vee (e_j \in b_k))}. \tag{11.4}$$

We employ k-medoid clustering algorithm [Duda 2000], [Hastie 2001] because the amount of necessary calculation is small. It is simple and efficient. It is an EM algorithm similar to k-means algorithm for numerical data. A medoid event $e_{\mathrm{med}(j)}$ is an event locating most centrally within a cluster c_j. They are initially selected at random. Other $|e| - |c|$ events are classified into the clusters based on the closeness to the medoids. Then, a new medoid is selected within the individual cluster so that the sum of closeness from events within the cluster to the modoid is maximal. The closeness is evaluated by the eq. (11.5). This is repeated until the medoid $e_{\mathrm{med}(j)}$ converges. Unsupervised learning techniques such as self-organizing map [Kohonen 1990] are also applied. The resulting clusters are denoted by c_j.

$$\text{Select } e_{\mathrm{med}(j)} \text{ to maximize } M(c_j) \equiv \sum_{e_i \neq e_{\mathrm{med}(j)} \wedge e_i \in c_j} J(e_{\mathrm{med}(j)}, e_i). \tag{11.5}$$

The predictor function $W(b_i)$ in eq. (11.6) is used to evaluate the likeliness of the individual baskets b_i as a candidate which should have included dark events. The dark events are the hidden participant to the basket, which is the origin of attraction in the empty space among clusters. The baskets ranked more highly are retrieved by the baskets.

$$W(b_i) \equiv \frac{1}{|c|} \sum_{j=0}^{|c|-1} \max_{e_k \in c_j} \frac{B(e_k \in b_i)}{\sum_l B(e_k \in b_l)}. \tag{11.6}$$

The gateway event $e_{\mathrm{gtw}(j)}$ in a cluster c_j is selected by eq. (11.7). The gateway event is a key event which is associated with the dark events.

$$e_{\mathrm{gtw}(j)} \equiv \arg \max_{e_k \in c_j} \frac{B(e_k \in b_i)}{\sum_l B(e_k \in b_l)}. \tag{11.7}$$

Finally, topology analysis comes. The dark events and the corresponding gateway events in each cluster are drawn as bridges between clusters. The force-direct placement technique [Sugiyama 2002], [Fruchterman 1991] is employed to determine the place of the nodes, where links are replaced by springs.

11.3 Node Discovery

We present a simulation experiment of the algorithm to discover a node in the test data generated from a scale-free random network [Barabasi 1999], [Watts 1998]. The scale-free random network is a commonly used model to describe human's communication, relationship or dependence in engineering, social, or economic problems. The scale-free random network tends to contain centrally located hub events. The hub events influence the way the network operates. However, random deletion of events has little effect on the network's connectivity and effectiveness.

11.3.1 Simulation Model

The event in eq. (11.1) is a node (edge) in a network. Fig. 11.4 and Fig. 11.5 shows a scale-free random network including 400 nodes. The occurrence frequency distribution of nodal degree is ruled by a power law ($P(d) \propto d^{-2.4}$). It includes a primary hub node (labeled 0-00) and 20 clusters (belonging nodes are labeled 1-xx, 2-xx, 3-xx, up to 20-xx). The clusters include secondary hub nodes (labeled 1-00, 2-00, 3-00, up to 20-00) and 380 (=19 × 20) nodes. The node is connected with node belonging

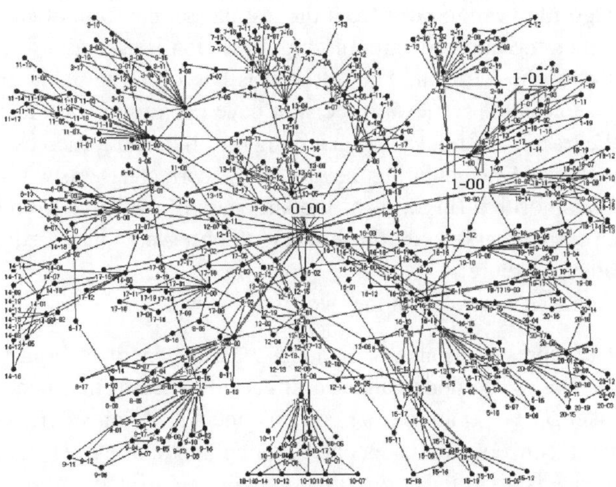

Fig 11.4. Scale-free network of 400 nodes (20 groups × 20 nodes/group) when the ratio of the links connecting nodes belonging to different groups to the whole links is $p_c = 0.057$. The nodes 0-00, 1-00, 1-01 are studied in the simulation. (Copyright, IEEE 2006)

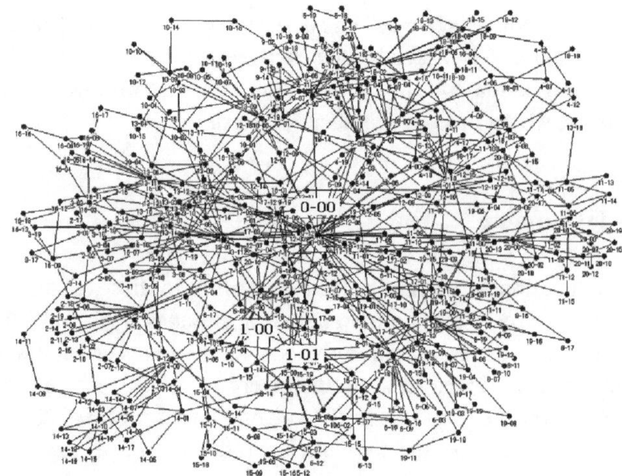

Fig 11.5. Scale-free network of 400 nodes (20 groups $\times$ 20 nodes/group) when the ratio of the links connecting nodes belonging to different groups to the whole links is $p_c = 0.19$. The nodes 0-00, 1-00, 1-01 are studied in the simulation. (Copyright, IEEE 2006)

to a different cluster by a given probability p_c. The overall ratio where a link connects nodes belonging to different clusters is $p_c = 0.057$ for Fig. 11.4. The boundary between clusters is not clear, but we can observe that the 20 clusters exist in the network. The overall ratio is $p_c = 0.19$ for Fig. 11.5. It is difficult to distinguish the 20 clusters in this network.

The objective is to evaluate how much information regarding a hub node in the network the algorithm can recover from the test data in the form of eq. (11.1) where the information on the hub node are missing. Precision is used as a measure to evaluate the quality of the retrieval. The retrieval is based on the predictor function in eq. (11.6). In information retrieval, precision have been used as evaluation criteria. Precision is the fraction of relevant data among the all data returned by search. Here, precision is evaluated by calculating the ratio of the correct baskets within the baskets retrieved by the algorithm (in terms of the largeness of the value from the predictor function). The correct baskets are those where unobserved nodes are hidden and related to the latent structure.

The test data was generated in the two steps below.

1. Basket data representing neighbor nodes β_i was generated from the scale-free random network. The nodes under a direct influence from a node are grouped into a basket β_i. For example, we can imagine a situation where a person starts talking and a conversation takes place among neighboring persons. The area of such influence is specified approximately with the distance from a node. In this evaluation, we made up 400 basket data (β_0 to β_{399}) consisting of nodes within 2 hops from the initiator node. One hop is as long as one link on the graph. An example basket is $\beta_0 = \{$0-00, 1-00, 1-01, ..., 2-00, 2-01, ..., 20-00, ...$\}$, whose nodes are within 2 hops from 0-00.

2. A latent structure regarding hub nodes was configured to the basket data β_i by deleting the hub nodes. From the basket data β_i, either the primary hub node (0-00), the secondary hub node (1-00 for the cluster 1-xx), or other node (1-01 within the cluster 1-xx) was deleted so that these node can be made invisible on the basket data. Resulting data is denoted by b_i. An example basket is $b_0 =$ {0-00, 1-01, ..., 2-00, 2-01, ..., 20-00, ...}, whose nodes are within 2 hops from 0-00, in case where the secondary hub node 1-00 becomes the hidden hub node. The deleted hub node and the links inter-connecting the hub node and the 20 clusters became a latent structure hidden behind the basket data b_i. The algorithm tries to retrieve b_i which is different from the corresponding β_i.

11.3.2 Dependence on the Node

We investigate the sensitivity of the algorithm to the nodes. Fig. 11.6 shows the calculated precision as a function of the number of the retrieved baskets. These results are under the conditions in Fig. 11.4. The 20 baskets (appearing at the top in the basket list sorted according to the predictor function) recover the deleted hub nodes with precision above 0.8. This result shows the algorithm works fine. Precision decreases as the number of the retrieved baskets increases because the number of baskets where the hub nodes were deleted is 50 or so. It is generally difficult to recover non-hub node such as the node 1-01 within the cluster 1-xx.

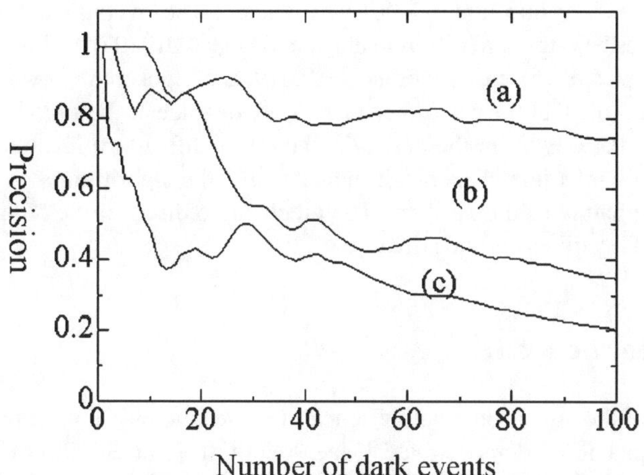

Fig 11.6. Precision to retrieve the baskets where a node was made invisible ($b_i \neq \beta_i$) as a function of the number of the retrieved baskets (corresponding dark events $\mathrm{DE}_i \in \beta_i$, $\mathrm{DE}_i \notin b_i$). The deleted node is (a) the primary hub node (0-00), (b) the secondary hub node (1-00), and (c) other node(1-01). (Copyright, IEEE 2006)

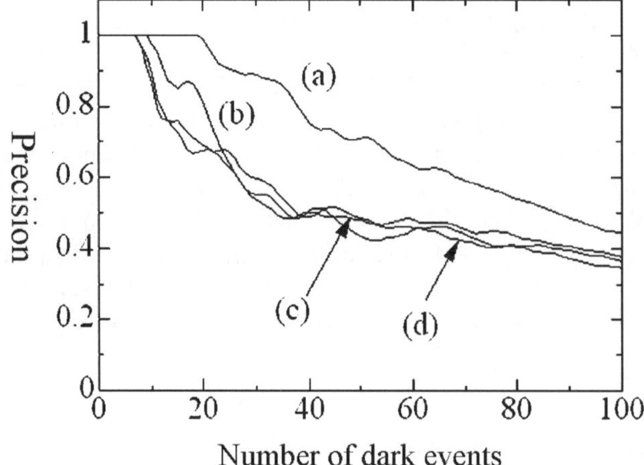

Fig 11.7. Precision to retrieve the baskets where a node was made invisible ($b_i \neq \beta_i$) as a function of the number of the retrieved baskets (corresponding dark events $\mathrm{DE}_i \in \beta_i$, $\mathrm{DE}_i \notin b_i$). The deleted node is the secondary hub node (1-00). The overall ratio where a link connects nodes belonging to different clusters is (a) $p_c = 0.019$, (b) $p_c = 0.057$ (Fig. 11.4), (c) $p_c = 0.095$, and (d) $p_c = 0.19$ (Fig. 11.5). (Copyright, IEEE 2006)

11.3.3 Dependence on the Network

We investigate the sensitivity of the algorithm to the network structure. Fig. 11.7 shows the calculated precision as a function of the number of the retrieved baskets when the secondary hub node (1-00) was deleted. The overall ratio where a link connects nodes belonging to different clusters is $p_c = 0.019$, 0.057 (Fig. 11.4), 0.095, and 0.19 (Fig. 11.5). In case where $p_c = 0.019$, the algorithm shows a very good performance. Note that the performance does not degrade as the ratio increases from $p_c = 0.057$ to 0.19. Even in the network where it is difficult to identify 20 clusters, the algorithm works fine. This result indicates that the algorithm is not sensitive to the clustering nature of the network. The precision to discover the hidden hub node remains good regardless of the structure.

11.4 Corporate R&D

Application of the human-interactive annealing to technology development based on analysis on patents is demonstrated [Maeno 2006b], [Horie 2006]. This is an application toward creative thinking [Hori 1994]. A patent is a set of exclusive rights granted by a state to a patentee for a fixed period of time in exchange for a disclosure of an invention. The rights are indicated by the claims defining the invention. Patents provide with technological elements representing a means to solve a specific engineering design problem.

11.4.1 Simple Demonstration

We try to identify an unknown, but significant technological element by analyzing published patents. It may be a technology hidden by a rival company like a submarine patent, an emerging technology from other field of expertise, a technology owned by a niche company or a small technician community, or a completely new technology. These latent structures are potential threat to corporate research and development. By becoming aware of them, we can turn the threat to opportunity. We also describe extension for integrated data from multilateral observation to help human understand and invent promising scenarios.

Twenty nine patents applied in Japan[1] are picked up as known technological expertise in the field of knowledge discovery. Here, we configure the input baskets so that the subjects of the baskets can be objective or preferred effect on the engineering design problems. Content of the baskets is a set of patent application numbers which is suitable for the subjects of the baskets. Thirteen baskets are configured. The result derived from the human-interactive annealing process at the second iteration is shown in Fig. 11.8. Seven baskets (and corresponding dark events) are retrieved and drawn on the scenario map. They are connected to the biggest clusters consisting of eighteen patents (cluster c_1), and smaller clusters (cluster c_2, c_3 etc.).

Fig. 11.8 includes the annotations put in human's interpretation. The annotation is based on the comments understood from the patent documents. Three new technological elements DE_1 (corresponding to b_1), DE_6 (b_6), and DE_7 (b_7) appeared between the biggest cluster and 2 smaller clusters.

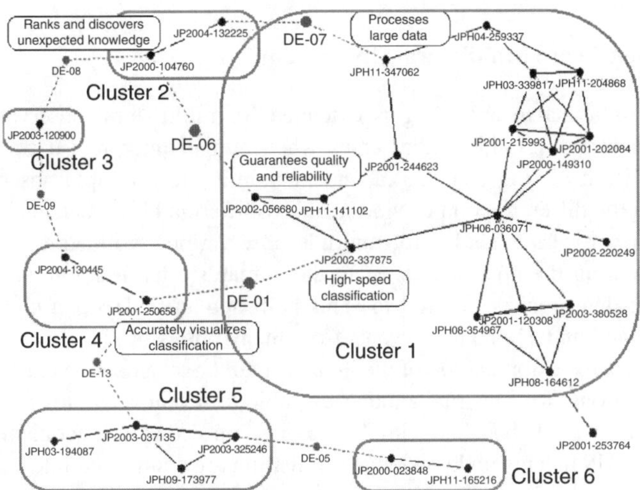

Fig 11.8. Scenario map drawn for the 29 Japanese patents on knowledge discovery.

The cluster c_1 corresponds to a set of conventional means developed for statistical analysis or data mining in knowledge discovery. Particularly, discovery of association rules in knowledge discovery has evolved along three performance criteria. The first criterion is speed. This is required in real-time and on-line applications such as a contact center for product support and services. The second criterion is the amount of data. This is required in batch processing applications such as long-term customer trend analysis. The third criterion is quality. It means that more precise and more accurate association rules are required. The two two-event clusters (clusters c_2 and c_4) incorporate technological elements for discovering unexpected knowledge and for visualizing knowledge respectively. Unexpected knowledge tends to be neglected in human's recognition, but significant for decision-making. In this sense, it could be related to the chance discovery methods. Visualization is an important technical expertise which has been employed in many fields in science and engineering. These are mentioned as annotations in the figure.

The 3 new technological elements between the clusters suggest a new and unknown technological element which combines these three clusters. The human-interactive annealing, about which you are reading, could be just such a technology. It could indicate unexpected threat by visualizing invisible dark events with use of the technical expertise in statistics and machine learning. The technological element could represent a technique to incorporate human cognitive factor into the process. The result recommends the technology analyst to investigate closely whether potential competitor companies are developing such technological element or not. For this purpose, extension for integrated data from multilateral observation is useful, where observation from many points of view is included.

11.4.2 Extension to Multilateral Observation

The human-interactive annealing is extended for multilateral observation. Multilateral observation provides more clues when human interprets dark events. The clues arise from mixing contexts of understanding with multiple observations obtained from the different point of views, and by different observers, analysts, or data sources. The process extended for multiple observations is illustrated in Fig. 11.9. After elaborating the area of interest in the human's interpretation, another observation from data pool is referred to. This help human understand clusters, invent hypothesis, and understand dark events from many points of view.

Toward realistic application of the human-interactive annealing extended to multiple observations, we illustrate another example to invent scenarios for technology development by published analyzing US patents[2] on knowledge acquisition. At first, twenty-three US patents on knowledge acquisition are configured into baskets based on the US class identifiers. The US class represents tree-structured categories of technologies. The result is shown in Fig. 11.10. This figure is similar to Fig. 11.8. Five dark events emerge. Here, we focus on the latent structure between the clusters c_2

---

[2] http://www.uspto.gov/ (Related information is available at the United States Patent and Trademark Office).

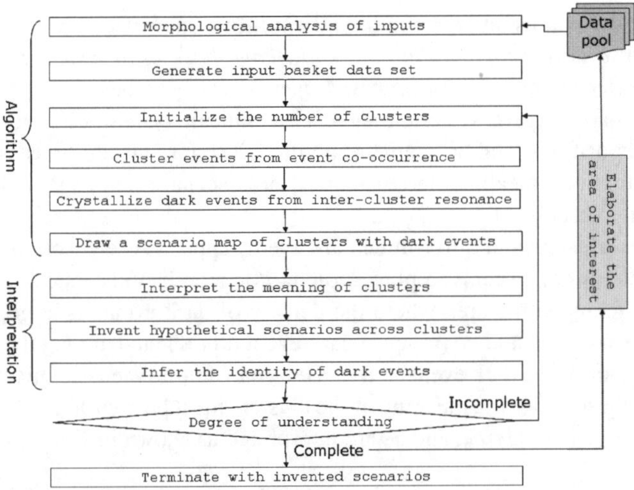

Fig 11.9. Human-interactive annealing process extended for integrated data from multilateral observation (extension of fig. 11.3 with data pool).

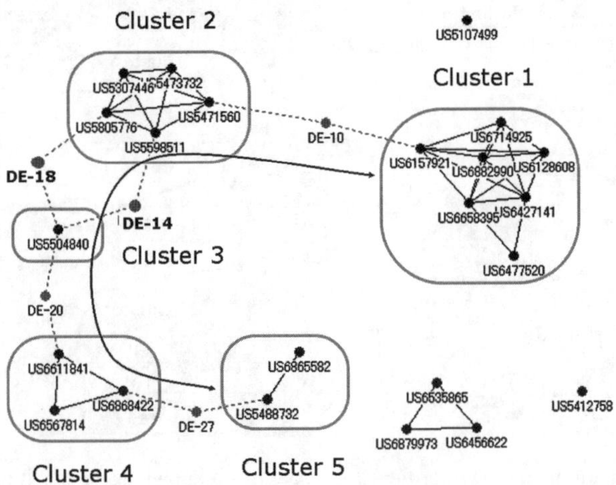

Fig 11.10. Scenario map drawn for the 23 US patents on knowledge acquisition from the data on patent identifiers and classes.

and c_3. The dark events DE_{14} and DE_{18} refer to the baskets b_{14} and b_{18}, whose subjects are US class = *706/50: knowledge processing system/having specific management of a knowledge,* and US class = *706/59: knowledge processing system/creation or modification.*

Then, we refer to another data to understand the dark events deeply. It is claims in the patents 5805776 and 5598511 in the cluster c_2, and 5504840 in the cluster c_3. Although the other portions of patent texts can be used, claims directly represents a means to solve a problem; technological elements of interest. Morphological analysis investigates the claims and decomposes them into a sequence of words. For example, a claim: *a device for building the knowledge base of an expert system, which diagnoses a technical system comprised of modules* becomes b_0 = {device, build, knowledge, base, expert, system, diagnose, technical, system, comprise, module}. Articles, prepositions, and pronoun are omitted. The baskets are processed with the human-interactive annealing, and visualized on a scenario map again. The result is shown in Fig. 11.11.

Ten word clusters and thirty-three dark events appear. A few isolated events also appear. The cluster c_5 seems to play a role to connect the clusters c_1, c_2, c_3, c_4, c_6, and c_8. We focus on the area where the density of dark events is large; among the clusters c_1, c_2, c_3, c_5, and c_8 (= dense dark event area 1), and among the clusters c_4, c_5, c_6, and c_7 (= dense dark event area 2). Before inventing scenario on technological elements in these areas, the meaning of the clusters must be interpreted. For example, the clusters c_2, c_3, c_4, c_5, c_6, and c_7 are interpreted as shown in Table 11.1.

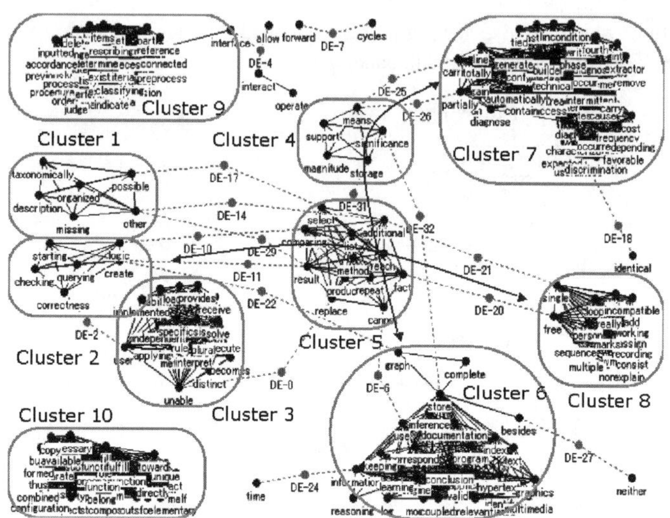

Fig 11.11. Scenario map drawn for the 23 US patents on knowledge acquisition from the data on words (morphological analysis of the sentences) in claims.

Table 11.1. Interpretation (keywords) of clusters in a scenario map.

Cluster	Interpretation (keywords)
c_2	Correct, check, logic, query
c_3	Interpret, execute, apply, implement, rule, user
c_4	Support, store, significance, magnitude
c_5	Replace, compare, select, produce, repeat, fact
c_6	Learn, infer, identify, reason, text, document, multimedia, graphics
c_7	Diagnose, extract, frequency, occurrence, rate, cost, knowledge base

From the interpretation of the clusters, we invented a few scenarios incorporating the identity of dark events in dense areas. Here we regard the dense area as a structure to represent a technical idea as a whole. Two examples are listed below, which are rich in terms of proposability, unnoticability, and growability. Expert knowledge is inferred toward the dense dark event area 1 as a new technological means. Cross-validation of knowledge is inferred toward the dense dark event area 2 as a new technological means.

- Enhance learning by replacing facts with the aid of interpretation of queries on *expert knowledge* (=dense dark event area 1).
- Assign significance to knowledge inferred from multiple information (texts, documents, multimedia, graphics, etc. based on diagnosis by *cross-validating knowledge* (=dense dark event area 2).

Although the analysis shown in this section is for a simple demonstration purpose, it indicates how we should proceed to get an insight into a scenario for harnessing threat from hidden technological property based on a latent structure. The usage of integrated data from multilateral observation is also illustrated. Popular approaches in the present technology research and development employ engineering design methods such as TRIZ[3] [Fey 2005], Value Engineering (VE), or Taguchi Methods. TRIZ has been developed by G. Altshuller and his colleagues since 1946. It is a collection of methodologies, tool sets, knowledge bases, and model-based technologies to obtain innovative ideas and solutions to solve engineering design problems. In TRIZ, systems, failures, and evolution patterns are analyzed of system evolution toward new systems and refinement of old systems. Value Engineering is a systematic method to improve the value of products and services by examining functions. Value is the ratio of function to cost. The value can be increased by enhancing the function or reducing the cost. This practice identifies and removes unnecessary expenditures. The Taguchi Methods are statistical methods for quality engineering developed by G. Taguchi. It improves the quality of manufactured goods, and can also be applied to marketing or advertising. This practice is a combination of the off-line quality engineering in the design phase and the on-line quality engineering in the manufacturing phase.

These methods mainly aim at utilizing precedent successful cases and optimizing combination of technological elements under cost and quality constraint. Identifying an invisible new technological element emerging as a niche is getting more important. The human-interactive annealing process is expected to shed a new light on such problems.

11.5 Summary

Unobserved events play an important role in the dynamics of observed events. They are either something hidden intentionally, or something new not yet recognized. Such invisible events are named dark events. A new method named human-interactive

---

[3] (abbrev.) Theory of Inventive Problem Solving in Russian.

annealing is presented to understand the threat arising from the dark events and to invent a scenario to turn the threat to opportunity. The method is extended for integrated data from multilateral observation. A scenario invention experiment is demonstrated using patent documents in the field of knowledge acquisition to design corporate R&D strategies. Multilateral observation provides more clues when the engineers and strategists obtain an idea on emerging technologies.

References

[Barabasi 1999] A. L. Barabasi, R. Albert, and H. Jeong: Mean-field theory for scale-free random networks, Physica A **272**, 173-187 (1999).

[Batallas 2006] D. A. Batallas, and A. A. Yassine: Information leaders in product development organizational networks: Social network analysis of the design structure matrix, IEEE Transactions on Engineering Management **53**, 570-582 (2006).

[Chakrabarti 1993] A. K. Chakrabarti, I. Dror, and N. Eakabuse: Interorganizational transfer of knowledge: an analysis of patent citations of a defense firm, IEEE Transactions on Engineering Management **40**, 91-94 (1993).

[Chakrabarti 1991] A. K. Chakrabarti: Competition in high technology: analysis of patents of US, Japan, UK, France, West Germany, and Canada, IEEE Transactions on Engineering Management **38**, 78-84 (1991).

[Duda 2000] R. O. Duda, P. E. Hart, and D. G. Stork: Pattern classification (2nd edition). Wiley-Interscience (2000).

[Fey 2005] V. Fey, and E. Rivin: Innovation on demand: New product development using TRIZ. Cambridge University Press (2005).

[Fruchterman 1991] T. M. J. Fruchterman and E. M. Reingold: Graph drawing by force-directed placement, Software - Practice and Experience **18**, 1129-1164 (1991).

[Hastie 2001] T. Hastie, R. Tibshirani, and J. Friedman: The elements of statistical learning: Data mining, inference, and prediction (Springer series in statistics). Springer-Verlag (2001).

[Hori 1994] Koichi Hori: A system for aiding creative concept formation, IEEE Transactions on Systems, Man, and Cybernetics **24**, 882-894 (1994).

[Horie 2006] K. Horie, Y. Maeno, and Y. Ohsawa: Human-interactive annealing process with pictogram for extracting new scenarios for patent technology, Proceedings of the CODATA (International Council for Science: Committee on Data for Science and Technology) International Conference, Beijing (2006).

[Klerks 2002] P. Klerks: The network paradigm applied to criminal organizations, Connections **24**, 53-65 (2002).

[Kohonen 1990] T. Kohonen: The self-organizing map, Proceedings of the IEEE **78**, 1464-1480 (1990).

[Krebs 2002] V. E. Krebs: Mapping networks of terrorist cells, Connections **24**, 43-52 (2002).

[Liben-Nowell 2004] D. Liben-Nowell, and J. Kleinberg: The link prediction problem for social networks, Proceedings of the Annual ACM International Conference on Information & Knowledge Management, New York (2004).

[Maeno 2007a] Y. Maeno, and Y. Ohsawa: Human-computer interactive annealing for discovering invisible dark events, IEEE Transactions on Industrial Electronics **54**, 1184-1192 (2007).

[Maeno 2007b] Y. Maeno, K. Ito, and Y. Ohsawa: Detecting invisible relevant persons in a homogeneous social network, Lecture Notes in Computer Science **4490**, 74-81, Springer-Verlag (2007).

[Maeno 2007c] Y. Maeno, Y. Ohsawa, and T. Ito: Catalyst personality for fostering communication among groups with opposing preference, Lecture Notes in Artificial Intelligence **4570**, 806-812, Springer-Verlag (2007).

[Maeno 2006a] Y. Maeno, and Y. Ohsawa: Stable deterministic crystallization for discovering hidden hubs, Proceedings of the IEEE International Conference on Systems, Man, and Cybernetics, Taipei, 1393-1398 (2006).

[Maeno 2006b] Y. Maeno, K. Ito, K. Horie, and Y. Ohsawa: Human-interactive annealing for turning threat to opportunity in technology development, Proceedings of the IEEE/WIC/ACM International Conference on Data Mining, Workshop on threat Mining, Hong Kong, 714-717 (2006).

[Ohsawa 2006] Y. Ohsawa eds.: Chance discovery in real world decision making. Springer-Verlag (2006).

[Ohsawa 2005] Y. Ohsawa: Data crystallization: chance discovery extended for dealing with unobservable events, New Mathematics and Natural Computation,**1**, 373-392 (2005).

[Sugiyama 2002] K. Sugiyama: Graph drawing and applications for software and knowledge engineers (Series on software engineering and knowledge engineering 11). World Scientific Publishing (2002).

[Suzuki 2005] E. Suzuki, and J. M. Zytkow: Unified algorithm for undirected discovery of exception rules, International Journal of Intelligent Systems **20**, 673-691 (2005).

[Watts 1998] D. J. Watts, and S. H. Strogatz: Collective dynamics of small-world networks, Nature **398**, 440-442 (1998).

[Weiss 1998] G. M. Weiss, and H. Hirsh: Learning to predict rare events in event sequences, Proceedings of the 4th International Conference on Knowledge Discovery and Data Mining, New York City, USA (1998).

Human-Interactive Annealing Process with Pictogram for Extracting New Scenarios for Patent Technology

Kenichi Horie[1], Yoshiharu Maeno[2], and Yukio Ohsawa[3]

[1] School of Engineering, the University of Tokyo, Bunkyo-ku, Tokyo 113-8563
`kenhorie1231@hotmail.com`
[2] Graduate School of System Management, the University of Tsukuba, Bunkyo-ku, Tokyo 113-856
`maeno.yoshiharu@nifty.com`
[3] School of Engineering, the University of Tokyo, Bunkyo-ku, Tokyo 113-8563
`ohsawa@q.t.u-tokyo.ac.jp`

Summary. *It is only observable part of the real world that can be stored in data. Latent structure behind observation often plays an important role in the dynamics of visible events. Such latent structure is composed of unobservable events. Data crystallizing aims at presenting the latent structure among events including unobservable events. Human-interactive annealing is developed to understand unobservable events by tuning the granularity level of structure to be visualized after data crystallization. This paper presents application of the data crystallization and human-interactive annealing for extracting new scenarios of product design from latent technology structure behind current patented technology. The results show its effect to industrial decision making.*

Keywords: chance discovery, data crystallization, unobservable events, Human machine interaction, Design, pictogram

12.1 Introduction

This study is dedicated to decision makers working in real domains where discovery of unobservable events are desired. We have been challenging to reveal events potentially important but never observed. Because unobservable events can not be included in given data, existing data mining methods hardly worked in identifying such events.

Data crystallization is the challenge to this hard problem from an extension of our challenge of chance discovery since 2000 ([1–3]). Chance discovery is to discover a chance, defined as an event significant for decision. This has been a challenge beyond data mining, in that the goal was the understanding of the meaning of rare events for human's decisions, rather than learning rules for predicting rare events as done in ([4, 5]). Methods of chance discovery show the relation of concepts behind

K. Horie et al.: *Human-Interactive Annealing Process with Pictogram for Extracting New Scenarios for Patent Technology*, Studies in Computational Intelligence (SCI) **123**, 205–219 (2008)
`www.springerlink.com`

the words, and aid human-computer interactions for finding influential events, ideas, and people in the market ([6–8]), with recognizing the uncertainty of the future. For aiding this creation of the future rather than the prediction, tools of data-visualization have been developed ([2, 9, 10], Matsuo, Ohsawa & Ishizuka, 2002).

Unobservable events which are not included in the given data, however, can be visualized by the breaking through new method, data crystallization ([11]), where artificially dummy nodes may potentially correspond to them. In addition, a new method, *human-interactive annealing* ([12]) is developed to reveal the latent structure along with a simplified stable crystallization algorithm.

In this paper, we address an issue to understand the data crystallization method with human-interactive annealing process. Then, we present our application of data crystallization with human-interactive annealing to patents documents for discovering hidden idea from latent structure and making a decision to design new products by creating new scenarios from them. In this study, we propose six steps in a process with a method of *Pictogram*, which is composed of picture, chart and text data in order to aid human understanding and creating new scenarios from a latent structure visualized by data crystallization and human-interactive annealing process.

12.2 Data Crystallization and its Limit

The objective of data crystallization is to understand the role of (not only rare but) unobservable significant events to the decision of user. In this paper, we show an approach of integrating two new methods for the breakthrough from the current state of arts in chance discovery.

The first is a method for visualizing data with inserting artificial dummy items. These dummy items mean unobservable events, of which the entity is totally unknown and are not includes in the given data. By this, the user can see the overview of the real world, with the potential existence of significant but hidden (unobservable/unknown) events. The second is the cognitive process of human for understanding the role of hidden events in the real world. Basically, the process of chance discovery so far has been following the Double Helix Process in ([13, 14]). The effect of this process, where the granularity of information about chances is tuned, enabled applications such as selling new products in marketing ([6, 7]), detecting earthquake signs ([15]), treatment opportunity of hepatitis ([16]), and designing products ([17], Maeno & Ohsawa, 2006). In order to create finer strategic scenarios in real business, data crystallization is expected to work for introducing unobservable events to the creation, where the previous methods of chance discovery introduced rare events included in the data. We present a new process of data crystallization with human-machine interaction in this paper, for enabling to apply data crystallization to the problem of designing new products.

12.2.1 KeyGraph: The Basic Tool for Visualizing Event Maps

KeyGraph is a tool we developed for visualizing relations among events. If the environment here means the existing activities a criminal group, KeyGraph shows the relation of members on their co-existing frequencies (See ([2] and [3]) for details). In Eq. (12.1), let data D1 express a set of meetings, putting a period (g.h) at each end of meeting.

Here, *member*1 in Eq. (12.1) can be regarded as "member1_attends" i.e., an event that a member appeared in a meeting place. Regarding each item in the data as an event rather than as an object is meaningful in interpreting KeyGraph as an event map, where the sequence of events should be grasped from the connections among nodes.

$$D1 = \begin{bmatrix} member1 \ member2 \ member3. \\ member1 \ member2 \ member3 \ member4. \\ member4 \ member5 \ member7 \ member8. \\ member5 \ member2 \ member3 \ member7 \ member8. \\ member1 \ member2 \ member7 \ member8 \ member9. \\ member5 \ member7 \ member8 \ member9. \end{bmatrix} \qquad (12.1)$$

KeyGraph ([2,11]), of the following steps, is then applied to D1. Then Fig. 12.1 is obtained as a result.

KeyGraph-Step 1: The $M1$ most frequent events in the data (e.g., *member*1 in Eq. (12.1)) are depicted with black nodes. The $M2$ most strongly co-occurring events-pairs get linked with black lines. Here, the co-occurrence is computed on the Jaccard equation in Eq. (12.2), where Freq(X) means the number of baskets (lines in Eq. (12.1)) including elements of X.

member1, member2, and member3 are connected with solid lines in Fig. 12.1. Each connected graph forms one *island*, implying a basic context shared by its members.

KeyGraph-Step 2: $M3$ events co-occurring with multiple islands the most strongly, e.g., member9, are obtained as *hubs*. A path of links connecting islands via hubs is called a *bridge*. If a hub is rarer than black nodes, it is colored in a different color (e.g. red or white) than black. We regard such a hub as a candidate of *chance*,

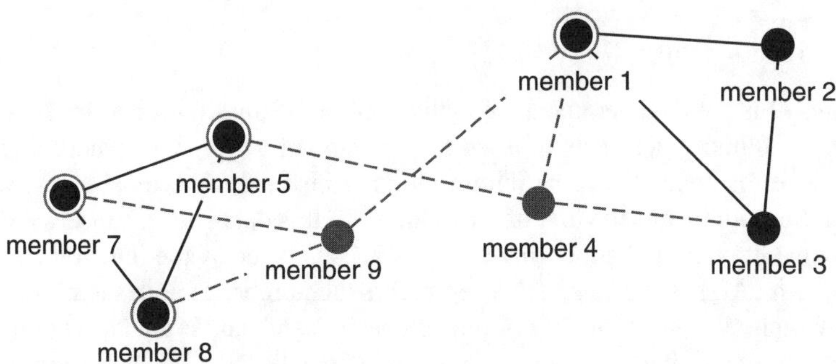

Fig 12.1. An output of KeyGraph: Islands are obtained from D1, including sets member1, member2, member3 and member5, member6, member7 respectively. The nodes in the islands show frequent events, and member4 and member9 show rare hubs bridging islands.

because it can be meaningful for a decision to jump from an island, corresponding to a context represented by the cluster of events, to another island.

$$Ja(e_i, e_j) = \frac{Freq(e_i \cap e_j)}{Freq(e_i \cup e_j)} \qquad (12.2)$$

Figure 12.1 helps in making a scenario of criminal behaviors, such as "member1, member2, and member3 are working together, and member5, member6, member7 form another group. When they meet member9, member9 gives commands to both groups from a higher level of organization," via recollecting information about the members from the memory of intelligence analysts.

The appearance of a bridging member can be a central topic to the analysts.

Discussion of analysts by looking at the output diagram of KeyGraph may resolve the uncertainty about which member4 or member9 is the real leader, because human can reflect their knowledge acquired from the real interaction with the external environment.

12.2.2 Data Crystallization: Extending KeyGraph for Analysis of Hidden Events

Data crystallizing aims at presenting the hidden structure among events including unobservable ones. This is realized with inserting dummy items, which may correspond to unobservable events, to the given data on past events. The existence of unobservable events and their relations with other events are then visualized by applying KeyGraph. The core of data crystallization is represented as follows:

```
[The algorithm of data crystallization]
Hidden_0 := {}; line_0 :={}; oM1 , M2p: given values
For all i, j ? {0, 1, c, N} such that  j G.E. i do
if line_i and line_j are same then Insert (D, i, j);
H: = KeyGraph (D, M1, M2, M3, = M1/2);
For j =1 to N do
If j ? H then Delete (D, j);
```

Here, D is the data-set given. N is the number of lines (co-occurrence units) in the data. A dummy item gets inserted to each line of D. If two or more lines have the same set of items, the same dummy item is inserted to all those lines, sufficed with the line-number of the first of those lines. To this data-set with inserted dummy nodes, KeyGraph is applied. Formally, D is to be analyzed by the function KeyGraph (D, M_1, M_2, M_3). The value of M_1 represents the number of nodes to be visualized by KeyGraph. M_2 is the number of links in each island, and is set larger (smaller) if user likes to see a small (large) number of large (small) islands. Then, dummy items which did not appear on the bridges of KeyGraph get deleted from D.

$Insert(D, i, j)$ means to insert $dummy\_j$, the dummy node for the j-th line, to the i-th line of data D and from data D. The second and the third lines of the procedure mean to insert $dummy\_i$ to the $dummy\_j$-th line, and, if there is a line (the i-th line) of the same set of items as the j-th line, $dummy\_j$ is inserted to all those lines. $Delete(D, j)$ means to delete $dummy\_j$, the dummy item for the j-th line, from all its appearances in data D. H represents the set of the line-numbers where the dummy items, which appeared on the bridges of KeyGraph, are positioned in the data.

Data crystallization works in the way like the crystallization of snow. A dummy item plays a role of a particle of dust connecting molecules of water in the air. The increase in M_2 corresponds to the decrease in temperature. In the case of snow, a well-structured crystal is made because the temperature is decreased gradually and water molecules are collected via dust participles.

For showing a simple example, let us take series of meetings in a team of 21 members, as the target data to analyze. In D_a, in Eq. (12.3), a part of data on the participants is listed, obtained in *Step 2*) for the concern with the real leader in the team. In Eq. (12.3), each line corresponds to a meeting.

$$D_a = \begin{bmatrix} Prof.U \;\; Prof.K \;\; Prof.O \;\; Prof.J \\ Prof.Q \;\; Prof.M \;\; Prof.A \;\; Prof.N \;\; Prof.I \\ Prof.U \;\; Prof.K \;\; Prof.J \;\; Prof.I \\ Prof.I \;\; Prof.J \;\; Prof.G \\ Prof.O \;\; Prof.A \;\; Prof.U \;\; Prof.I \;\; Prof.N \\ Prof.N \;\; Prof.L \;\; Prof.U \;\; Prof.M \\ Prof.F \;\; Prof.G \\ \cdots \end{bmatrix} \tag{12.3}$$

Figure 12.2 is the result of KeyGraph, for $M_1 = 20$, $M_2 = 20$, and $M_3 = 20$, from D_a. Even though KeyGraph search 20 hubs bridging among islands, we find all islands separated i.e., no hubs among them. That is, the team looks like a set of groups irrelevant to each other. Thus, we should investigate hidden levels. The dummy nodes are inserted, denoted 1_x for the x-th line, to obtain D_b below.

$$D_b = \begin{bmatrix} Prof.U \;\; Prof.K \;\; Prof.O \;\; Prof.J \;\; dummy1\_1 \\ Prof.Q \;\; Prof.M \;\; Prof.A \;\; Prof.N \;\; Prof.I \;\; dummy1\_2 \\ Prof.U \;\; Prof.K \;\; Prof.J \;\; Prof.I \;\; dummy1\_3 \\ Prof.I \;\; Prof.J \;\; Prof.G \;\; dummy1\_4 \\ Prof.O \;\; Prof.A \;\; Prof.U \;\; Prof.I \;\; Prof.N \;\; dummy1\_5 \\ Prof.N \;\; Prof.L \;\; Prof.U \;\; Prof.M \;\; dummy1\_6 \\ Prof.F \;\; Prof.G \;\; dummy1\_7 \\ \cdots \end{bmatrix} \tag{12.4}$$

Figure 12.3 is the output of KeyGraph for D_b in Eq. (12.4). Some dummy nodes appear in the graph, bridging among islands. For example, dummy1_5 between Prof.A and Prof.O means some hidden thing relevant to the fifth meeting (the fifth line in Eq. (12.4)) made a significant bridge for the structure.

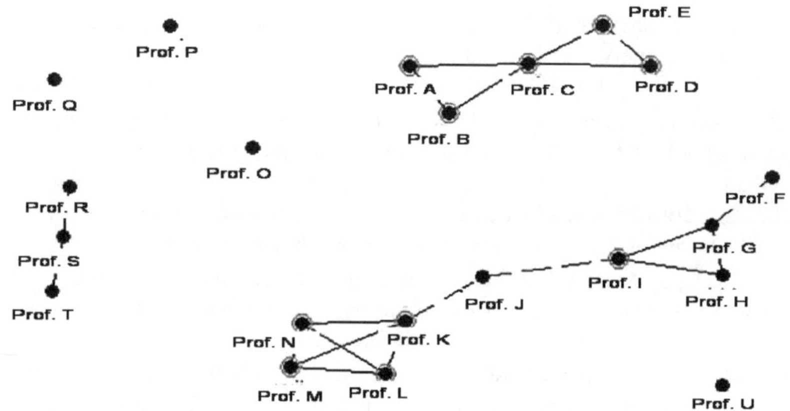

Fig 12.2. The original KeyGraph for members of a group.

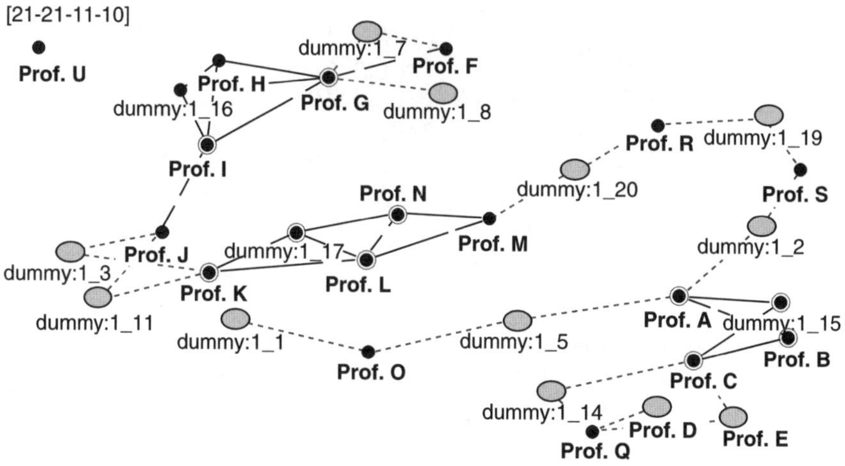

Fig 12.3. The output for data with first-order dummies (1_x).

From the new figure obtained by data crystallization, we can obtain newer findings. For example, dummy1_1 means there might have been some powerful leader who just sent a command such as "do meeting for this problem!" to the members of the first meeting. He/she may not have appeared in the meetings, but his command can be regarded as the first voice of the meeting i.e., he/she was the hidden leader.

Obtaining the diagram in Figure 12.3, we obtain the new dataset as in Eq. (12.5), by leaving dummy items appearing in Figure 12.3. The two dummies as dummy1_4 and dummy1_6 are discarded, because they do not appear in Figure 12.2.

$$
D_c =
\begin{bmatrix}
Prof.U & Prof.K & Prof.O & Prof.J & dummy1\_1 \\
Prof.Q & Prof.M & Prof.A & Prof.N & Prof.I & dummy1\_2 \\
Prof.U & Prof.K & Prof.J & Prof.I & dummy1\_3 \\
Prof.I & Prof.J & Prof.G \\
Prof.O & Prof.A & Prof.U & Prof.I & Prof.N & dummy1\_5 \\
Prof.N & Prof.L & Prof.U & Prof.M \\
Prof.F & Prof.G & dummy1\_7 \\
\cdots
\end{bmatrix}
\tag{12.5}
$$

In Ohsawa ([11]), Ohsawa presented a method to automatically decrease M2, corresponding to the temperature@in winter. The expected result is that various granularity of crystallized structure can be obtained. For example, in the application of Ohsawa's data crystallization to the data on peoples' meetings, relations were shown between lower-level leader such as the leader of a small subgroup and top-leader such as the dean of a school faculty.

However, according to the experiments, his method worked only for a small number of items in the data as the example in ([11]). For a larger number of items, meaningless dummy nodes appear in the output of graph of data crystallization.

Although Ohsawa proposed to combine the talent of human to the data crystallization algorithm, it was a sheer extension of existing double helix process of chance discovery. For real application of data crystallization, because dummy nodes do not have names corresponding to real entities in the real world, human demands to see more simplified structures.

12.3 Refinement by Human-Interactive Annealing

The human-interactive annealing is a new technique of which the basic procedure has been presented by Maeno and Ohsawa ([12, 18]). The process is similar to the annealing in materials science and simulated annealing. A graph, i.e., event map as mentioned above, is used to represent crystallized unobservable events. The annealing process is a combination of two complementary elements; interpretation by human and the crystallization by computer. The two elements are combined as in Figure 12.4.

Here, the temperature is the single control parameter representing the depth of human's understanding of the structure of the real world. When the understanding should to be richer, the temperature shall be set higher, resulting in more complex hidden structures among many islands (co-occurrence clusters of observable events) to be visualized. This leads to the discovery of novel and unexpected scenarios. On the other hand, when the understanding should become simplified, the temperature shall be set lower. Then the user tried to understand the basic structures from the graph. This iteration in the annealing process is continued until human is satisfied by the posterior understanding.

As the temperature increases, the following three structural changes occur on the graph. These are embedded in the annealing process and independent of the stable deterministic crystallization algorithm.

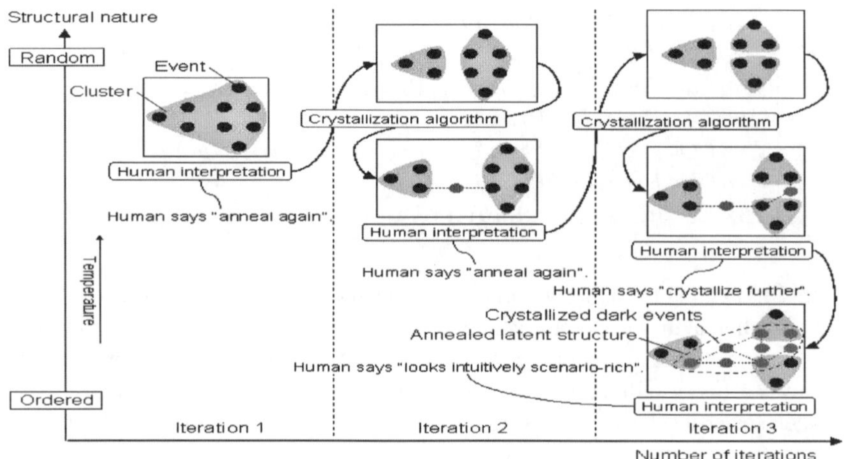

Fig 12.4. Human-interactive annealing; iteration of human's interpretation and computer's data crystallization

1. Weaker inter-cluster links, i.e., connecting dummy events and clusters, are @destroyed.
2. Weaker intra-cluster links, i.e., connecting events within a cluster, are destroyed.
3. The events are divided into larger number of clusters.

In this annealing-based crystallization, the computer basically analyzes the occurrence frequencies and the co-occurrence of events. As in Figure 12.4: In the heating step, up to the specified peak temperature, the number of edges between visible events decreases i.e., weak associations are destroyed and crystallized unobservable events disappear. Then, a cooling step comes after the heating step, where event relations are solidified as temperature goes down. The number of unobservable events between clusters of visible events, corresponding to islands in KeyGraph, increases in the graph. As a result, the clusters are connected to each other to form a single large structure. The crystallization is followed by the interpretation by human, where it is checked whether the termination condition is fulfilled, i.e., if the user is satisfied with his/her understanding.

In Maeno & Ohsawa ([18]), it has been shown that human-interactive annealing successfully obtains hidden leaders even if they do not appear in the given dataset. The method has been applied to a large-sized data-set of meetings, produced from an artificial human network produced by a community simulator. The results showed people who meet many people can be restored by this method, even though they had been deleted from the original dataset.

The scalability of the method has been thus evaluated, because the network included 100 nodes, which was much larger than the 20 nodes they had been testing. Yet, this was a sheer toy problem: The application to real-world problems has been still open to data crystallization and human-interactive annealing.

12.4 Application of Human-interactive Annealing to Patent Documents

12.4.1 Preliminary Study and Tasks

We executed a preliminary test based on text data made by combining documents of six Japanese patents about surface defects inspection system with couple charged device (CCD) and marking systems for defects on works. All claims in these patents were used for text data and processed by data crystallization with human-interactive annealing. After finishing the iterations in the human-interactive annealing process, an event map was shown by KeyGraph on Polaris ([9]) (Figure 12.5 & Figure 12.7).

We executed the test for two hours, with four examinees, which are composed of one sales manager, two sales engineers and one engineer. These people were instructed to communicate looking at the output graph of KeyGraph as a result of data crystallization, where each opinion was instructed to be a scenario for developing products.

However, as a result, only one engineer understood the meaning of all clusters in the diagram and creates scenarios of each cluster. Neither new words corresponding to the hidden events nor new scenarios were communicated or created. We interviewed all examinees after the preliminary test and found the major problems as listed below:

1. Too many hidden events to predict and difficult to think of suitable words to express the hidden events connecting other words in clusters.
2. The word-structure is complex, reflecting the complexity of patent documents, composed of multiple contents such as purposes, implementations, technologies, etc.

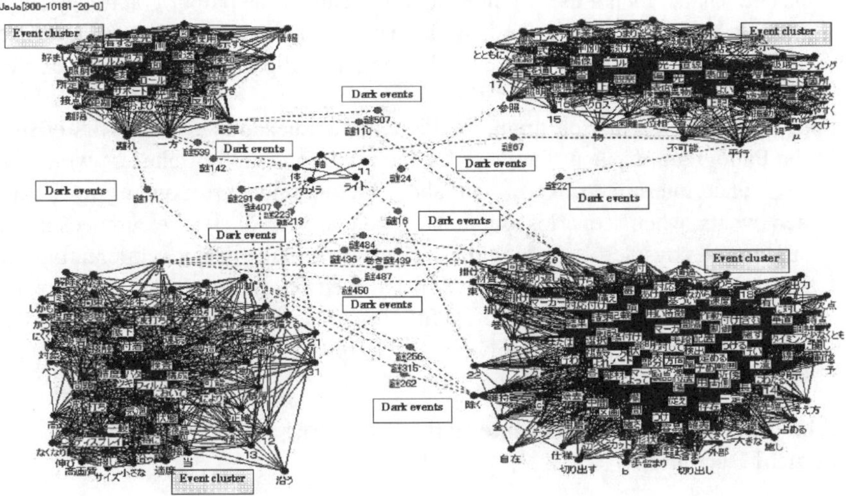

Fig 12.5. Event map after human-interactive annealing

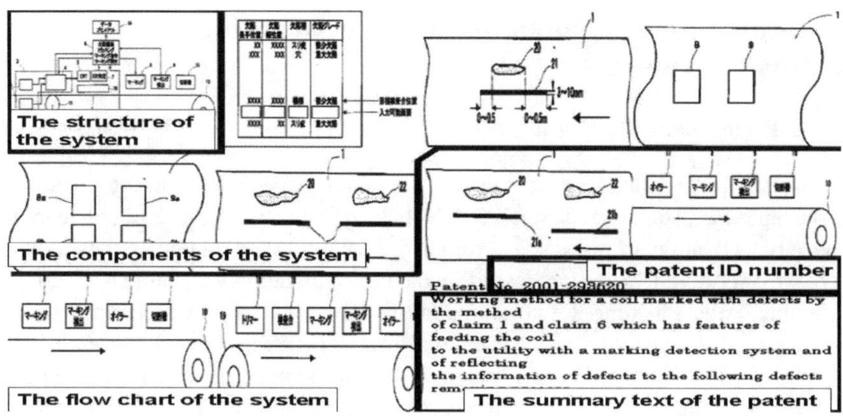

Fig 12.6. A sample of Pictogram

12.4.2 Improvement of Tasks

The improvements listed below were obtained, heuristically, for solving the problems above in the preliminary tests:

Impr.1) Choose topics from patent claims for text data so as to focus on limited @topics.

Impr.2) Add each patent number to the end of corresponding claim, as if they were words to be dealt with by KeyGraph.

Impr.3) Mark each island (cluster) with a keyword summarizing the cluster.

Impr.4) Make Pictograms (Figure 12.6) for all patents. A Pictogram is a visual summary of each patent, composed of the patent number, the flow chart of the patented system, drawings of components, corresponding to the claims in the patent. Then, when a user is interested in a part of the output graph KeyGraph, Pictograms including the words in the part are extracted and shown to the user. Then the user chooses the suitable Pictogram and pastes onto the part of KeyGraph (Figure 12.7)

Impr.5) Set the presentation timing of Pictograms to examinees, as to firstly show the Pictogram of patent numbers for the reinterpretation of clusters which had been once interpreted. And, then, show Pictograms corresponding to the hidden events, when scenarios are about to be considered. Here, examinees are instructed to consider as many hidden events to connect islands, in creating new scenarios. This order, of presentation is based on the results of ([17]), where users who successfully created useful scenarios first paid attention to clusters and then to the bridges, in applying KeyGraph to product designs.

12.4.3 Application of Human-interactive Annealing with Pictogram to Patent Documents

We adopted the new proposals for the application of data crystallization and human-interactive annealing process to 106 Japanese patents including both gmarking

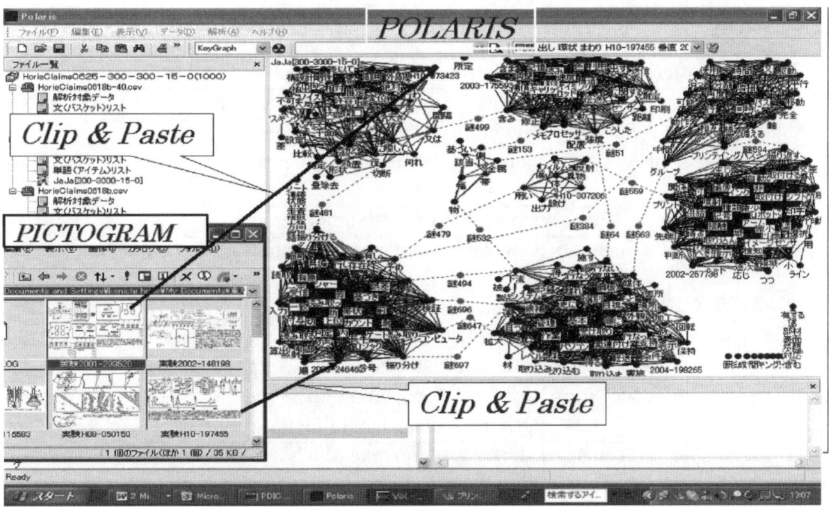

Fig 12.7. Pasting Pictogram on event map after human-interactive annealing

systemsh and ginspectionh and executed the real test by the six steps below for two hours with five examinees: one sales manager, two sales engineers and two engineers working in a company (call this Company A hereafter) to develop and sell SIS machines. Following the steps below, reflecting the improvements (Impr.1 through Impr.6), and the examinees did a design communication with looking at the result of KeyGraph with data crystallization:

[Phase of presenting the scenario map]

Step1: See the event map visualized by KeyGraph with data crystallization, on the text data obtained by combining the documents of 106 patents.

[Phase of discussing about islands]

Step2: Interpret events underlying each cluster in the graph, and write the title corresponding to the meaning of each cluster, on the presented graph (Table 12.1).

Step3: Talk about scenarios using words in each cluster and write them on the white board in the room during the discussion (Table 12.1).

Step4: Find suitable Pictograms (Figure 12.7) to be pasted to patent numbers in clusters depicted as circles in Figure 12.8, and reinterpret the cluster referring to them.

[Phase of discussing about bridges]

Step5: Past Pictograms corresponding to dummy nodes in the graph, considering hidden events, e.g., "If events in the cluster of Hidden events1 occurs, then events in the cluster of Hidden events2 will also occur" and so on, to connect the clusters in the graph (Figure 12.9). Then, write created new scenarios in the white board. For example, subjects may create a scenario *change the speed of conveyer with the progress of the*

Table 12.1. Cluster Title and Scenarios

Cluster	Item	Scenarios
1	Title	Technology for the marking system
	Scenario	System to remove specific defects from unmasked planate area by irradiating laser beam
2	Title	Controller of marking system
	Scenario	Control system which transmits a position of defects on the film by measuring the distance of between defects
	Correction of scenarios	Control system to transmit the positional information to the back end equipment, which identify the position of defects.
3	Title	Driving mechanism of marking system
	Scenario	Marking equipment which moves in parallel with the travel direction of film on roll and is able to mark multiple defects
4	Title	Ink jet marking system
	Scenario	Robotic arms, by which ink jet heads are automatically moved to a position of defects on the product and mark the defects by ink
5	Title	Back end inspection system
	Scenario	Computational system to detect the position of defects and separate the controversial defects for aiding the visual inspection at the back end
6	Title	Post process of CCD inspection
	Scenario	System to allocate the number to each of defects on the film by measuring the distance
	Correction of scenarios	Control system to transmit the positional information to the back end equipment, which identify the position of defects.

film on the belt - then the back end checking may be realized by combining cluster No.5 of "Technology for back end inspection system" and cluster No.6 of "Technology for control of CCD inspection" via the four hidden events (a double headed arrow) at "Hidden events1."

Step6: Select new scenarios which can be agreed by all participants and evaluate them in the view points of feasibility in development and marketing, and of the product's novelty.

12.4.4 Experimental Results

We evaluated the results of the procedure on the following criteria. It should be noted that the purpose of this evaluation is to see if the proposed procedure in 4.3 supplied users with the ability to obtain novel and acceptable to their forthcoming team-work

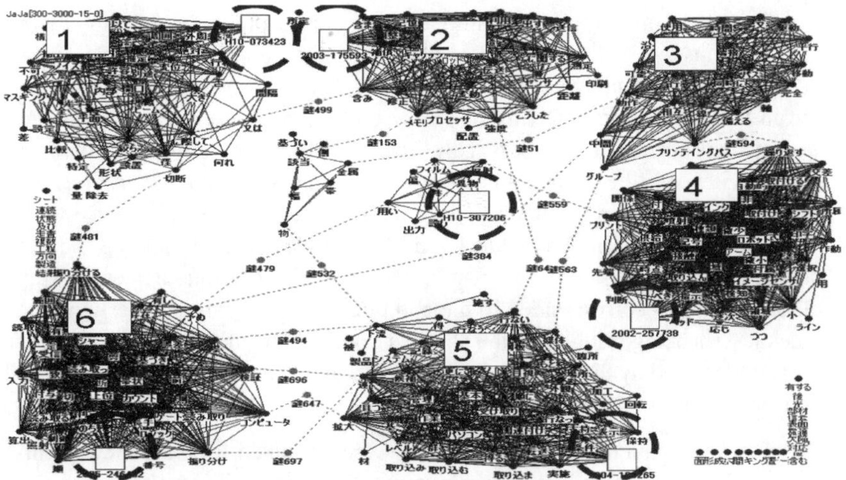

Fig 12.8. Pasting Pictogram on each cluster (Step 1 to Step 4)

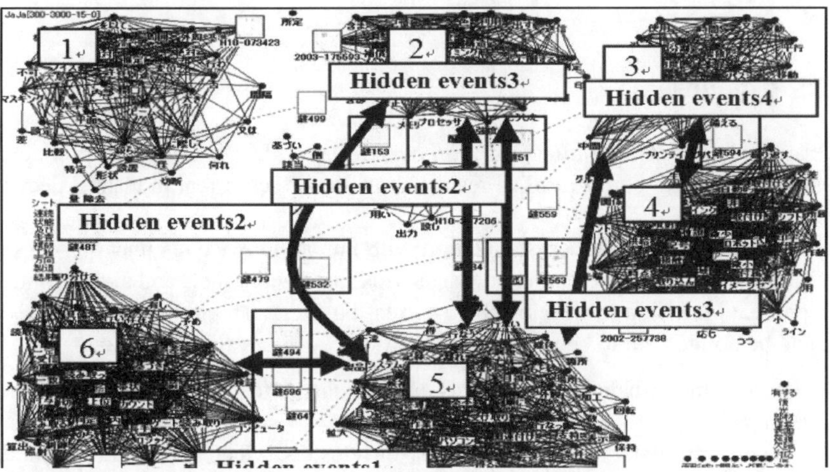

Fig 12.9. Pasting Pictogram on hidden events (Step5 to Step6)

in business. Thus, we do not include a comparison of KeyGraph with other methods, but compare the scenarios obtained for the various bridging hidden events with scenarios, and check if the problems in the difficulties stated in 4.1 were solved. By this, we can investigate if the hidden events, which play significant role in the structure of the graph, which really contributed to creating scenarios.

1. *Creativity in reinterpreting clusters and correcting scenarios*: Each scenario was reinterpreted finely with Pictograms on patent number in each cluster at Step.4. Here, the two scenarios of clusters No.2 and No.6 were corrected (Table 12.1). The drawings and the charts on each Pictogram aided examinees in their different

expertise to reinterpret the meaning of each cluster precisely. In comparison to the previous test in 4.1., we find an apparent improvement here.

2. *Creating new scenarios via bridging dummy nodes*: New scenarios emerged from the six clusters referring to Pictograms of hidden events numbers on the event map at Step 5 of 4.3. All the five examinees initially paid attention to "Hidden events1", which connected No.5 cluster titled as *back end inspection system* and No.6 cluster titled as *post process of CCD inspection*. Then, they paid attention also to "Hidden events2" and "Hidden events3" following the procedures described in Step5. As a result, new scenarios continuously emerged, and those scenarios corresponded to the combination of scenarios underlying multiple clusters being bridged by the "hidden events". The meaning of hidden events was understood by the Pictograms, and these understanding were reflected to the created scenarios here.

After the experiment, the new scenarios were evaluated from the view points of novelty of ideas and feasibility of development. Scenarios 4, and 6 after correction in Table 12.1, were selected and currently introduced into Company A's new products respectively.

12.5 Conclusion

We applied the *data crystallization with human-interactive annealing* to product re-designing in a real company. The results show the effect to real industrial decision making.

In this paper, the data crystallization with human-interactive annealing process was performed well for creating new scenarios for new products and aiding to make a significant decision to develop them in real business. The tasks below, however, should be pointed out to be solved for the efficiency of creating scenarios.

1. Improve the visualization of words on each cluster to show the relation of them easily.
2. Prepare multilateral datum of nodes on clusters which connected to hidden events to narrow the degree of ambiguity of hidden events.

Modifying and improving these tasks of the data crystallization with human-interactive annealing process, the other application for real business can be expanded to patent analysis, analysis of consumer behaviors in marketing, and the analysis of disciplinary boundaries in science.

References

1. Chance discovery Consortium. [Online] Available: *http://www.chancediscovery.com/english/index.php*.
2. Ohsawa, Y., (2003). KeyGraph: Visualized Structure Among Event Clusters, in Y. Ohsawa and P. McBurney. Eds., *Chance Discovery*, Springer Verlag, pp.262-275.

3. Ohswa, Y., (2006). Scenario Maps on Situational Switch Model, Applied to Blood-Test Data from Hepatitis C Patients, Ohsawa, Y., and Tsumoto, S. (eds) *Chance Discoveries in Real World Decision Making*, Springer, pp.69-80.
4. Joshi, M., Kumar, V., Agarwal, R., (2001). Evaluating Boosting Algorithms to Classify Rare Classes: Comparison and Improvements, *In Proceedings of the First IEEE International Conference on Data Mining*.
5. Weiss, GM., and Hirsh, H (1998). "Learning to Predict Rare Events in Event Sequences", *In Proceedings of the Fourth International Conference on Knowledge Discovery and Data Mining (KDD-98)*, AAAI Press, Menlo Park, CA, 359-363.
6. Ohsawa, Y., Usui, M., (2006). Creative Marketing as Application of Chance Discovery, Ohsawa, Y., and Tsumoto, S., *Chance Discoveries in Real World Decision Making*, Computational Intelligence, Springer-Verlag, pp.253-272.
7. Ohsawa, Y., and Fukuda, H., (2002). Chance Discovery by Stimulated Group of People - An Application to Understanding Rare Consumption of Food, *Journal of Contingencies and Crisis Management*, Vol.10, No.3, pp.129-138.
8. Yada, K., Motoda, H., Washio, T., and Miyawaki, A., (2005). Consumer Behavior Analysis by Graph Mining Technique, *New Mathematics and Natural Computation*, Vol.2, No.1, pp.59-68.
9. Okazaki, N., Ohsawa, Y., (2003). *Polaris: An Integrated Data Miner for Chance Discovery*, In Proc. of The 3rd International Workshop on Chance Discovery and Its Management, Greece.
10. Matsumura, N., et al, (2002). Discovering Emerging Topics from WWW, *Journal of Contingencies and Crisis Management*, Vol.10, No.2, pp.73-81.
11. Ohsawa, Y., (2005). Data crystallization: chance discovery extended for dealing with unobservable events, *New Mathematics and Natural Science*, vol. 1, pp.373-392.
12. Maeno, Y., Ohsawa, Y., (2006A). Understanding of Dark Events for Harnessing risk, Ohsawa, Y., and Tsumoto, S. (eds) in *Chance Discoveries in Real World Decision Making*, Springer-Verlag.
13. Ohsawa, Y., and McBurney, P. eds., (2003). *Chance Discovery*, Advanced Information Processing, Springer-Verlag, pp.2-15.
14. Ohsawa, Y., and Nara, Y., (2003). Understanding Internet Users on Double Helical Model of Chance-Discovery Process, *New Generation Computing*, Vol.21 No.2, pp.109-122.
15. Ohsawa, Y., (2002). KeyGraph as Risk Explorer from Earthquake Sequence, *Journal of Contingencies and Crisis Management*, Vol.10, No.3, pp.119-128.
16. Ohsawa Y, Fujie H, Saiura A, Okazaki N, and Matsumura N, (2004). Process to Discovering Iron Decrease as Chance to Use Interferon to Hepatitis B, in Paton, R. (ed) *Multidisciplinary Approaches to Theory in Medicine*.
17. Horie, K., Ohsawa, Y., Okazaki, N., (2006). Products Designed on Scenario Maps using Pictorial KeyGraph: *WSEAS Transitions on Information Science and Applications* Vol.3 pp.1324-1331., Issue 7, ISSN 1790-0832.
18. Maeno, Y., and Ohsawa, Y., (2006B). *Human-computer interactive annealing for crystallization of invisible dark events*, IEEE Transactions on Industrial Electronics, to appear.
19. Ohsawa, Y., and Tsumoto, S., *Chance Discoveries in Real World Decision Making*, Series on Computational Intelligence, Springer-Verlag (2006).

Pharmaceutical Drug Design Using Dynamic Connectionist Ensemble Networks

Ajith Abraham[1], Crina Grosan[2] and Stefan Tigan[3]

[1] Norwegian Center of Excellence, Center of Excellence for Quantifiable Quality of Service, Norwegian University of Science and Technology, O.S. Bragstads plass 2E, Trondheim, Norway
ajith.abraham@ieee.org
[2] Department of Computer Science, Babes-Bolyai University, Cluj-Napoca, 3400, Romania
cgrosan@cs.ubbcluj.ro
[3] University Iuliu Hatieganu, Faculty of Medicine, Department of Biostatistics and Medical Informatics, Cluj-Napoca, Romania
stigan@umfcluj.ro

Summary. This article presents a dynamic ensemble neural network model for a pharmaceutical drug design problem. Designing drugs is a current problem in the pharmaceutical research domain. By designing a drug, we mean to choose some variables of drug formulation (inputs), for obtaining optimal characteristics of drug (outputs). To solve such a problem, we propose a dynamic ensemble neural network model and the performance is compared with several neural network architectures and learning approaches. The idea is to build a dynamic ensemble neural network depicting the dependence between inputs and outputs for the drug design problem. Bootstrap techniques were used to generate more samples of data since the number of experimental data is reduced due to the costs and time durations of experimentations. We obtain in this way a better estimation of some drug parameters. Experiment results indicate that the proposed method is efficient.

13.1 Introduction

This article presents a dynamic neural network ensemble for modeling the situations that interfere in the process of designing drugs. By designing a drug, we mean to choose some variables of drug formulation, for obtaining optimal characteristics of drug [2, 3]. Our application is made on a particular class of drugs, namely retard drugs. We approach this problem with a bootstrap simulation which is suitable in some particular situations [1, 4].

The problem comes from the pharmaceutical research activity. It refers to a specific class of drugs that has delayed action called generically *retard drugs*. The pharmaceutical experimental situation leads to a mathematical optimization problem [12, 14, 15]. The pharmacist researcher must take into account several variables

A. Abraham et al.: *Pharmaceutical Drug Design Using Dynamic Connectionist Ensemble Networks*, Studies in Computational Intelligence (SCI) **123**, 221–231 (2008)
www.springerlink.com

Table 13.1. Sample data showing the *inputs* and *outputs*

Variables of formulation: Inputs						Responses: Outputs					
ExpNo	X_1	X_2	X_3	X_4	X_5	Y_1	Y_2	Y_3	Y_4	Y_5	Y_6
1	20	2	3	5	1	84.0	973.8	4.2	1.043	7.85	1.165
2	40	2	3	0	0	71.9	1150.0	1.6	1.016	8.2	2.264
3	20	8	3	0	1	92.5	1121.4	4.2	1.044	8.83	0.700
4	40	8	3	5	0	88.1	1200.0	3.7	1.038	8.87	1.205
5	20	2	9	5	0	99.2	910.0	5.8	1.061	8.3	1.914
6	40	2	9	0	1	68.2	985.1	4.1	1.043	7.9	2.550
7	20	8	9	0	0	99.1	1010.0	5.3	1.056	9.05	1.160
8	40	8	9	5	1	83.9	925.4	5.5	1.058	8.5	1.265
9	30	5	6	2.5	0.5	85.0	1055.8	3.8	1.036	8.3	1.535
10	30	5	6	2.5	0.5	81.2	1030.0	4.1	1.042	8.37	1.490
11	30	5	6	2.5	0.5	85.0	1060.0	4.1	1.042	8.4	1.535

of formulation of the drug such as: the speed of mixing turbine, the concentration of the binder, addition speed, the proportion of talc, the proportion of *lauril sulfate Na*.

We name these variables *inputs* and denoted as X_i, $i = \overline{1,n}$. For the problem considered, we have five inputs, X_1, X_2, X_3, X_4 *and* X_5. With those variables of formulation, for each combination taken into account, the researcher obtains a variant of drug with certain characteristics. For each obtained variant, some parameters are measured, called *responses* that characterizes the drug: charging performance, average diameter of the pill, Carr value, Hausner value, the flow time and the brittleness. We consider these responses as *outputs* denoted by Y_j, $j = \overline{1,m}$. For our problem, we have 6 outputs denoted by Y_1, Y_2, Y_3, Y_4, Y_5 andY_6.

The costs of experimentations are high and it is necessary to devote a long time to determine all responses for each variant of drug. So, the research group realized only 11 experiments. With the formalization proposed above, we grouped the experimental data as depicted in Table 13.1.

The aim is to determine a combination of variables $(x_1, x_2, ..., x_5)$ such that the responses $(y_1, y_2, ..., y_6)$ are optimal. By optimal we mean that outputs respect some conditions and also by taking into account some restraints for outputs.

1. The first response, output Y_1, must be maximized, so the goal is to obtain a value as close as possible to 100 %.
2. Te second output Y_2 must not outrun some values determined by the fact it is a value representing tablet's diameter. So, the requirement is that $y_2 \in [800\mu\text{m}, 1000\mu\text{m}]$. A value around the average $900\mu\text{m}$ is suitable.
3. The third output Y_3 has also an admissible interval for its value, $[1, 20]$, but we must determine it as close as possible to 1.
4. The fourth response, output Y_4, has a narrower interval for its values, $[1, 1.2]$, but it also has to be closest to 1.
5. The fifth output Y_5, representing a time quantity, must be as small as possible, but positive.
6. For the last output Y_6, the goal is to minimize it, with positive values, so the 0 value is considered desirable.

We search for the values of X_i $i = \overline{1,5}$ for which we obtain a drug formulation with optimal characteristics Y_j $j = \overline{1,6}$. Variables are chosen by the researcher from a continuous domain. Not all values are accepted. So we must consider domains of definition, real intervals, for each input variable. Accepted variation intervals for inputs, for our problem, are: $X_1 \in [0,50]$, $X_2 \in [1,8]$, $X_3 \in [3,9]$, $X_4 \in [0,5]$, and $X_5 \in [0,1]$.

13.2 Dynamic Ensemble of Neural Networks

Consider a population of n networks trained on a set $A = (x_m, y_m)$ of labeled instances of a binary classification problem. A simple approach to combining network outputs is to simply average them together [11, 13, 16]. The basic ensemble method (BEM) output is defined by:

$$f_{BEM} = \frac{1}{n} \sum_{i=1}^{n} f_i(x) \tag{13.1}$$

This approach can lead to improved performance, but does not take into account the fact that some networks may be more accurate than others. It has the advantage of being easy to understand and implement and can be shown not to increase the expected error. A generalization to the BEM method is to find weights for each output that minimizes the mean squared error (MSE) of the ensemble. The general ensemble model (GEM) is defined by:

$$f_{GEM} = \sum_{i=1}^{n} \alpha_i f_i(x) \tag{13.2}$$

where α_i is chosen to minimize the MSE with respect to the target function, f (estimated using the cross validation set), and sum to 1. Define the error, $\varepsilon_i(x)$, of a network, f_i, as:

$$\varepsilon_i(x) = f(x) - f_i(x) \tag{13.3}$$

If the correlation matrix, $C_{ij} = E[\varepsilon_i(x) \, e_j(x)]$, then the task is to find weights that minimize the following:

$$MSE_{[f_{GEM}]} = \sum_{i=1}^{n} \sum_{j=1}^{n} \alpha_i \alpha_j C_{ij} \tag{13.4}$$

It is shown that the optimal choice could be derived as follows:

$$\alpha_i = \frac{\displaystyle\sum_{j=1}^{n} C_{ij}^{-1}}{\displaystyle\sum_{k=1}^{n} \sum_{j=1}^{n} C_{kj}^{-1}} \tag{13.5}$$

The performance of GEM depends on a reliable estimate of C and the fact that it is non-singular so that it can be easily inverted. In practice, errors are often highly correlated, thus, the rows of C are nearly linearly dependent so that inverting C leads to significant round-off errors.

If the output of a neural network, $y = f_i(x)$, can be interpreted as the probability that an instance, x, is in a class, then, as y approaches 1, we feel more certain that the instance is in the class. As y approaches 0, we become more certain that the instance is not in the class. To quantify this notion; we define the certainty, $c(y)$, of a neural network output as:

$$c(y) = \begin{cases} y & if\, y \geq 0.5 \\ 1 - y & otherwise \end{cases} \quad (13.6)$$

The certainty rises for output y less than 0.5 as y falls, and for outputs $y \geq 0.5$ as y rises. It is possible to depict that one network output, y_1, is less certain than another, y_2, if $c(y_1)$ is less than $c(y_2)$. Note that the certainty behaves symmetrically with respect to positive and negative decisions; the certainty of an output of 0.1 is the same as that of an output of 0.9, but the decision they are certain about is different.

In a dynamic ensemble network (DEN), instead of choosing static weights derived from f_i performance on a sample of the input space, the weights are adjusted proportional to the certainties of the respective network outputs [6, 7]. The dynamically averaged network (DAN) is defined as:

$$f_{DAN} = \sum_{i=1}^{n} w_i f_i(x) \quad (13.7)$$

where $w_i = \dfrac{c(f_i(x))}{\sum\limits_{i=1}^{n} c(f_i(x))}$ and $\sum\limits_{i=1}^{n} w_i = 1$

The weight vector is calculated each time the ensemble output is evaluated, to try to give the best decision for the particular instance under consideration, instead of statically choosing weights that give an optimal decision with respect to a cross validation set. Each network's contribution to the sum is proportional to its certainty. A value close to 0.5, for instance, would contribute very little to the sum while a very certain value of 0.99 (or 0.01) among many less certain values would dominate the sum. Each fully connected neural network in the ensemble is generated with random initial weights. Then, each neural network is trained partially with training data and tested with the validation data.

13.2.1 Artificial Neural Networks

Artificial neural networks (ANNs) were designed to mimic the characteristics of the biological neurons in the human brain and nervous system. An artificial neural network creates a model of neurons and the connections between them, and trains it to associate output neurons with input neurons. The network learns by adjusting the interconnections (called weights) between layers. When the network is adequately trained, it is able to generate relevant output for a set of input data. A valuable property of neural networks is that of generalization, whereby a trained neural network is able to provide a correct matching in the form of output data for a set of previously unseen input data.

Multi-Layered Perceptron Networks (MLPN)

Typical MLPN is arranged in layers of neurons (nodes), where every neuron in a layer computes the sum of its inputs and passes this sum through a nonlinear function (an activation function) as its output. Each neuron has only one output, but this output is multiplied by a weighting factor if it is to be used as an input to another neuron (in a next higher layer). There are no connections among neurons in the same layer.

Activation functions for the hidden layers are required to introduce nonlinearity into the network. Without nonlinearity, hidden layers would not make networks more powerful. The training of a network by backpropagation (BP) involves three stages: the forward propagation of the input training pattern(s), the calculation and back-propagation of the associated error, and the adjustment of the weights. After training, application of the network involves only the computation of the feedforward phase.

Basically, BP is a gradient descent technique to minimize the error E for a particular training pattern. For adjusting the weight (w_k),in the batched mode variant the descent is based on the gradient $\nabla E(\frac{\delta E}{\delta w_k})$ for the total training set:

$$\triangle w_k(n) = -\epsilon \cdot \frac{\delta E}{\delta w_k} + \triangle w_k(n-1) \qquad (13.8)$$

The gradient gives the direction of error E. The parameters ε and α are the learning rate and momentum respectively. A good choice of both the parameters is required for training success and speed of the ANN.

In the conjugate gradient algorithm (CGA) a search is performed along conjugate directions, which produces generally faster convergence than steepest descent directions. A search is made along the conjugate gradient direction to determine the step size, which will minimize the performance function along that line. A line search is performed to determine the optimal distance to move along the current search direction. Then the next search direction is determined so that it is conjugate to previous search direction. The general procedure for determining the new search direction is to combine the new steepest descent direction with the previous search direction. An important feature of the CGA is that the minimization performed in one step is not partially undone by the next, as it is the case with gradient descent methods. An important drawback of CGA is the requirement of a line search, which is computationally expensive. Moller [5] introduced the scaled conjugate gradient algorithm (SCGA) as a way of avoiding the complicated line search procedure of conventional CGA. According to the SCGA, the Hessian matrix is approximated by:

$$E''(w_k)p_k = \frac{E'(w_k + \sigma_k p_k) - E'(w_k)}{\sigma_k} + \lambda_k p_k \qquad (13.9)$$

where E' and E'' are the first and second derivative information of global error function $E(w_k)$. The other terms p_k, σ_k and λ_k represent the weights, search direction, parameter controlling the change in weight for second derivative approximation and parameter for regulating the indefiniteness of the Hessian. In order to get a good quadratic approximation of E, a mechanism to raise and lower λ_k is needed when the Hessian is positive definite. Detailed step-by-step description can be found in the literature [5]. We used the MLPN trained using SCGA.

Elman Recurrent Neural Networks (ERNN)

ERNN, also known as partially recurrent neural network, are a subclass of recurrent networks [9]. They are multilayer perceptron networks augmented with one or more additional context layers storing output values of one of the layers delayed by one step and used for activating this or some other layer in the next time step. The ERNN has context units, which store delayed hidden layer values and present these as additional inputs to the network. The ERNN can learn sequences that cannot be learned with other recurrent neural network e.g. with Jordan recurrent neural network, since networks with only output memory cannot recall inputs that are not reflected in the output. Several training algorithms for calculation of error gradient in general recurrent networks exist. Usually, both hidden and output units have nonlinear activation functions. Note that external input at time t does not influence the output of any unit until time $t + 1$. The network is thus a discrete dynamical system.

Radial Basis Function Network (RBFN)

RBFN network consists of 3-layers: input layer, hidden layer, and output layer. The neurons in hidden layer are of local response to its input and known as RBF neurons, while the neurons of the output layer only sum their inputs and are called linear neurons. It is well known that neural network training can result in producing weights in undesirable local minima of the criterion function [10]. This problem is particularly serious in recurrent neural networks as well as for MLPN with highly nonlinear activation functions, because of their highly nonlinear structure, and it gets worse as the network size increases. This difficulty has motivated many researchers to search for a structure where the output dependence on network weights is less nonlinear. The RBFN has a linear dependence on the output layer weights, and the nonlinearity is introduced only by the cost function for training, which helps to address the problem of local minima. There are two basic methods to train an RBFN in the context of neural networks. One is to jointly optimize all parameters of the network similarly to the training of the MLPN. This method usually results in good quality of approximation but also has some drawbacks such as a large amount of computation and a large number of adjustable parameters. Another method is to divide the learning of an RBFN into two steps. The first step is to select all the centers μ in terms of an unsupervised clustering algorithm such as the K-means algorithm proposed by Linde et al., and choose the radii σ by the k-nearest neighbor rule. The second step is to update the weights B of the output layer, while keeping the μ and σ fixed. The two-step algorithm has fast convergence rate and small computational burden. We used a two-step learning algorithm to speed up the learning process of the RBFN.

The selection of the centers and radii of RBF neurons can be done naturally in an unsupervised manner, which makes this structure intrinsically well suited for weather prediction. As a result, we adopt below a self-organized learning algorithm for selection of the centers and radii of the RBF in the hidden layer, and a stochastic gradient descent of the contrast function for updating the weights in the output layer. For the selection of the centers of the hidden units, we may use the standard k-means clustering algorithm. This algorithm classifies an input vector x by assigning it the label most frequently represented among the k-nearest neighbor samples. Specifically, it places the centers of RBF neurons in only those regions of the input space, where significant data are present. Once the centers and radii are established, we can make use of the minimization of the contrast function to update the weights of the RBFN.

Hopfield Model

This network is a single layer network with symmetric weight matrices in which the diagonal elements are all zero. The diagonal elements need not be zero, but we assume that is the case since the performance is improved when taken to be zero. Thus, for a Hopfield network with weight matrix W, $w_{ij} = w_{ji}$ and $w_{ii} = 0$ for all $i, j = 1$, $2, \ldots, n$. Inputs are applied simultaneously to all neurons, which then output to each other and the process continues until a stable state is reached, which represents the network output. The feedback loops involve the use of particular branches composed of unit-delay elements (denoted by z^{-1}), which result in a nonlinear dynamical behavior by virtue of the nonlinear nature of the neurons.

13.2.2 Ensemble of Neural Networks

Figure 13.1 represents the architecture of the ensemble neural network [13]. The ensemble is formulated using combinations of MLPN, ERNN, RBFN and HFM architectures using the dynamic ensemble network strategy discussed above.

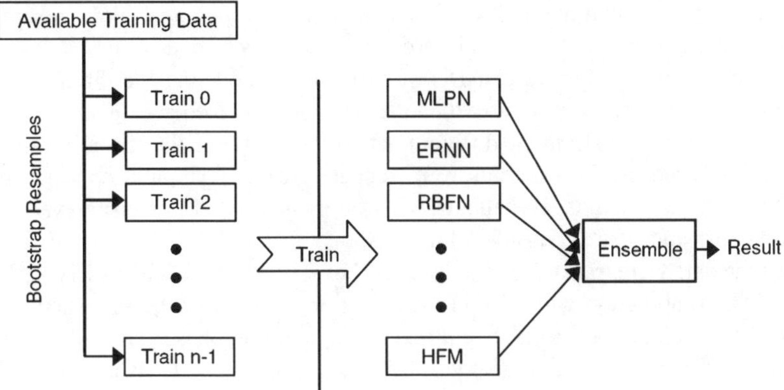

Fig 13.1. Dynamic ensemble architecture

13.3 Experiment Setup and Results

In this kind of pharmaceutical drug design problem, the researcher has to deal with two major aspects of the experimentation:

1. The relation between inputs and outputs is unknown, so we must find a way to make the best approximation
2. Data are difficult to obtain due to the restraints of costs and time. So, a method to simulate similar data is desirable

13.3.1 Bootstrap Re-sampling of Data

If the proposed solution is not acceptable or the system developed is not compatible and the process of evaluating an optimal solution is not convergent, we can conclude that the sample of data is too poor and the regression functions are not suitable to describe the links between variables. We resort, in this case, to a resampling method that allow us to manage uncertainty. The aim is to improve our sample of data with pseudo data and to evaluate some statistical parameters. We use a bootstrap method of resampling data. We group each input variable of formulation X_i, $i = \overline{1,5}$ with each output Y_j, $j = \overline{1,6}$. We obtain 30 vectors with bivariate data (X_i, Y_j). For those vectors we apply a bivariate bootstrap resampling procedure. Finally, among the bootstrap simulated sets of X_i variables we select a combination of inputs, a set of values x_i, $i = \overline{1,5}$, that correspond to best situated values of Y_j. We mean by this to observe the resampling process of (X_i, Y_j) and select a proper combination of y_j, $j = \overline{1,6}$. In our application a such a combination is

$$(y_1 = 99.1, y_2 = 910, y_3 = 1.6, y_4 = 1.036, y_5 = 7.85, y_6 = 0.7)$$

We look in the bivariate bootstrap resampling process and extract the input combination $(x_1, x_2, x_3, x_4, x_5)$ for which we obtain the above combination of y_j, $j = \overline{1,6}$.

From other point of view, such a resampling method is useful to obtain new data and to improve the function approximation of the dependence $y = f(X)$. So, we can say that no matter what method we choose to approach such a problem, resampling bootstrap methods are desirable to improve the accuracy of data sets.

The experimental system consists of two stages: modeling the different neural network models (and constructing the ensemble) and performance evaluation. 70% of the data was used to train the different network models and the ensemble and 30% for testing purposes. Experiments were repeated three times and the worst errors are reported. The test data is then passed through the trained models to evaluate the learning efficiency of the considered models.

All the data were transformed into values between -1 and 1. The main goal of this scaling, in combination with weight initialization, is to allow the squashed activity function to work at least at the beginning of the learning phase. Thus, the gradient, which is a function of the derivative of the non-linearity, will always be different from zero. At the end of each algorithm, the outputs were re-scaled into the original data format for achieving the desired result.

Table 13.2. Test results and performance comparison of drug design system

Output	Y_1	Y_2	Y_3	Y_4	Y_5	Y_6
MLPN						
RMSE	0.0487	0.039	0.043	0.040	0.038	0.039
CC	0.979	0.968	0.967	0.956	0.966	0.975
RBFN						
RMSE	0.0322	0.0391	0.0390	0.0421	0.0412	0.0381
CC	0.972	0.981	0.972	0.964	0.956	0.973
ERNN						
RMSE	0.0532	0.0456	0.0489	0.0534	0.0456	0.0453
CC	0.954	0.956	0.966	0.945	0.946	0.947
HFM						
RMSE	0.0675	0.0756	0.0453	0.0563	0.0645	0.0756
CC	0.923	0.091	0.0967	0.0961	0.0933	0.0921
Dynamic Ensemble Network						
RMSE	0.0192	0.0223	0.0213	0.0310	0.0202	0.0276
CC	0.988	0.991	0.987	0.976	0.978	0.983

The configuration of the neural network depends highly on the problem. To decide on the architectures of the MLPN, ERNN and HFM, a trial and error approach was used. Networks were trained for a fixed number of epochs, and the error gradient was observed over these epochs. Performance of the MLPN, ERNN and HFM networks were evaluated by increasing or decreasing the number of hidden nodes. The activation functions for the MLPN and ERNN models were chosen to be log-sigmoid and hyperbolic-tangent-sigmoid for hidden units, respectively, and linear for the output units.

Since there is no exact rule for fixing the number of hidden neurons and hidden layers to avoid under-fitting or over-fitting in the MLPN and ERNN networks, therefore, the RBFN model is investigated to address this difficulty. In RBFN, the Gaussian activation function was chosen for the hidden units, and linear for the output units.

The obtained results indicate that satisfactory accuracy has been achieved using the MLPN, ERNN, HFM and RBFN models. The performance is evaluated using root mean squared error (RMSE) and correlation coefficient (CC).

Compared to the HFM performance, MLPN exhibited lower errors. It is capable of modeling the problem considered better than HFM. However, the learning process of the MLPN algorithm is time-consuming and its performance is heavily dependent on the network learning parameters. The ERNN model, compared to MLPN, could efficiently capture the dynamics of the model, resulting in a more compact and natural representation of the temporal information contained in the data. The RMSE of the ERNN model was much lower than that of the HFM method. The RBFN performed well in terms of accuracy. Since RBFN has unsupervised learning characteristics and a modular network structure, these properties make it more effective for fast and robust problem modeling. It is indicated that the HFM model overestimated most of the predicted values. Overall, the performance of HFM is reasonable. However, compared to the other models, it is less accurate for the drug design problem.

The optimal network is the one that has a lower error and the highest correlation coefficient. The experimental comparisons of the MLPN, ERNN, RBFN and HFM methods pointed out that no single algorithm can be regarded as the best to model all variables. Thus, the use of ensemble of neural networks as an alternative approach is considered and the empirical performance clearly illustrates the importance of the approach. Dynamic ensemble-averaging tends to cancel out the noise part as it varies among the different ensemble members, and tends to retain the fitting of the regularities of the data.

13.4 Conclusions

This article introduced a dynamic ensemble neural network model for a pharmaceutical drug design problem. Test results reveal that the proposed connectionist models are capable of modeling all the outputs accurately. Compared to the different artificial neural network approaches, the dynamic ensemble model performed better in terms of RMSE and correlation coefficient values.

Performance could have been improved by providing more training data. The most important achievement of this result is that it gives to the researcher a new starting point of experimentation in stead of making other 20 - 30 experiments and to arrive to the same conclusion as the ensemble model recommends.

Acknowledgements

Authors would also like to thank the colleagues of the Department of Maxillofacial Surgery, University of Medicine and Pharmacy, Iuliu Hatieganu Cluj-Napoca, for the initial contributions of this research.

References

1. Abdelhak M. Zoubir, D. Robert Iskander, (1998), *Bootstrap MATLAB Toolbox*, Software Reference Manual.
2. Remus Câmpean, A. Prodan, (2003), *Biomatematică – aplicatii în Excel*, Editura Medicală Universitară "Iuliu Hatieganu", Cluj-Napoca, ISBN: 973-693-016-5.
3. Remus Câmpean, Augustin Prodan, (2003) *A Rating Model Applied for Designing Drugs*, Proceedings of the 12-th IASTED International Conference on Applied Simulation and Modelling, Marbella, Spain, pg 557-561, ACTA press, ISBN: 0-88986-384-9, ISSN: 1021-8181
4. T. Hesterberg, S. Monaghan, D. S. Moore, A. Clipson, R. Epstein, (1993), *Bootstrap Methods and Permutation Tests*, W. H. Freeman and Company, New York, 2003

5. Moller, A. F., A Scaled Conjugate Gradient Algorithm for Fast Supervised Learning, Neural Networks. 6:525-533.
6. Hansen LK, Salamon P. (1990), Neural network ensembles. IEEE Transactions on Pattern Analysis 12(10):993-1001.
7. Jimenez D, Walsh N (1998) Dynamically weighted ensemble neural networks for classi.cation. In: Proceedings of the international joint conference on neural networks (IJCNN98), Anchorage, Alaska, May 1998, pp 753-756.
8. Sharkey AJC (1999) Combining artificial neural nets: ensemble and modular multi-net systems. Springer, Berlin Heidelberg New York.
9. Elman J.L. (1991), Distributed representations, simple recurrent networks and grammatical structure, Machine Learning, Vol. 7, No. 2/3, pp. 195-226.
10. Orr, M. J.(1995), "Regularization in the selection of radial basis function centers", Neural Computation, Vol. 7, No. 3, pp. 606-623.
11. Maqsood I, Khan MR and Abraham, A. (2004), Neural Network Ensemble Method for Weather Forecasting, Neural Computing & Applications, Springer Verlag London Ltd., Volume 13, No. 2, pp. 112-122.
12. Abraham, A., Grosan C. and Tigan S. (2007), Ensemble of Hybrid Neural Network Learning Approaches for Designing Pharmaceutical Drugs, Neural Computing & Applications, Springer Verlag London Ltd., Volume 16, No. 3. pp. 307-316.
13. Maqsood I and Abraham, A. (2007), Weather Analysis Using an Ensemble of Connectionist Learning Paradigms, Applied Soft Computing Journal, Elsevier Science, Volume 7, Issue 3, pp. 995-1004.
14. Grosan C., Abraham, A. and Tigan S.(2006), Engineering Drug Design Using a Multi-Input Multi-Output Neuro-Fuzzy System, 8th International Symposium on Symbolic and Numeric Algorithms for Scientific Computing (SYNASC'06), Timisoara, Romania, IEEE CS Press, pp. 365-371.
15. Grosan C., Abraham, A., Tigan S., Chang TG and Kim, DH (2006), Evolving Neural Networks for Pharmaceutical Research, International Conference on Hybrid Information Technology (ICHIT'06), IEEE Press, Korea, pp. 13-19.
16. Maqsood I, Khan MR and Abraham, A. (2004), Neural Network Ensemble Method for Weather Forecasting, Neural Computing and Applications, Springer Verlag London Ltd., Volume 13, No. 2, pp. 112-122.

14

A Framework of Knowledge Management Platform for Middle and Small Business

Xingsen Li[1,2], Jun Li[1,2], Yuejin Zhang[1,2] and Yong Shi[2,3]

[1] Management School of Graduate University of Chinese Academy of Sciences, 100080, Beijing, China
vc6a@sohu.com
[2] Chinese Academy of Sciences Research Center on Data Technology and Knowledge Economy, 100080, Beijing, China
[3] Graduate University of Chinese Academy of Sciences, 100080, Beijing, China

Summary. "Ruled by orders instead of rules" is a severe problem that restricts the development of the most middle and small businesses in China. It overemphasizes the influence of some key employee's ability ("able person") and ignores the rule system. Based on the study of knowledge management and its implementations, we analyzed the reasons of the "able person" problem, and found that the key factor is lacking of an effective knowledge management system to share skills and experiences in the company. Referred from the current research on knowledge management and a consulting company's solutions, a simple and effective knowledge management platform has been constructed, it consists of knowledge collection, transformation and accumulation and runs together with business operation systems. It can collect knowledge both from employees and from open data, the Web or information systems by data mining. Its application in a Chinese company shows that the framework of knowledge management platform can be applied efficiently and can solve "able person" problem well.

Keywords: Knowledge Management, Middle and Small Businesses, Knowledge Management Platform, Business-Oriented Knowledge Management, "Able Person" Problem

14.1 Introduction

In the past few years, corporate management skill has been improving, but most Chinese middle and small businesses are not able to undertake big projects due to lack of resources. They have to hire "able person" on short-term to fulfill important tasks. These caused a severe problem called "able person" problem which means "ruled by men instead of rules" that restricts the development of the company, it overemphasizes the influence of "able person" and ignores the rule system. "Able person" is usually not reliable and the companies that rely on "able person" often make the function of their company system weaker and weaker.

X. Li et al.: *A Framework of Knowledge Management Platform for Middle and Small Business*, Studies in Computational Intelligence (SCI) **123**, 233–248 (2008)
www.springerlink.com

In recent years, many companies have been aware of the weakness of "able person" culture and its negative impacts on a company's development. However, this problem is very difficult to solve because companies are often lack of system management knowledge and skilled persons to fulfill a rule-based system.

In order to solve the above problems in middle and small businesses, there has been a growing interest in treating knowledge management as a significant organizational resource [1] and emphasize the important role of knowledge management systems on strategic management, innovation practice, organizational learning [2], process reengineering [3] and other fields. Almost all these literatures' solutions are based on big companies which have a good management information system or with a standard rules to run for the operations. But they neglected the need for knowledge management in middle and small sized businesses and therefore can't find a simple method for those middle and small sized businesses to implement knowledge management systems. Meanwhile, with the development of world wide trade, knowledge management is getting more and more a key performance indicators for sustainable development especially for those middle and small sized businesses.

The World Wide Web has become a huge data source. Billions of pages are publicly available, and it is still growing dramatically. Thus, the internet has become an important infrastructure of information for most people [19]. Web Mining [19] that can discover knowledge from huge amounts of web pages, including web documents, hyperlinks between documents, usage logs of web sites, etc, has become an important knowledge source for business.

The open-access movement is the worldwide effort to provide free online access to scientific and scholarly research literature, especially peer-reviewed journal articles and their preprints [20]. Open-access (OA) literature is digital, online, free of charge, and free of most copyright and licensing restrictions. The 'Berlin Declaration' was published in 2003 as a guideline to policy makers to promote the Internet as a functional instrument for a global scientific knowledge base. Because knowledge is derived from data, the principles of the 'Berlin Declaration' should apply to data as well. Gradually, access to scientific data will be widely available and allowing further use of the data by the scientific community [21].

Data mining refers to extracting or "mining" knowledge hidden from large amounts of data [22]. The importance of collecting data that reflect your business or scientific activities to achieve competitive advantage is widely recognized now. Data might be one of the most valuable assets of your corporation - but only if you know how to reveal valuable knowledge hidden in raw data. Data mining can help you reveal knowledge hidden in data and turn this knowledge into a crucial competitive advantage. Data mining allows you to extract diamonds of knowledge from your historical data and predict outcomes of future situations. It will help you optimize your business decisions, increase the value of each customer and communication, and improve satisfaction of customer with your services. Today increasingly more companies acknowledge the value of this new opportunity and turn to invest on data mining projects.

The web, the open access movement, knowledge management system and data mining tools make it possible for companies to overcome the "able person" problems.

The purpose of this paper is to propose a combined knowledge management platform for middle and small businesses to solve "able person" problems gradually. The rest of our paper is organized as follows. In Section 2, we analyze the disadvantages of "able person" system and its reasons. In Section 3, we propose a solution of knowledge management platform for middle and small businesses. The Section 4 introduces design outlines and how to develop software of the knowledge management system. The fifth section described the knowledge collection process in detail. Section 6 applied our model to a real application to verify the effectiveness and got a satisfactory answer. The paper is summarized in Section 7.

14.2 Analysis on Able Person Problem

"Able person" corporation culture emphasizes on individual ability, neglects the rules and followed by random management. This kind of culture causes many negative influences. Particularly, with the development and expansion of company, "able person" management has become a major obstacle for the company's development. (See Table 14.1)

First, the decision-making process often ignores scientific demonstration and depends overconfidently on able person's own insights. Due to lacking of a normative decision-making process, many important decisions concerning future development are made only by able persons briefly. Second, on the senior management level, owner of the company always sets up special business or position for a particular person just because of his/her excellence, sometimes even violate the company's regulations. Management system becomes useless. In this situation, it is able person who controls management system, but not the other way around. This can lead to able person crisis, such as division or group completely dependence on able person. These crisis can occur when able person decide to leave the company. A 1998's survey of European firms by KPMG Peat Marwick found that almost half of the companies reported having suffered a significant setback from losing key staff with 43% experiencing impaired client or supplier relations and 13% facing a loss of income because of the departure of a single employee [1]. Third, on Executive level, "able person" corporation culture emphasizes on able person's contribution and ignores teamwork

Table 14.1. Characteristics of Able Person Corporation Culture

	To do Tasks	To manage human resource	
Decisions Level	Decision only by able person or owners, No Normative Decision Process	Year-end Performance Evaluation	No Normative Scoring Process, mostly by owner's sense
Management Level	Excessive Depending on Able Person to do tasks	Daily Performance Evaluation	Scoring only on short-term Results
Executive Level	Responsible only for leaders but not for results	Training Plan for employees	More chance for senior leaders and Ignore junior managers

and training for junior managers. Because of this behindhand lower management development system, junior manager's enthusiasm is disrupted.

For each former successful company, there must be a skilled manager, who, with his keen insight, led his company from small to large, from puny to strong, even to a multi-billion asset business. However, its development process was not on rules but on random orders, and the effect depends on the ability of the leader who gave orders. In a word, "Able person" may lead company success, but the success is difficult to sustain. Overemphasizing on able person will impact on system and professionalism. In the long term, it is harmful for company's development and its effectiveness.

By now, many companies have realized the weakness of able person culture and its restriction on company development. It is not that these businesses are unwilling to set up a new system, but due to lack of knowledge about system management and a platform to reproduce able person. It has been shown that many Chinese companies do not know how to reform "able person" system and establish a management system because they lack under-mentioned managerial elements and technical methods [4].

(1) Lack of real business objective plan/budge; managed by boss' emotional orders, not by logical rules. Almost 99.9% companies' business plan and budgeting have not in real use but faked numbers even became a number game for the boss.

(2) Lack of specific duty and responsibility; choosing personnel according to boss's authority and affection but not by system interviews or skills.

(3) Lack of systematic management for operation process; managing by boss's senses but not by scientific method.

(4) Lack of a scientific and effective performance evaluation system, scoring employees by boss's liking but not by performance.

By surveying existing research results on corporate life cycle theory [5, 6], we find that the key reason comes from the birth phase. All the problems mentioned above are due to lack of a business & knowledge management system at birth phase and growth phase, such a system is not built in a day, and the first step is to collect knowledge in processes, products, data and/or services. Then transfer existing knowledge into other parts of the organization. Knowledge management includes organizational learning, personal management, cultural, etc [7], and includes models, support tools and environments on the technical aspect [8].

To solve the above problems of able person system, the key is to establish a knowledge management system within a company so that everyone works within the system, it is this system that will support company's strategy and cultivate skilled persons.

14.3 Knowledge Based Solutions

Despite growing interest about a strategic perspective on knowledge management (KM) [12,13], there is still a lack of a procedure and methods to guide the implementation of KM strategies. Ronald Maier (2003) [23] identified four scenarios for potentially successful KM initiatives. The majority of organizations can be described as being a knowledge management starter. In order to improve these KM initiatives and link them to business strategy, a process-oriented knowledge management approach had been used as a step to bridge the gap between human- and technology-oriented KM. This approach is outlined with the help of the four levels of intervention: (1) strategy, (2) KM organization and processes, (3) topics/content, and (4) instruments/systems [23].

In order to utilize the free resources and skilled persons in society for middle and small companies, paper [25] gives a knowledge management model which consists of project disassembling, task issuing, bidding, signing, monitoring and assembly testing etc. It makes good use of the available human resources in society and makes knowledge management a workflow. This idea can also be used to collect human's knowledge inside the company.

As we have analyzed in Section 2, the knowledge management platform for middle and small businesses should be simple and can be used easily in daily works, so the framework of knowledge management platform must combine corporation strategy [9] with all employees and with business operation system, it should include: (1) each month's specific strategy and work target within a year; (2) designated responsibility for each person with key tasks; (3) performance monitoring and improvements; (4) knowledge sharing and innovation. At the same time, the system should join three administrative levels together: strategic planning, operation management and business operations [10, 11]:

1) strategic management level

According to the development prospects and internal and external competitive elements, company will set down its strategic goal, determine strategic executing approach and divide the goal into each specific plan. It also obtains real time feedback from the strategic execution process of management control and business operation, and then accordingly optimizes strategy.

2) operation management level

Translate action plan into program budget, real-time monitoring and inquiry of performance operation process, adjust action plan according to performance analyzing.

3) Business operation level

Using business information systems, execute specific business operation, report results and instruct upgrade to improve the business.

To fulfill these functions, the company needs the preconditions of establishing system:

(i) The management knowledge of business workflow process;

(ii) A platform to accumulate knowledge, so it can generate dormant knowledge and each employee can acquire needed knowledge easily;

(iii) A platform of knowledge application which can bring the knowledge into play and produce values for both employees and corporation.

Accordingly, we designed a business process-oriented model of knowledge management platform shown in Figure 14.1.

It is closely combined with company's operation management and makes all experience, work flow process and business system together into a uniform knowledge management system. Joining with operation system such as plan and action

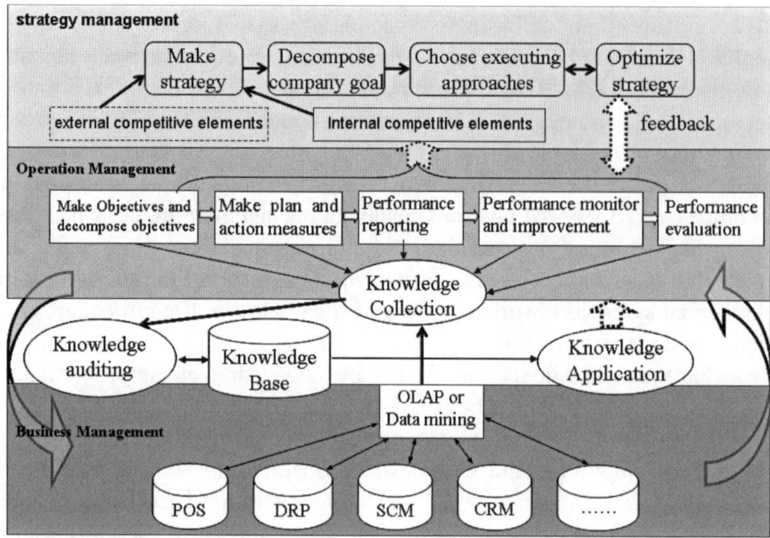

Fig 14.1. A Business Process-Oriented Knowledge Management Model

measure, key responsibility, achievement tracking, performance evaluation, etc., this knowledge management platform will collect knowledge from information system and process and then apply the knowledge to standardize the systematic management in company [7, 8]. This includes six steps:

(1) In strategy management level, record external and internal competitive elements by which company strategy was made. Decompose company goal to operation level and choose executing approaches.

(2) In operation management level, make department objectives and decompose objectives to sub-units, all units make their plan and propose action measures up to their leader, make employees involved in company operation and realize promised goal. Instruct how to translate strategic plan into executable program, and put into effect in relevant departments and setup key action measures.

(3) Define responsibility of each position, set up an inquiry platform, report performance results and periodically inquire leading officials for the achievement based on facts and data, propose improvement method to ensure the implementation of yearly operation objective.

(4) Performance evaluation: check and evaluate each employee's achievement and performance include knowledge contribution rate. Then grade evaluation results by defined criterions, and fulfill rewards or punishment.

(5) Knowledge management: Accumulate management and business operation knowledge in all the process and from data in business management systems by OLAP or data mining. the knowledge will be saved in knowledge base after been audited, then applied to daily works, such as share experiences in the process of operation, find answers from knowledge base, refine a flow process and use the knowledge platform to make employees growth.

(6) The knowledge and the operation performance will feedback to strategy level and optimize the strategy, make operation a work flow, gradually most work process will be standardized and optimized.

This platform will regard the company as an organic entity. By budgeting and planning, defining key responsibility, tracking achievement, and performing performance evaluation, etc., everyone knows his/her authority and responsibility clearly, and "automatically" working towards company's objective.

14.4 Design of Knowledge Management Platform

In order to use the platform easily for middle and small businesses, we designed a software system to help its application.

14.4.1 Main Functions

System contents and their explanations are showing in Table 14.2:

14.4.2 Software Operation Flow Chart and its Explaination

Figure 14.2 is the flow chart of software operation management.

The arrows in the chart show the work sequence of the functions. The software makes knowledge management a workflow process. Main functional modules of system are given as Table 14.3 in detail.

14.4.3 Network Construction of System

The system does not require high level hardware configuration, general PC is enough for a server. System network is displayed as following Figure 14.3:

14.5 Knowledge Collection from Working Process

Collect knowledge from working process have 5 main steps:

Step1. Business target setting

According to the development prospects and internal and external competitive elements, company will set down its strategic goal, and designate to each month, each business unit; then designated responsibility for each person with key tasks.

Step 2. KPI definition and assignation

Define KPI including KPI code, name, its unit code and name, formula for computing and its description in detail according to unit's specific strategy and work target within a year, and then assign KPIs as part of duty to employees. The interface of KPI definition is showing in Figure 14.4.

Table 14.2. System Contents and their Explainations

Business topic	Content	Explanation
1. Budget planning: Translate strategic plan into executable action program	Make a yearly operation /budget plan and divide into business units. the operation /budget plan should also include key measures, time schedule and responsible persons.	Instruct how to correctly translate strategic plan into executable program, and put into effect in specific departments and key action measures.
2. Position responsibility: Positioning each job's role and design performance goal	Analyze business position, define each position's authority and responsibility; on the basis of yearly operation /budget plan, make sure responsibility specifically.	Undefined position responsibility will cause management chaos and inefficiency, buck-passing and conflict cannot be avoided. So define and fix position power and responsibility by job description, positioning job role, and by quantified financial and organizational targets, provide a evaluation criteria for employee's performance
3. Achievement tracking: Performance monitor and improvement	Periodically put forward general manager's supervision and instruction, find problems occurred in operation process and their causes, improve action and optimize management.	Without an achievement tracking system, the plan made at the beginning of year will be a "fake" one-just a format, the fulfillment of the goal cannot be measured. So periodically inquire achievement of relevant leading officials based on facts and data, at the same time propose improvement method to ensure the implementation of year operation goal.
4. Performance evaluation: Fulfill rewards or punishment according to individual's achievement	Evaluate according to the fulfillment of year operation/budget and their knowledge contribution.	If payment does not assort with achievement, certainly it will cause inefficiency. By evaluation of performance and achievement, award employees with good performance and punish those bad ones.
5. Knowledge management: Manage knowledge systematically, and utilize them comprehensively.	Collect knowledge according to plan, achievement report form and action improvement, audited and save them in knowledge database, and indexed for application.	If employee cannot get value from knowledge platform, the system will be abandoned soon. So by knowledge application, provide value to employees, and make employee's ability grow up with platform together.

Step 3. Knowledge collection from work planning

Month plan or week plan has to be made according to the targets of the month assigned from company goal. Key measures, important actions, resource plan and risk analysis were recorded in system. Leaders can check the employee's plan and will give suggestions if their plan were failed. The interface is showing in Figure 14.5.

Step 4. Knowledge collection from working reports

At the end of the month or week, working results will be reported in the system, such as KPI value finished, experiences or reason analysis, key action analysis. Leaders can monitor at any time and give opinions if necessary. The interface is showing in Figure 14.6.

Step 5. Knowledge collection from action improving process

If the working results are not good as what have planed, the leader can send a query instruction to the person who in charged. Then improvement actions and its key measures will be planed and the leader also can guide him/her. The interface is showing in Figure 14.7.

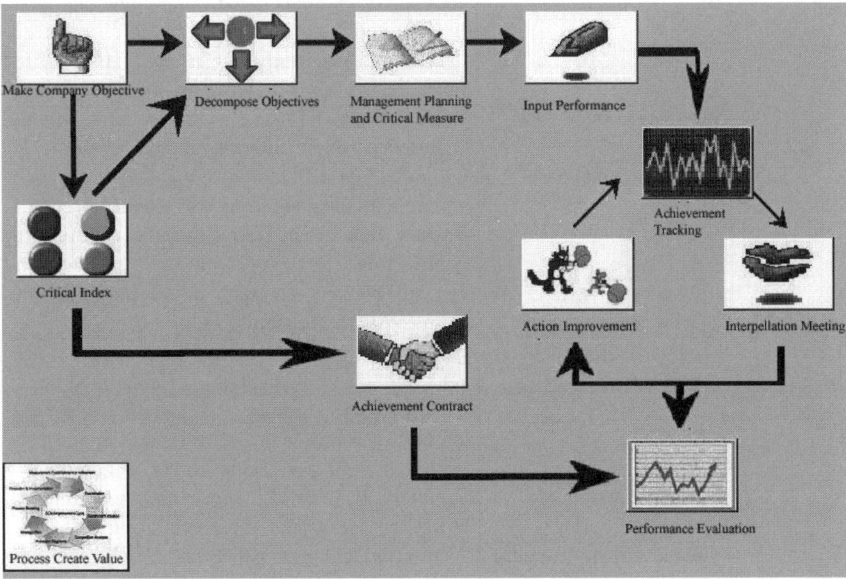

Fig 14.2. Software Operation Flow Chart

Table 14.3. Function modules of knowledge management system

Primary menu	Sub menu	Function description
planning budget	set down yearly objective	Making yearly operation objective for head office
	division of year objective	divide year operation objective into departments and individuals one by one
	planning (include action measures, resource demand, risk analysis)	Make year, quarter, month and week plan, include quantified target, topic, key measure, risk analysis & scheme, supervisor can examine the planning target value of different departments, and accordingly make his own plan.
	plan auditing	Supervisor can audit plan, auditing reports will be shown in task list, and unqualified plan should be reworked.
	plan checking	check information of plan, and send notice
	plan file management	can save as other formats file and send with plan together
		save plans in server computer as modules

Table 14.3. Continued

Primary menu	Sub menu	Function description
position responsebility	set head office KPIs	Set KPI and its target value. Can be sorted into four aspects: finance, customer, inside operation, study and growth, etc.; can setup collecting type. Free definition, so can be applied to different industries and departments. Can also be sorted by finance, operation, organization, etc.
	set department KPIs	set department KPI and its target value, briefly explain department authority and responsibility
	set individual KPIs	Set individual KPI and its target value; briefly explain individual authority and responsibility. individual target can be empty, then it accede department KPI directly
	responsibility and achievement contract	sign achievement contract, define target weight and position responsibility in it
result feedback	input achievement report form	Type in achievement of year, quarter, month and week, include implementation of quantified target and experience summary; supervisor can review and collect achievement of different departments one by one, and input achievement of their own.
	achievement query	can query achievement information according to department, personnel, time periods, etc.
	achievement monitoring	Customized color of gap between real achievement and plan value; can send inquiry instruction in monitor window to review plan and achievement results.
	inquiry instruction	send inquiry instruction, employee can receive inquiry information in task list and learn from it
	send meeting notice	employee can receive meeting information in task list, and can check receiving status of meeting notice
	inquiry feedback	fill in action improvement measure, share experience
	meeting bulletin	Note important meeting content, issue bulletin.
	action measure auditing	supervisor makes instruction to the action measure by his experience
	Action report	fill in actual result of improved action, summarize experience
achievement/ performance evaluation	performance evaluation	according to defined algorithm, calculate achievement score, and ranking it
	print report form	choose department, time period to review and print performance reports

knowledge management	Working memo	add, inquire, save and send working memo
	module management	add module, module inquiry, auditing, downloading for application
	daily report	share daily experience
	BBS forum	issue topic, reply, discuss, BBS system management
system maintenance	log in system	login via LAN or telnet by user name and password
	registration	register new user, authorizing by administrator
	update password	change login password
	authority management	setup user's operation limitations
	department setup	setup and maintain department information of a company
	personnel setup	setup personnel information
	position setup	setup position information
	membership management	setup membership of the organization
	basic information setup	setup plan type, bulletin sort, module type, measure unit, warning rule, default display page
	server setup	setup server FTP address

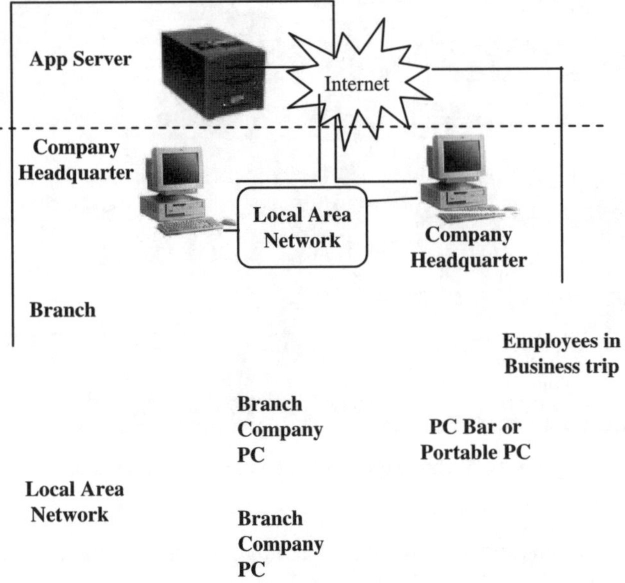

Fig 14.3. Knowledge Management System Network

All the knowledge can be from the Web, from data mining, or from one's experience in whole process and will be saved in the system. The knowledge can be

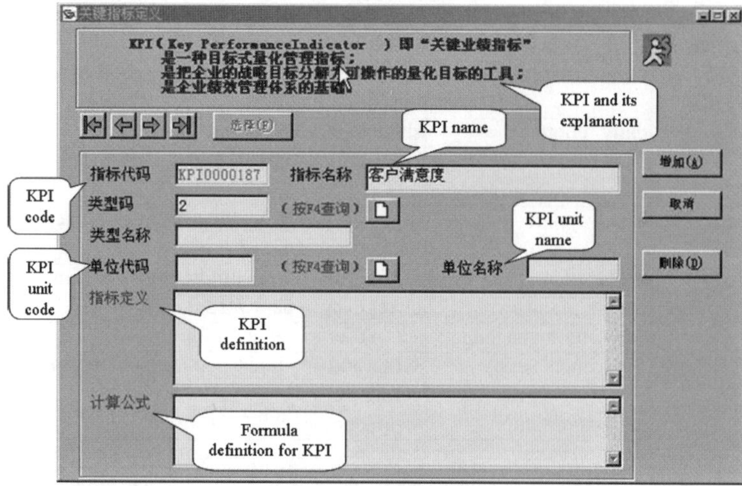

Fig 14.4. KPI definition

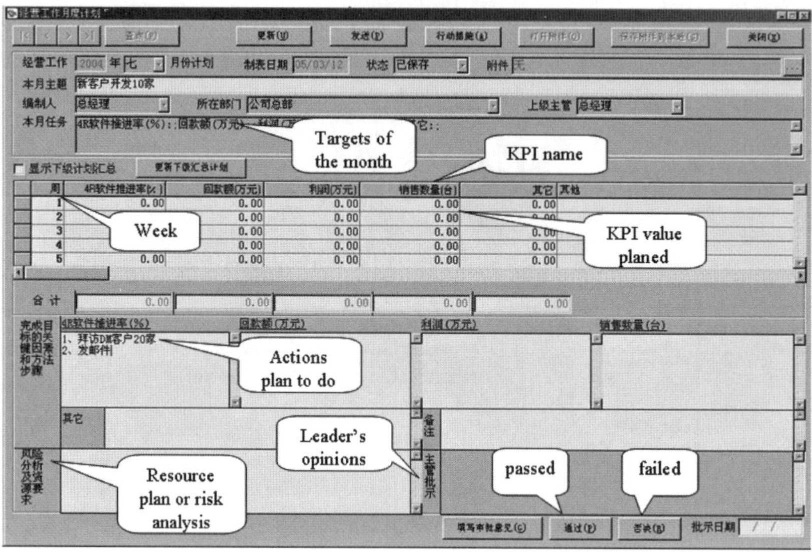

Fig 14.5. Knowledge collection from working plan

indexed, queried, marked, updated or deleted by knowledge officials through a management process.

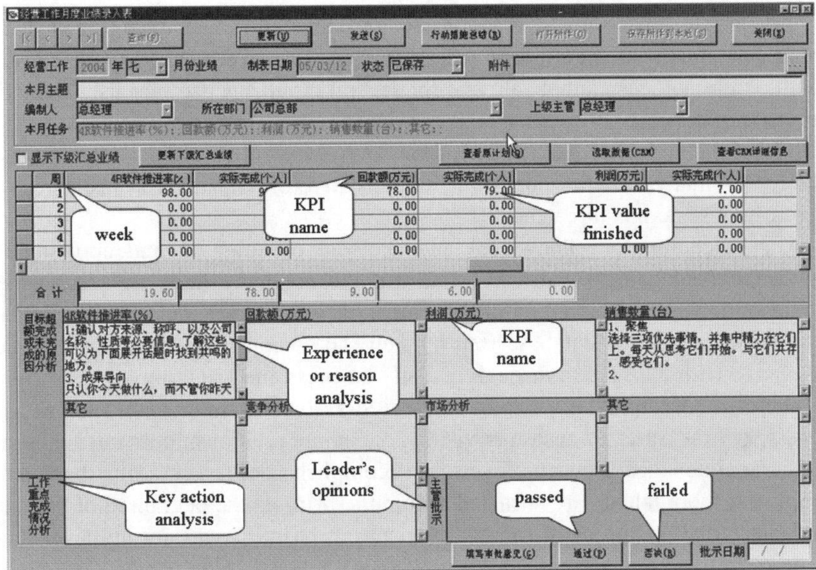

Fig 14.6. Knowledge collection from working reports

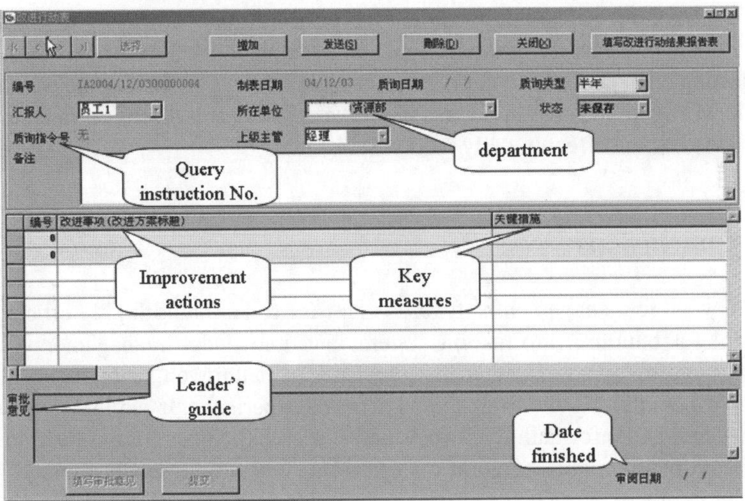

Fig 14.7. Knowledge collection from action improving

14.6 Case Study

We have applied the basic concepts of the proposed platform to a national-wide engineering machine agent company, which has six subsidiaries in six provinces, acting as agents for engineering firms of Korean Daewoo and Japanese Shinko, the agent company achieves yearly sales value is 300 million RMB. Its business includes product distribution, sales, maintaining service, spare parts supply and technical support.

The problem of this company is that there is a noticeable achievement gap between sales representatives; experienced salesman can sell more than 10 machines each month, while novices cannot sale one for several months. Customer lists, bank and supplier relationship, etc., all of these resources are grasped by several persons. Therefore, the company has to depend on several executive officials and excellent sales representatives. In this condition, it is hard to reform managerial structure; even the payment system is threatened.

After accepting our knowledge management consultation and applying this knowledge management software platform, the company built an achievement-oriented culture, emphasized on improvement of performance and achievement, and established knowledge database step by step. The company is able to accumulate knowledge with the system through its daily business and train employees by using the platform. Gradually, a series of rule-based systems is setting up, such as sales process, key measures of customer visiting, customer relationship maintenance, and supplier maintenance. New employee can accomplish sales target during the first few month, and more talents are attracted to join the firm. At the beginning of this year, they adjusted all positions of subsidiary manager, business was not influenced, and on the contrary, it has achieved a 7% growth. As a result, "able person" system was changed, and gradually forms a new corporation culture that values performance by truth data and facts.

It also met a lot of problems during the project, one is unwilling to make plans and write down measures before action, some people think it's extra-work. Another is collecting all experience in the system for compilation of books; some people think there is wrong knowledge in books. We solve them by full discuss in meetings and show its value and effect in daily works.

14.7 Conclusions

This knowledge management software is an easy to learn and highly efficient, low cost tool for building information systems and knowledge management systems. It collects international management experience from the Web, from the information system by data mining and from skilled person's experience. It can integrate existing systems, build an information network, and extract KPI status for executive officials in decision-making and operation management. It sets up a result-oriented performance system with KPI. It sets up quantify performance target and evaluation criteria, and sets up a performance system based on facts and data. Via network software platform, real-time supervise the achievement situation, and by inquiry, team study to ensure the implementation of strategic objective and finally establishes achievement-oriented corporation culture.

The other advantages of this system include: 1) Interactivity which makes all employees' action to be strategy-oriented. Set down objective, and implement it effectively by planning and checking. 2) High efficiency: set KPI target, real-time monitor implementation situation via Internet and make early warning for target that exceeds defined limit. 3) Economy: automatically extract target value that executive officials want to know from business data. Managers can spend more time in analysis since the quantities of low quality reports are reduced. 4) Systematicality: establish a systematic, standard and continuous business driven platform that does not rely on able person, meanwhile accumulate company's knowledge database, and cultivate talent within the system.

In summary, although there many work left to do, such as how to make people share their experience more happier, how to use the Web efficiently as a knowledge source, This knowledge management platform has helped small and middle businesses to build a knowledge-based management system with low costs. Company can accumulate knowledge in operation process gradually, and transfer "able person" system into institutional operation system.

Acknowledgments

This research has been partially supported by a grant from National Natural Science Foundation of China (#70621001, #70531040, #70501030#70472074), 973 Project #2004CB720103, Ministry of Science and Technology, China, and BHP Billiton Co., Australia. The authors would like to thank Prof. Han Yongsheng, Institute of Software Chinese Academy of Sciences; Ph.D. Jiang Ruxiang, general manager of Beijing Zion Consulting Company; Wang Haihong, software engineer of Youngor group Ltd., who finished most software developing works; vice Prof. Zhang Lingling, Chinese Academy of Sciences Research Center on Data Technology and Knowledge Economy and anonymous referees for their help and very constructive comments. Corresponding author: Jun Li, email: Lijun@huatai-serv.com

References

1. Maryam Alavi, Dorothy E. Leidner, Review: Knowledge management and knowledge management systems: Conceptual foundations and research issues, MIS Quarterly; Vol. 25 No. 1 Mar 2001; pp. 107-136
2. Garry L. Adams and Bruce T. Lamont, Knowledge management systems and developing sustainable competitive advantage, Journal of knowledge management; Vol. 7 No. 2, 2003; pp. 142-154
3. Ricardo Aler, Daniel Borrajo, etc., Knowledge-based approach for business process reengineering, SHAMASH, Knowledge-Based Systems, 15(2002), pp. 473-483
4. Zion Consulting Beijing Company <4R consulting plan of business management >
5. Neil C. Churchill, Virginia L. Lewis, The five stages of small business growth, Harvard Business Review, 1983(5-6), pp. 30-50
6. Chen Xuezhong, Wang Bin, etc. The Cycle Theory of Enterprises' Life and the Growth Strategies of Small-Medium Businesses, Research On Development, 2005 No. 6, pp. 78-81
7. Dianne J. Hall and David Paradice, Philosophical foundations for a learning-oriented knowledge management system for decision support. Decision Support Systems, Volume 39, Issue 3, May 2005, pp. 445-461

8. Claire R. McInerney, Trust in Knowledge Management and Systems in Organizations. By Maija-Leena Huotari and Mirja Iivonen (Eds.), Idea Group Publishing, 2004, Information Processing & Management, Vol 41, Issue 3, May 2005, pp. 720-722

9. Chinho Lin and Shu-Mei Tseng, Riding the implementation gaps in the knowledge management system for enhancing corporate performance. Expert Systems with Applications, Volume 29, Issue 1, July 2005, pp. 163-173

10. M. du Plessis, Drivers of knowledge management in the corporate environment. International Journal of Information Management, Volume 25, Issue 3, June 2005, pp. 193-202

11. Edward W. Rogers, Maija-Leena Huotari and Mirja Iivonen, Trust in Knowledge Management and Systems in Organizations, Idea Group Publishing, Hershey, Library & Information Science Research, Available online 24 May 2005.

12. Mohammed Quaddus and Jun Xu, Adoption and diffusion of knowledge management systems: field studies of factors and variables. Knowledge-Based Systems, Volume 18, Issues 2-3, April 2005, pp. 107-115

13. Karina Gibert, Miquel Sànchez-Marrè and Ignasi Rodríguez-Roda, GESCONDA: An intelligent data analysis system for knowledge discovery and management in environmental databases. Environmental Modelling & Software, 2006, 21(1), pp. 115-120

14. Bergeron B., Essentials of Knowledge Management. Hoboken: John Wiley & Sons, 2003

15. Grover V., Thomas H. Davenport. General perspectives on knowledge management: Fostering a research agenda. Journal of Management Information Systems. Armonk: 2001, 18(1): 5-22

16. Maier R. and Remus U. Defining Process-oriented Knowledge Management Strategies. Knowledge and Process Management, 2002(2): 103-118

17. Ti wana, A., The Knowledge Management Toolkit: Practical Techniques for Building a Knowledge Management System, New Jersey: Prentice-Hall, Inc., 2000

18. Jiang Rifu, Huo Guoqing and Guo Chuanjie, Research on schools of modern knowledge management, Management Review, 2006(10): 23-29

19. J. Srivastava, P. Desikan and V. Kumar, Web Mining- Concepts, Applications and Research Directions, In Data Mining: Next Generation Challenges and Future Directions. AAAI/MIT Press, Boston, MA, 2003.

20. Peter Suber, Timeline of the Open Access Movement, available online: http://www.earlham.edu/~peters/fos/timeline.htm, Dec, 2006.

21. Jens Klump, Roland Bertelmann, Jan Brase etc. Data publication in the open access initiative, Data Science Journal, Vol. 5 (2006) pp. 79-83

22. J. Han, K. Micheline, Data Mining: Concepts and Techniques, 2nd edition, Morgan Kaufmann, 2006.

23. Ronald Maier, Ulrich Remus, Implementing process-oriented knowledge management strategies, Journal of Knowledge Management, 2003, 7 (4): 62-74

24. Stefan Jablonski, Stefan Horn, Michael Schlundt, Process Oriented Knowledge Management, Eleventh International Workshop on Research Issues in Data Engineering on Document Management for Data Intensive Business and Scientific Applications, April 01-02, 2001, pp. 77-84

25. Xingsen Li, Ying Liu, Jun Li etc. A Knowledge Management Model for Middle and Small Enterprises, 2006 International Conference on Distributed Computing and Applications for Business, Engineering and Sciences DCABES2006 proceedings, Hangzhou, China, Oct. 2006, pp. 929-934

Part IV

Mining Risks from Multidisciplinary Data

Discovery of Clusters from Proximity Data: An Approach Using Iterative Adjustment of Binary Classifications

Shoji Hirano and Shusaku Tsumoto

Department of Medical Informatics, Shimane University, School of Medicine
89-1 Enya-cho, Izumo, Shimane 693-8501, Japan
hirano@ieee.org, tsumoto@computer.org

15.1 Introduction

Clustering is a task of forming groups of similar objects based on the predefined proximity (similarity/dissimilarity) measure and grouping criteria. A lot of approaches, for example, agglomerative/divisive hierarchical clustering, k-means and EM algorithms, have been proposed in the literature [1, 2] and widely used for exploratory analysis of real-world data. In order to find the best partition of objects that maximizes both inter-cluster homogeneity and between-clusters isolation, clustering methods often employ geometric measures such as the variance of distances. However, it becomes difficult to form appropriate clusters if only a proximity matrix is available as intrinsic information for analysis and the raw attribute values of data are unavailable or inaccessible. This is because the lack of attribute-value information may bring a difficulty in computing the global properties of groups such as centroids. Additionally, the choice of global coherence/isolation measures is limited if the proximity is defined as a subjective or relative measure, because such a measure may not satisfy the triangular inequality for any triplet of objects. Although conventional hierarchical clusterings are known to be able to deal with relative or subjective measures, they involve other problems such as erosion or expansion of data space by intermediate objects between large clusters and the results are dependent on the orders of object handling [1].

In order to deal with this problem, we propose a novel clustering method based on the indiscernibility of objects. It groups objects according to the colligation of N binary classifications where N denotes the number of objects. First, an equivalence relation that performs binary classification according to the local proximity is independently assigned to each object. Next, global assessment of N binary classifications is done according to a parameter called indiscernibility degree. The binary classifications are iteratively adjusted so that the resultant classification of objects meets the given level of indiscernibility. The main benefits of this method are twofold. Firstly, it can be applied to proximity measures that do not satisfy the

S. Hirano and S. Tsumoto: *Discovery of Clusters from Proximity Data: An Approach Using Iterative Adjustment of Binary Classifications*, Studies in Computational Intelligence (SCI) **123**, 251–268 (2008)
www.springerlink.com © Springer-Verlag Berlin Heidelberg 2008

triangular inequality. Secondly, it works on a proximity matrix and does not require direct access to original data values.

After introducing its methodology, we empirically investigate the fundamental properties of the method including effects of iterative adjustment of binary classifications, capability of handling relative proximity, and relationships between the indiscernibility degree and the resultant clusters. Through the experiments, we demonstrate that (1) clusters become coarser on each step of adjustment, (2) interpretable clusters can be formed even if elements in a dissimilarity matrix are locally disturbed and violate the triangular inequality, (3) the relationships between indiscernibility degree and the number of clusters draw a globally convex but multi-modal curve, and the range of indiscernibility degree that yields best cluster validity may exist on a local minimum around the global one which generates single cluster.

The remainder of this chapter is organized as follows. Section 15.2 introduces the basic definitions relevant to this work. Section 15.3 explains the methodology and computational procedures with simple examples. Section 15.4 provides experimental results on artificial data sets, and Section 15.5 presents a conclusion of the technical results.

15.2 Preliminaries

This section briefly introduces fundamental definitions, mainly came from rough sets [3], that are relevant to this work. Let $U \neq \phi$ be a universe of discourse and X be a subset of U. An equivalence relation R classifies U into a set of subsets $U/R = \{X_1, X_2, ...X_n\}$ that have the following properties: (1) $X_i \subseteq U, X_i \neq \phi$ for any i, (2) $X_i \cap X_j = \phi$ for any $i, j, i \neq j$, (3) $\cup_{i=1,2,...n} X_i = U$. Any subset X_i is called a category and represents an equivalence class of R. A category in R containing an object $x \in U$ is denoted by $[x]_R$. Objects x_i and x_j in U are *indiscernible on R* if $(x_i, x_j) \in P$ where $P \in U/R$. For a family of equivalence relations $\mathbf{P} \subseteq \mathbf{R}$, an indiscernibility relation over $\mathbf{P}$ is defined as the intersection of individual relations $Q \in \mathbf{P}$. Note that in the following sections we use the term 'category' interchangeably to 'cluster'.

15.3 Method

15.3.1 The Concept and Overall Procedure

A straightforward approach for incorporating indiscernibility into clustering is to define equivalence relations on the attribute domain and group the objects according to the derived indiscernibility relations [4]. Table 15.1 provides a simple example. Suppose we have four objects $U = \{x_1, x_2, x_3, x_4\}$ in two-dimensional attribute domain $A = \{a_1, a_2\}$ as shown in Table 15.1(a). First, in order to assess the indiscernibility of objects, we discretize continuous attribute values according to proper criteria such

Table 15.1. An example of conventional indiscernibility-based clustering

<table>
<tr><td colspan="3">(a) sample objects</td><td colspan="6">(b) classifications</td></tr>
<tr><td></td><td>a_1</td><td>a_2</td><td></td><td>a_1</td><td>a_2</td><td>U/R_{a_1}</td><td>U/R_{a_2}</td><td>C</td></tr>
<tr><td>x_1</td><td>0.1</td><td>0.1</td><td>x_1</td><td>S</td><td>S</td><td>0</td><td>0</td><td>c_1</td></tr>
<tr><td>x_2</td><td>0.2</td><td>0.2</td><td>x_2</td><td>S</td><td>S</td><td>0</td><td>0</td><td>c_1</td></tr>
<tr><td>x_3</td><td>0.7</td><td>0.5</td><td>x_3</td><td>L</td><td>M</td><td>1</td><td>1</td><td>c_2</td></tr>
<tr><td>x_4</td><td>0.8</td><td>0.8</td><td>x_4</td><td>L</td><td>L</td><td>1</td><td>2</td><td>c_3</td></tr>
</table>

as discerptibility in their distribution or based upon some human decisions. Let us assume that values of a_1 and a_2 are discretized into three levels S(small), M(medium), and L(large) as shown in a_1 and a_2 in Table 15.1(b). Then using equivalence relations R_{a_1} and R_{a_1} the following classifications can be defined.

$$U/R_{a_1} = \{\{x_1, x_2\}, \{x_3, x_4\}\},$$
$$U/R_{a_2} = \{\{x_1, x_2\}, \{x_3\}, \{x_4\}\}. \tag{15.1}$$

Using a family of equivalence relations $\mathbf{R} = \{R_{a_1}, R_{a_2}\}$ we obtain the classification of U by $\mathbf{R}$ as

$$U/\mathbf{R} = \{\{x_1, x_2\}, \{x_3\}, \{x_4\}\}. \tag{15.2}$$

Using a set of categories C, this can be rewritten as $U/\mathbf{R} = C = \{c_1, c_2, c_3\}$, where $c_1 = \{x_1, x_2\}$, $c_2 = \{x_2\}$, and $c_3 = \{x_4\}$. If we discretize a_2 into two levels, S(small) and L(large) with a threshold value of 0.4, we obtain

$$U/R_{a_2} = \{\{x_1, x_2\}, \{x_3, x_4\}\}, \tag{15.3}$$

and hence,

$$U/\mathbf{R} = \{\{x_1, x_2\}, \{x_3, x_4\}\}. \tag{15.4}$$

As seen in this example, indiscernibility-based clustering can be regarded as a process of determining a family of equivalence relations on attribute domain that classifies objects with an appropriate level of granularity. If the classification knowledge that the family provides is too fine, a lot of meaningless small categories would be obtained. On the contrary, if it is too coarse, only a small number of categories with uninteresting information would be obtained.

One of the main difference between such indiscernibility-based clustering techniques and our method is that our method defines an equivalence relation not on each attribute but on each object. That is to say, for N objects, we have N equivalence relations regardless of the number of attributes. Each of the N equivalence relations independently performs binary classification based on the relative proximity between the object to which the relation is assigned and other objects. Namely, we have N independently defined binary classifications.

Let us show an example using objects in Table 15.1(a). First, we calculate the dissimilarity matrix of the objects as shown in Table 15.2(a). For simplicity, we employ

Table 15.2. An example of the proposed indiscernibility-based clustering. Note: in (b) 0 represents class 'similar' and 1 represents class 'dissimilar'

<table>
<tr><td colspan="5" align="center">(a) dissimilarity matrix</td><td colspan="6" align="center">(b) classifications</td></tr>
<tr><td></td><td>x_1</td><td>x_2</td><td>x_3</td><td>x_4</td><td>U/R_1</td><td>U/R_2</td><td>U/R_3</td><td>U/R_4</td><td>C</td></tr>
<tr><td>x_1</td><td>0.00</td><td>0.14</td><td>0.72</td><td>0.99</td><td>x_1 0</td><td>0</td><td>1</td><td>1</td><td>c_1</td></tr>
<tr><td>x_2</td><td>0.14</td><td>0.00</td><td>0.58</td><td>0.84</td><td>x_2 0</td><td>0</td><td>1</td><td>1</td><td>c_1</td></tr>
<tr><td>x_3</td><td>0.72</td><td>0.58</td><td>0.00</td><td>0.31</td><td>x_3 1</td><td>1</td><td>0</td><td>0</td><td>c_2</td></tr>
<tr><td>x_4</td><td>0.99</td><td>0.84</td><td>0.31</td><td>0.00</td><td>x_4 1</td><td>1</td><td>0</td><td>0</td><td>c_2</td></tr>
</table>

the Euclidean distance as a measure of dissimilarity. Next, for object x_1, we assign an equivalence relation R_1 that classifies objects into two categories. One category contains objects similar to x_1, and another category contains objects dissimilar to x_1.

$$U/R_1 = \{\{x_1, x_2\}, \{x_3, x_4\}\}, \tag{15.5}$$

Here we regard that object $x_j (1 \leq j \leq 4)$ is similar to x_1 if the dissimilarity of x_1 to x_j, say $d(x_1, x_j)$ is smaller than 0.5. Note that R_1 evaluates only dissimilarities from x_1 to each of other objects and does not evaluate dissimilarities between other objects, e.g., $d(x_2, x_3)$ between x_2 and x_3. In the same way, we obtain R_2, R_3, and R_4, as shown in Table 15.2(b). Finally, according to the four binary classifications U/R_1, U/R_2, U/R_3 and U/R_4, we obtain a set of categories $C = \{c_1, c_2\}$ where $c_1 = \{x_1, x_2\}$ and $c_2 = \{x_3, x_4\}$. Similarly to the previous case, this can be rewritten with $\mathbf{R} = \{R_1, R_2, R_3, R_4\}$ and $U/\mathbf{R} = \{\{x_1, x_2\}, \{x_3, x_4\}\}$.

The main advantage of this approach is that it requires only a proximity matrix of objects and elements in the matrix do not need to satisfy the triangular inequality. However, the family of these 'initial' equivalence relations usually tends to split the objects into many equivalence classes, i.e., it produces a lot of fine categories. Since equivalence relations are determined independently without taking their relationships into accout, slightly different equivalence relations may discriminate globally similar objects.

In order to control the coarseness of the clusters, we adjust the initial binary classifications by accessing the parameter named *indiscernibility degree* of objects, which reflects the global consistency of classifications. An indiscernibility degree of two objects represents the ratio of binary classifications that regarded them as indiscernible objects. When this degree is larger than a predefined threshold value, the two objects are considered to be indiscernible, and all the binary classifications are adjusted to include them into the same category. The adjustment process is iterated several times because new candidates that should be considered as indiscernible objects may appear in the adjusted set of equivalence relations.

There are two parameters that control the behavior of this clustering method: the threshold value T_h for adjustment of binary classifications and the number N_r of iteration of adjustment. As shown in the experiments, N_r can be determined automatically, because the classifications will be stable after several cycles of adjustment.

The adjustment process can be terminated when no candidates for adjustment appear. We will discuss the relationships between T_h and resultant clusters in section 15.4.3.

In the remainder of this section, we formalize the procedures for assignment of initial binary classifications and their iterative adjustment.

15.3.2 Assignment of Initial Binary Classifications

Let $U = \{x_1, x_2, ..., x_N\}$ be a set of objects we are interested in. First, we define a binary classification for each object x_i $(1 \leq i \leq n)$ using an equivalence relation R_i as follows.

$$U/R_i = \{P_i,\ U - P_i\}, \tag{15.6}$$

where

$$P_i = \{x_j|\ d(x_i, x_j) \leq Th_{di}\},\quad \forall x_j \in U. \tag{15.7}$$

$d(x_i, x_j)$ denotes the dissimilarity between objects x_i and x_j, and Th_{di} denotes an upper threshold value of dissimilarity for object x_i. The equivalence relation R_i classifies U into two categories: P_i, which contains objects similar to x_i, and $U - P_i$, which contains objects dissimilar to x_i. When $d(x_i, x_j)$ is smaller than Th_{di}, object x_j is considered to be indiscernible to x_i. U/R_i can be alternatively written as $U/R_i = \{\{[x_i]_{R_i}\}, \{\overline{[x_i]_{R_i}}\}\}$, where $[x_i]_{R_i} \cap \overline{[x_i]_{R_i}} = \phi$ and $[x_i]_{R_i} \cup \overline{[x_i]_{R_i}} = U$ hold.

Methods for determining equivalence relations, including the choice of dissimilarity measure, is arbitrary under the condition that each of them provides binary classification of U. As an example, we describe in Appendix A a method for constructing initial binary classifications based on the denseness of the objects; however, one may use another approach for this purpose.

15.3.3 Global Adjustment of Initial Binary Classifications

Now let us assume $U = \{x_1, x_2, x_3, x_4, x_5\}$ and binary classifications of U by the equivalence relations R_1 to R_5 are given as follows.

$$
\begin{aligned}
U/R_1 &= \{\{x_1, x_2, x_3\}, \{x_4, x_5\}\}, \\
U/R_2 &= \{\{x_1, x_2, x_3\}, \{x_4, x_5\}\}, \\
U/R_3 &= \{\{x_2, x_3, x_4\}, \{x_1, x_5\}\}, \\
U/R_4 &= \{\{x_1, x_2, x_3, x_4\}, \{x_5\}\}, \\
U/R_5 &= \{\{x_4, x_5\}, \{x_1, x_2, x_3\}\}.
\end{aligned}
\tag{15.8}
$$

This example contains three types of equivalence relations: R_1 $(= R_2 = R_5)$, R_3 and R_4. Since each of them classifies U slightly differently, classification of U by the family of equivalence relations $\mathbf{R}$, $U/\mathbf{R}$, contains four very small, almost independent categories.

$$U/\mathbf{R} = \{\{x_1\}, \{x_2, x_3\}, \{x_4\}, \{x_5\}\}. \tag{15.9}$$

In this section, we present the global adjustment process of binary classifications. Through this process, binary classifications will be globally adjusted so that they produce adequately coarse categories that meet the user-specified level of indiscernibility. The global similarity of objects is represented by a newly introduced measure, the *indiscernibility degree*. Given a threshold value by a user, we repeat adjustment of the initial binary classifications until the indiscernibility degrees for all pairs of objects become larger than the threshold value.

First, we define an *indiscernibility degree*, $\gamma(x_i, x_j)$, for two objects x_i and x_j as follows.

$$\gamma(x_i, x_j) = \frac{\sum_{k=1}^{|U|} \delta_k^{indis}(x_i, x_j)}{\sum_{k=1}^{|U|} \delta_k^{indis}(x_i, x_j) + \sum_{k=1}^{|U|} \delta_k^{dis}(x_i, x_j)}, \tag{15.10}$$

where

$$\delta_k^{indis}(x_i, x_j) = \begin{cases} 1, & \text{if } (x_i \in [x_k]_{R_k} \wedge x_j \in [x_k]_{R_k}) \\ 0, & \text{otherwise.} \end{cases} \tag{15.11}$$

and

$$\delta_k^{dis}(x_i, x_j) = \begin{cases} 1, & \text{if } (x_i \in [x_k]_{R_k} \wedge x_j \notin [x_k]_{R_k}) \text{ or} \\ & \quad \text{if } (x_i \notin [x_k]_{R_k} \wedge x_j \in [x_k]_{R_k}) \\ 0, & \text{otherwise.} \end{cases} \tag{15.12}$$

Equation (15.11) shows that $\delta_k^{indis}(x_i, x_j)$ takes 1 only when x_i and x_j are indiscernible under U/R_k, with a condition that they belong to the same class of x_k. Equation (15.12) shows that $\delta_k^{dis}(x_i, x_j)$ takes 1 only when x_i and x_j are discernible under U/R_k, with a condition that either of them belongs to the same class of x_k. By summing up $\delta_k^{indis}(x_i, x_j)$ and $\delta_k^{dis}(x_i, x_j)$ for all $k(1 \leq k \leq |U|)$ as in Equation (15.10), we obtain the percentage of binary classifications under which x_i and x_j are indiscernible. Note that in Equation (15.11), we excluded the case in which x_i and x_j are indiscernible but not in the same class as x_k. R_k does not take into account the dissimilarity between x_i and x_j. Therefore, it does not positively access the discernibility between them if neither of them does not belong to $[x_k]_{R_k}$.

For example, the indiscernibility degree $\gamma(x_1, x_2)$ of objects x_1 and x_2 in the above case is calculated as follows.

$$\begin{aligned} \gamma(x_1, x_2) &= \frac{\sum_{k=1}^{5} \delta_k^{indis}(x_1, x_2)}{\sum_{k=1}^{5} \delta_k^{indis}(x_1, x_2) + \sum_{k=1}^{5} \delta_k^{dis}(x_1, x_2)} \\ &= \frac{1+1+0+1+0}{(1+1+0+1+0) + (0+0+1+0+0)} = \frac{3}{4}. \end{aligned} \tag{15.13}$$

Let us explain this example with the calculation of the numerator (1+1+0+1+0). The first value 1 is for $\delta_1^{indis}(x_1, x_2)$. Since x_1 and x_2 belong to the same class under U/R_1 and obviously, they belong to the same class to x_1, $\delta_1^{indis}(x_1, x_2) = 1$ holds. The second value is for $\delta_2^{indis}(x_1, x_2)$, and analogously, it becomes 1. The third value is for $\delta_3^{indis}(x_1, x_2)$. Since x_1 and x_2 belong to the different classes under U/R_3, it becomes 0. The fourth value is for $\delta_4^{indis}(x_1, x_2)$ and it obviously, becomes 1.

The last value is for $\delta_5^{indis}(x_1, x_2)$. Although x_1 and x_2 belong to the same class under U/R_5, their class is different to that of x_5. Thus $\delta_5^{indis}(x_1, x_2)$ equals 0.

Indiscernibility degrees for all pairs of objects in U are tabulated in Table 15.3. Note that the indiscernibility degree of object x_i to itself, $\gamma(x_i, x_i)$, will always be 1.

From its definition, a larger $\gamma(x_i, x_j)$ represents that x_i and x_j are commonly regarded as indiscernible objects under a large number of independent binary classifications. Therefore, if a binary classification U/R_l discerns the objects that have high γ value, we consider that it provides excessively fine classification knowledge, and adjust it according to the following procedure (note that R_l is rewritten as R_i below for the purpose of generalization).

Let $R_i \in \mathbf{R}$ be an initial equivalence relation on U. Now we define an adjusted version of binary classification U/R_i, denoted by U/R_i', as follows.

$$U/R_i' = \{P_i', \, U - P_i'\}, \tag{15.14}$$

where P_i' denotes a set of objects given by

$$P_i' = \{x_j | \gamma(x_i, x_j) \geq T_h\}, \quad \forall x_j \in U. \tag{15.15}$$

and T_h denotes the lower threshold value of the indiscernibility degree given by a user. If $\gamma(x_i, x_j)$ is larger than T_h, R_i is adjusted to include x_j into the class of x_i.

Suppose we set $Th = 3/5$ for the case in Equation (15.8). For U/R_1 we obtain the adjusted classifications U/R_1' as

$$U/R_1' = \{\{x_1, x_2, x_3\}, \{x_4, x_5\}\}, \tag{15.16}$$

because, according to Table 15.3, $\gamma(x_1, x_1) = 1 \geq T_h = 3/5$, $\gamma(x_1, x_2) = 3/4 \geq 3/5$, $\gamma(x_1, x_3) = 3/4 \geq 3/5$, $\gamma(x_1, x_4) = 1/5 \leq 3/5$, $\gamma(x_1, x_5) = 0/5 \leq 3/5$ hold. In the same way, the rest of adjusted binary classifications are obtained as follows.

$$\begin{aligned}
U/R_2' &= \{\{x_1, x_2, x_3\}, \{x_4, x_5\}\}, \\
U/R_3' &= \{\{x_1, x_2, x_3\}, \{x_4, x_5\}\}, \\
U/R_4' &= \{\{x_4\}, \{x_1, x_2, x_3, x_5\}\}, \\
U/R_5' &= \{\{x_5\}, \{x_1, x_2, x_3, x_4\}\}.
\end{aligned} \tag{15.17}$$

Then we obtain classification of U by the new family of binary classifications relations, $U/\mathbf{R}'$, as follows.

$$U/\mathbf{R}' = \{\{x_1, x_2, x_3\}, \{x_4\}, \{x_5\}\}. \tag{15.18}$$

In the above example, U/R_3, U/R_4 and U/R_5 are modified so that they include similar objects into the equivalence class of x_3, x_4 and x_5, respectively. The types of binary classifications remain three, however, the categories become coarser than those in Equation (15.9) by the adjustment.

15.3.4 Iterative Adjustment of Binary Classifications

It should be noted that the state of the indiscernibility degrees could also be changed after adjustment of the binary classifications, since the degrees are recalculated using the new family of binary classifications $U/\mathbf{R}'$.

Suppose we are given another threshold value $T_h = 2/5$ for the case in Equation (15.8). According to Table 15.3, we obtain $\mathbf{R}'$ after the first adjustment, as follows.

$$
\begin{aligned}
U/R_1' &= \{\{x_1, x_2, x_3\}, \{x_4, x_5\}\}, \\
U/R_2' &= \{\{x_1, x_2, x_3, x_4\}, \{x_5\}\}, \\
U/R_3' &= \{\{x_1, x_2, x_3, x_4\}, \{x_5\}\}, \\
U/R_4' &= \{\{x_2, x_3, x_4\}, \{x_1, x_5\}\}, \\
U/R_5' &= \{\{x_5\}, \{x_1, x_2, x_3, x_4\}\}.
\end{aligned}
\tag{15.19}
$$

Hence

$$
U/\mathbf{R}' = \{\{x_1\}, \{x_2, x_3\}, \{x_4\}, \{x_5\}\}.
\tag{15.20}
$$

The categories in $U/\mathbf{R}'$ are exactly the same as those in Equation (15.9). However, the state of the indiscernibility degrees are not the same because the binary classifications in $U/\mathbf{R}'$ are different from those in $U/\mathbf{R}$. Table 15.4 summarizes the indiscernibility degrees, recalculated using $U/\mathbf{R}'$. In Table 15.4, it can be observed that the indiscernibility degrees of some pairs of objects, for example $\gamma(x_1, x_4)$, increased after the adjustment, and now they exceed the threshold $th = 2/5$. Thus we perform adjustment of binary classifications again using the same T_h and the recalculated γ. Then we obtain

$$
\begin{aligned}
U/R_1' &= \{\{x_1, x_2, x_3, x_4\}, \{x_5\}\}, \\
U/R_2' &= \{\{x_1, x_2, x_3, x_4\}, \{x_5\}\}, \\
U/R_3' &= \{\{x_1, x_2, x_3, x_4\}, \{x_5\}\}, \\
U/R_4' &= \{\{x_1, x_2, x_3, x_4\}, \{x_5\}\}, \\
U/R_5' &= \{\{x_5\}, \{x_1, x_2, x_3, x_4\}\}.
\end{aligned}
\tag{15.21}
$$

Hence

$$
U/\mathbf{R}' = \{\{x_1, x_2, x_3, x_4\}, \{x_5\}\}.
\tag{15.22}
$$

After the second adjustment, the number of the binary classifications in $U/\mathbf{R}'$ are reduced from 3 to 2, and the number of categories are also reduced from 4 to 2. We further update the state of the indiscernibility degrees according to the binary classifications after the second adjustment. The results are shown in Table 15.5. Since no new pairs, whose indiscernibility degree exceeds the given threshold appear, adjustment process may be terminated and the stable categories may be obtained, as in Equation (15.22).

As shown in this example, adjustment of the binary classifications may change the indiscernibility degree of objects. Thus we iterate the adjustment process using the same T_h until the categories become stable. Note that each adjustment process

Table 15.3. Degree γ for objects in Eq. (15.8)

	x_1	x_2	x_3	x_4	x_5
x_1	3/3	3/4	3/4	1/5	0/4
x_2		4/4	4/4	2/5	0/5
x_3			4/4	2/5	0/5
x_4				3/3	1/3
x_5					1/1

Table 15.4. Degree γ after the 1st adjustment

	x_1	x_2	x_3	x_4	x_5
x_1	3/3	3/4	3/4	2/4	1/5
x_2		4/4	4/4	3/4	0/5
x_3			4/4	3/4	0/5
x_4				3/3	1/5
x_5					1/1

Table 15.5. Degree γ after the 2nd adjustment

	x_1	x_2	x_3	x_4	x_5
x_1	4/4	4/4	4/4	4/4	0/5
x_2		4/4	4/4	4/4	0/5
x_3			4/4	4/4	0/5
x_4				4/4	0/5
x_5					1/1

is performed using the previously 'adjusted' set of binary classifications. We will demonstrate, in experiments, that categories become stable after several cycles of adjustment.

15.4 Experimental Results

15.4.1 Effects of Iterative Adjustment

We first examined the effects of adjustment of the initial binary classifications. A two-dimensional numerical dataset was artificially created using Neyman-Scott method [5]. The number of seed points was set to 5. Each of the five clusters contained approximately 100 objects, and a total of 491 objects were included in the data. We evaluated validity of the clustering result based on the following measure:

$$\text{Validity } v_{\mathbf{R}}(C) = \min\left(\frac{|X_{\mathbf{R}} \cap C|}{|X_{\mathbf{R}}|}, \frac{|X_{\mathbf{R}} \cap C|}{|C|}\right), \tag{15.23}$$

where $X_{\mathbf{R}}$ and C denote the clusters obtained by the proposed method and the expected clusters, respectively. Initial binary classifications were determined by the density-based method described in Appendix A. The threshold value for adjustment T_h was set to 0.2, meaning that if two objects were commonly regarded as indiscernible by 20% of objects in the data, all the binary classifications were modified to regard them as indiscernible objects.

Without adjustment, the method produced 461 small clusters. Validity of the result was 0.011, which was the smallest value observed in this dataset. This was because the small size of the clusters produced very low coverage, namely, amount of overlap between the generated clusters and their corresponding expected clusters was very small compared with the size of the expected clusters.

By performing adjustment one time, the number of clusters was reduced to 429, improving validity to 0.013. As the adjustment proceeds, the small clusters merged as shown in Figures 15.1 and 15.2. Validity of the results continued to increase. Finally, clusters became stable at the 6th adjustment, where 10 clusters were formed as shown in Figure 15.2. Validity of the clusters was 0.927. One can observe that a few small clusters, for example, clusters 5 and 6, were formed between the large clusters. These objects were classified into independent clusters because of the competition of the

260   S. Hirano and S. Tsumoto

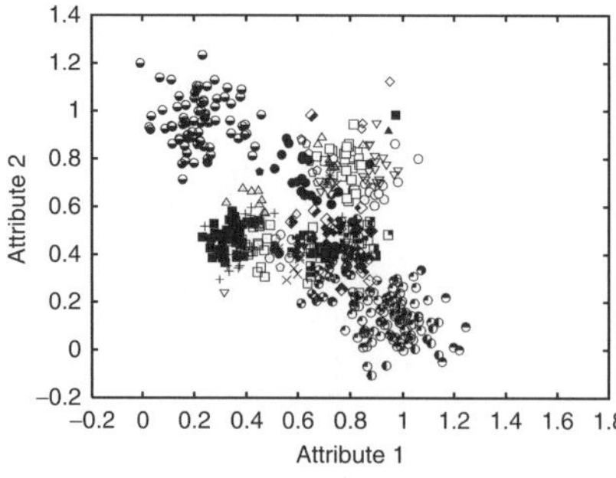

Fig 15.1. Clusters after 4th adjustment.

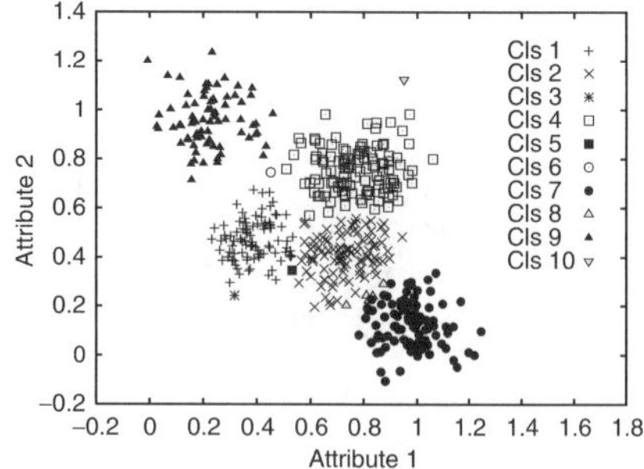

Fig 15.2. Clusters after 6th adjustment.

large clusters containing almost the same populations. Aside from this, the results revealed that the proposed method automatically produced good clusters that have high correspondence to the original ones.

The computational complexity almost followed $O(n^3)$, where n denotes the number of objects. This was mainly due to the high computational costs in calculating indiscernibility degree of objects. On a PC (Pentium 4, 1.7GHz), the method required about 5 seconds to process this dataset.

15.4.2 Capability of Handling Relative Proximity

In order to validate the method's capability of handling relative proximity, we performed clustering experiments using another dataset. The data was originally generated on the two-dimensional Euclidean space likewise the previous dataset; however, in this case we randomly modified distances between data points so that they locally violates the triangular inequality.

The dataset was prepared as follows. First, we created a two-dimensional data set by using the Neyman-Scott method similarly as in the previous case. The number of seed points was set to three for simplicity, and a total of 310 points were included in the dataset. Next, we calculated the Euclidean distances between the data points and constructed a 310×310 proximity matrix. Then we randomly selected some elements of the proximity matrix and mutated them to zero. The ratio of mutated elements was changed from 0% to 50% with 10% intervals. For each mutation ratio, we created 10 proximity matrices in order to include enough randomness. Consequently, we obtained a total of 60 proximity matrices.

We took each of the proximity matrices as an input and performed clustering of the dataset. Parameters used in the proposed method were manually determined through pilot experiments as $\sigma = 15.0$ and $T_h = 0.3$.

For comparison, we employed average-linkage and complete-linkage agglomerative hierarchical clustering methods (for short, AL-AHC and CL-AHC respectively) [1]. Note that we disregarded the original data values and took the mutated proximity matrix as input of the clustering methods. Therefore, we did not employ clustering methods that require direct access to the data value.

Table 15.6. Comparison of the clustering results

Mutation Ratio[%]	0	10	20	30	40	50
AL-AHC	0.990	0.688±0.011	0.670±0.011	0.660±0.011	0.633±0.013	0.633±0.018
CL-AHC	0.990	0.874±0.076	0.792±0.093	0.760±0.095	0.707±0.098	0.729±0.082
Our method	0.981	0.980±0.002	0.979±0.003	0.980±0.003	0.977±0.003	0.966±0.040

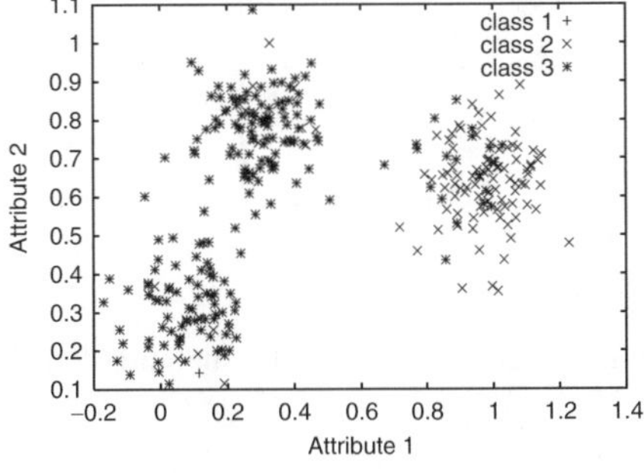

Fig 15.3. Clustering results by AL-AHC. Mutation ratio = 40%. Linkage was terminated when three clusters were formed.

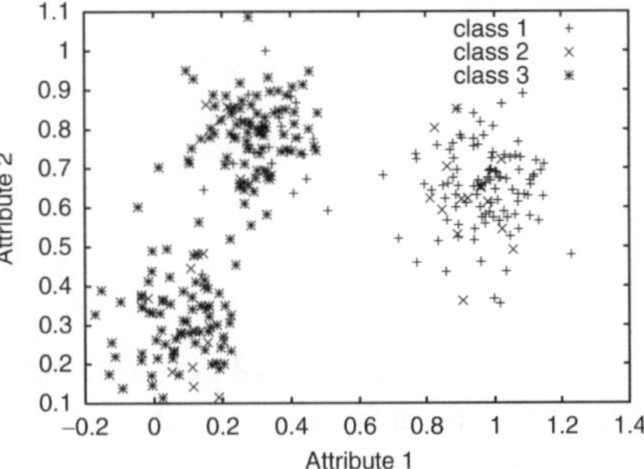

Fig 15.4. Clustering results by CL-AHC. Mutation ratio = 40%. Linkage was terminated when three clusters were formed.

We evaluated validity of the clustering results using the same measures as in the previous case. Table 15.6 provides the comparison results. The first row of the table represents the ratio of mutation. For example, 30 represents that 30% of the elements in the proximity matrix were mutated to zero. The next three rows contain the validity obtained by AL-AHC, CL-AHC and the proposed method, respectively. Except for the cases in zero mutation ratio, validity is represented in the form of 'mean ± SD', summarized from the 10 randomly mutated proximity matrices.

Without mutation, the proximity matrix was the exactly the same as the one obtained by using the Euclidean distance. Therefore, both of AL-AHC and CL-AHC could produce high validity over 0.99. The proposed method also produced the high validity over 0.98.

However, when mutation had occurred, the validity of clusters obtained by AL-AHC and CL-AHC largely decreased to 0.688 and 0.874, respectively. They kept moderately decreasing following the increase of mutation ratio. The primary reason for the decrease of validity was considered as follows. When the dissimilarity between two objects was forced to be mutated into zero, it brought a kind of local warp to the proximity of the objects. Thus the two objects could become candidates of the first linkage. If the two objects were originally belonged to the different clusters, these clusters were merged at an early stage of the merging process. Since both of AL-AHC and CL-AHC do not allow inverse of the cluster hierarchy, these clusters would never be separated. Consequently, inappropriately bridged clusters were obtained as shown in Figures 15.3 and 15.4.

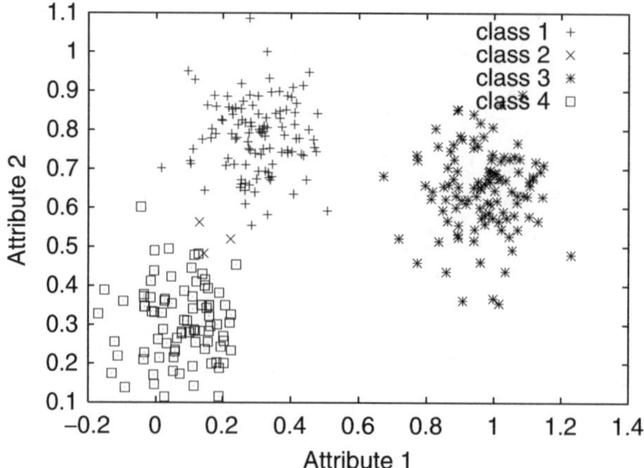

Fig 15.5. Clustering results by the proposed method. Mutation ratio = 40%. Iteration terminated at the fourth cycle.

On the contrary, the proposed method produced high validity even when the mutation ratio approached to 50%. In this method, influence of a mutation was very limited, i.e., localized. The two concerning objects would consider themselves as indiscernible objects, however, the majority of other objects never change their classifications. Although the categories obtained by the initial equivalence relations could be distorted, they could be globally adjusted through iterative adjustment of the binary classifications. Consequently, good clusters were obtained as shown in Figure 15.5. This demonstrates the capability of the method for handling locally distorted proximity matrix that do not satisfy the triangular inequality.

15.4.3 Relationships Between T_h and Clusters

Result of our clustering method is mainly affected by two factors: (1) initial binary classifications that form the basic partition of objects, and (2) threshold value of indiscernibility degrees for iterative adjustment of initial binary classifications. The first factor is basically data dependent; therefore, we investigated the second factor, i.e., relationships between the threshold value T_h of indiscernibility degree γ and resultant clusters.

In order to seclude the influence of determination method for initial binary classifications, we employed a special set of classifications named perfect initial binary classifications, that were derived from the class labels of objects. Based on the perfect binary classifications, we first examined the relationships between the threshold value of indiscernibility degree and resultant clusters. After that, we applied random disturbance to the perfect classifications, and examined how the result changes.

Perfect Binary Classifications

A perfect binary classification U/R_i for object x_i is denoted as follows.

$$U/R_i = \{P_i,\ U - P_i\}, \tag{15.24}$$

where

$$P_i = \{x_j|\ c[x_i] = c[x_j]\}, \quad \forall x_j \in U. \tag{15.25}$$

where $c[x_i]$ denotes the class label of x_i assigned when creating the dataset. Obviously, the types of perfect binary classifications in $U/\mathbf{R}$ are equal to the number of classes in the dataset, because if objects x_i and x_j belong to the same class, R_i and R_j become identical. The family of perfect binary classifications $U/\mathbf{R}$ certainly produces the original groups.

Datasets

We artificially created a total of four numerical datasets named c3-1, c3-2, c5-1, and c5-2 shown in Table 15.7. Datasets c3-1 and c3-2 contain three clusters, and c5-1 and c5-2 contain five clusters respectively. The number of data points in each cluster was controlled to be substantially different and imbalanced, because the balanced data may induce special effect of T_h on a specific range. The data points were generated to follow a two-dimensional normal distribution for easy visualization; however, note that the geometric distribution of data points is not significant in this experiment because we used only their class labels for creating the perfect equivalence relations.

Procedures

The following procedure was applied to each dataset.

1. Form perfect initial binary classifications: according to Eqs. (15.24) and (15.25), assign a perfect initial binary classification for each $x_i \in U$.
2. Disturb the initial classifications: Select one of the following disturbance operation randomly at each time and apply it to initial classifications. This process is repeated $card(P_i) \times \rho$ times, where ρ denotes disturbance ratio (from 0.0 to 1.0, with interval of 0.2).

    a) Delete: Randomly select one element in P_i and remove it from P_i.
    b) Add: Randomly select one element from U and add it to P_i.

Table 15.7. Number of data points in datasets

Dataset	Cls 1	Cls 2	Cls 3	Cls 4	Cls 5	total
c3-1	52	40	93	–	–	185
c3-2	224	31	177	–	–	432
c5-1	52	171	148	215	55	641
c5-2	64	164	126	58	155	567

    c) Replace: Randomly select one element from P_i and replace it with randomly selected element in U.

3. Clustering: Apply the iterative adjustment process to the disturbed initial binary classifications and obtain clusters. This process is repeated by changing T_h (from 0 to 1.0, interval 0.05) . For each T_h, calculate the validity of clustering result according to $v_{\mathbf{R}}(C)$ defined previously.

Results and Discussions

Figures 15.6-15.9 show the results on the four datasets respectively. Each of the figures consists of two sub-figures: Th-Number of clusters curves (left) and Th-Cluster Validity curves (right). The horizontal axis corresponds to the threshold value T_h of indiscernibility degree γ. The vertical axis corresponds to the number of clusters or cluster validity for the left or right figure, respectively. Each figure contains six curves indexed by ρ, which corresponds to the ratio of disturbance of the perfect initial binary classifications described previously.

Let us first see the global characteristics the curves. At $T_h = 0$, every equivalence relation was modified to include all objects. This means that, regardless of the characteristics of initial equivalence relations, all objects would be grouped into the

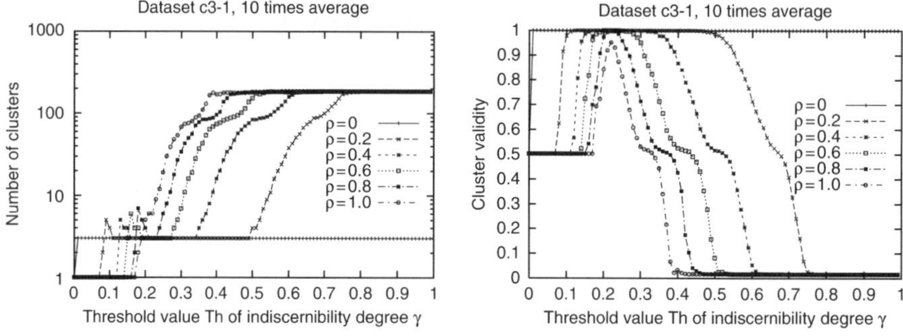

Fig 15.6. Results for dataset c3-1. Left: Number of clusters. Right: Cluster Validity.

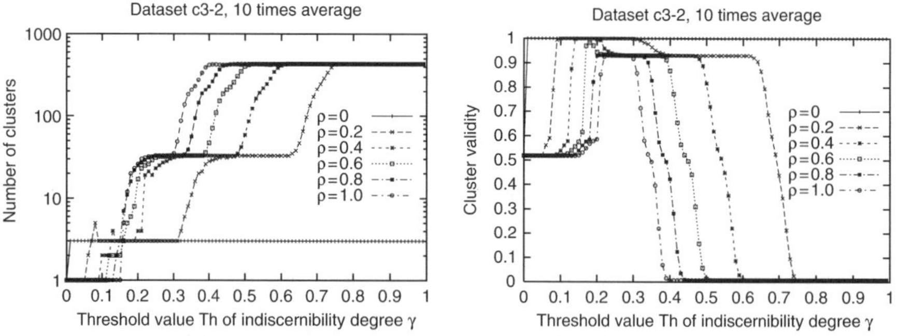

Fig 15.7. Results for dataset c3-2. Left: Number of clusters. Right: Cluster Validity.

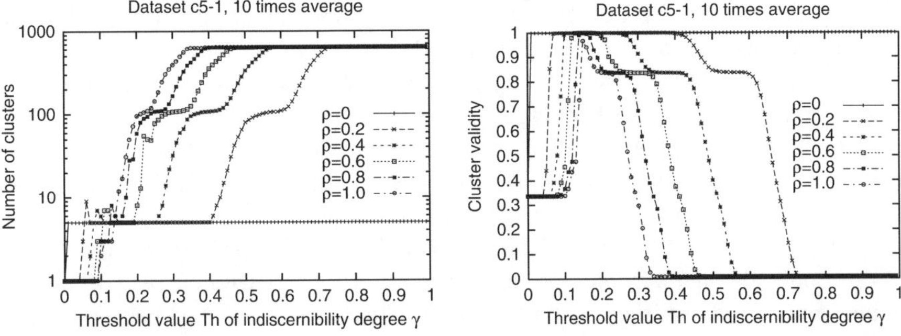

Fig 15.8. Results for Dataset c5-1. Left: Number of clusters. Right: Cluster Validity.

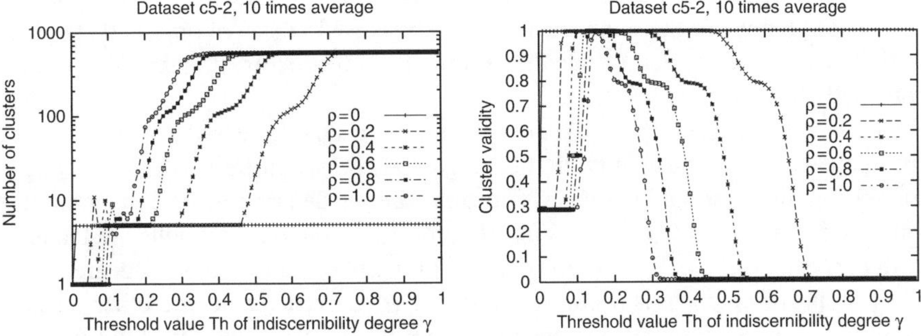

Fig 15.9. Results for Dataset c5-2. Left: Number of clusters. Right: Cluster Validity.

same cluster. Therefore the number of clusters was always 1 at $T_h = 0$. The cluster validity took a constant value which was dependent only to the class distribution of the dataset (around 0.5 or 0.3 for the datasets used here).

When $\rho = 0$, initial binary classifications were identical to the perfect relations since no disturbance was applied. In this case the indiscernibility degrees were 0 for all pairs of objects belonging to different clusters, and 1 for those belonging to the same cluster. Therefore, correct clusters of validity $= 1$ were formed for all values of $T_h > 0$ without any adjustment.

If $\rho > 0$, situations become close to those of real-world datasets. The variety of initial equivalence relations drastically increase because of disturbance. Even a small difference of binary classifications results in producing fine clusters due to the increase of total discrimination ability. Hence, without adjustment of the classifications, excessively large number of fine clusters were produced. Let us first see the case of dataset c3-1 in Figure 15.6. For large values of $T_h > 0.8$, only a few binary classifications satisfied the condition for adjustment in Eq. (15.15). As most of the relations remained unchanged, the number of induced clusters kept high value - almost equal to the number of objects in the dataset. When T_h became smaller, the number of binary classifications to be adjusted increased. The adjustment made classification coarser and made the number of clusters smaller, inducing the increase of cluster validity. The level of T_h for starting this improvement was higher if ρ was smaller, because at smaller ρ initial binary classifications were only slightly and locally modified from the perfect binary classifications. Therefore, the indiscernibility degree of each object pairs kept high value, while the types of binary classifications were quite large. As ρ becomes larger, more severe and global disturbance could occur. Since it induced the decrease of average level of the indiscernibility, the values of T_h should be smaller to do the necessary adjustment.

For $0.5 > T_h > 0.1$, the number of clusters kept 3 with the highest validity of 1. In this range, the method could produce the correct clusters with the help of iterative adjustment of the disturbed initial binary classifications. The range became narrow as ρ became small. For example, when $\rho = 0.6$ the range was about $2.7 > T_h > 1.8$ and when $\rho = 1.0$, there was no range of T_h that could generate the correct clusters. If there exists too much disturbance, the level of indiscernibility degrees for objects that should belong to the same cluster would be close to those of objects that belong to different clusters. Hence it would be difficult to form correct clusters, especially for small clusters.

For small values of $T_h < 0.1$, the number of clusters decreased to 1, followed by the decrease of cluster validity. In this range, too coarse cluster was obtained due to too much adjustment of binary classifications. Let us denote by min_γ the minimum value of indiscernibility degrees. Actually, for $T_h < min_\gamma$, the results are identical with the case of $T_h = 0$, due to the discrete property of indiscernibility degree.

The above characteristics were commonly observed for all the other datasets used in this experiment. It demonstrated that, by changing the threshold value of indiscernibility degree, we could control the roughness of classification knowledge, namely, granularity of the data.

Furthermore, an interesting feature about the number of clusters was observed on all datasets. Around $T_h = 0.1 - 0.2$, there existed a short spike at the left end of the range for yielding the correct number of clusters. Although it could disappears on extremely disturbed cases, the convex features of the curve may be used for determining the best range of T_h semi-automatically.

15.5 Conclusions

In this chapter, we have presented an indiscernibility-based clustering scheme which groups objects according to the iterative adjustment of binary classifications. We introduced the concept of the indiscernibility degree, represented by the number of binary classifications that commonly regard two objects as indiscernible, as a measure of global similarity between the two. By the use of the indiscernibility degree and iterative adjustment of binary classifications, the method classified objects according to their global characteristics, regardless of small and local difference that might produce unacceptably fine categories. Experimental results from the artificially created numerical data showed that objects were classified into the expected clusters if adjustment was performed, whereas they were classified into many small categories without adjustment. This indicated that iterative adjustment of binary classifications based on indiscernibility degree served well as a successful classification protocol. The experiments also demonstrated that this method could produce good clusters for proximity data that do not satisfy the triangular inequality.

We have also empirically investigated the relationships between indiscernibility degree and clustering results by using perfect binary classifications. The results demonstrated that the threshold parameter might be associated with roughness of knowledge, which also controls the granularity of dataset. Additionally, although it still requires exploratory approach, the convex shape of th-AC curve suggested the possibility of guiding appropriate range of the thresholds.

References

1. B. S. Everitt, S. Landau, and M. Leese (2001): Cluster Analysis Fourth Edition. Arnold Publishers.
2. P. Berkhin (2002): Survey of Clustering Data Mining Techniques. Accrue Software Research Paper. URL: http://www.accrue.com/products/researchpapers.html.
3. Z. Pawlak (1991): Rough Sets, Theoretical Aspects of Reasoning about Data. Kluwer Academic Publishers, Dordrecht.
4. J. W. Grzymala-Busse and M. Noordeen (1988): "CRS – A Program for Clustering Based on Rough Set Theory," Research report, Department of Computer Science, University of Kansas, TR-88-3, 13.
5. J. Neyman and E. L. Scott (1958): "Statistical Approach to Problems of Cosmology," *Journal of the Royal Statistical Society*, Series B20: 1–43.

Appendix A: Denseness-based Determination of Initial Binary Classifications

This appendix provides a density-based method for determining initial binary classifications, which was mentioned in Section 15.3.2.

Let $U = \{x_1, x_2, ..., x_N\}$. A binary classification U/R_i for object x_i is defined as

$$U/R_i = \{P_i, \, U - P_i\}, \tag{15.26}$$

$$P_i = \{x_j \mid d(x_i, x_j) \leq Th_{di}\}, \; \forall x_j \in U. \tag{15.27}$$

where $d(x_i, x_j)$ denotes dissimilarity between objects x_i and x_j. The threshold of dissimilarity, Th_{di}, for object x_i is determined based on the spatial density of objects. The procedure is summarized as follows.

1. Sort $d(x_i, x_j)$ in ascending order. For simplicity, we denote the sorted dissimilarity using the same representation $d(x_i, x_s)$, $1 \leq s \leq N$.
2. Generate a function $f(d)$ that represents the cumulative distribution of d. For a given dissimilarity d, function f returns the number of objects whose dissimilarity to x_i is smaller than d. Figure 15.10 shows an example. Function $f(d)$ can be generated by linearly interpolating $f(d(x_i, x_s)) = n$, where n corresponds to the index of x_s in the sorted dissimilarity list.

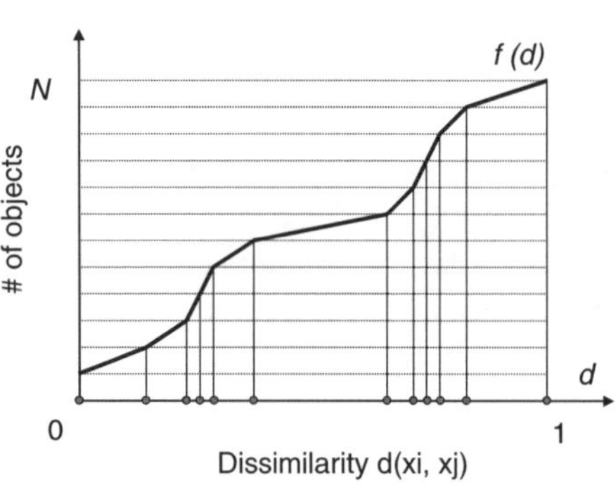

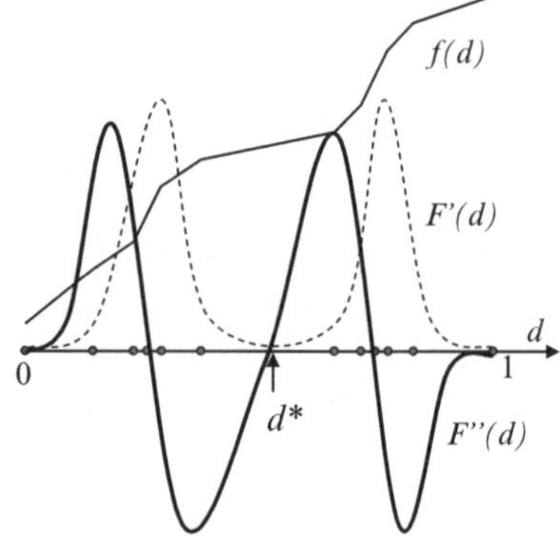

Fig 15.10. An example of function $f(d)$ generated by $d(x_i, x_s)$.

Fig 15.11. Relations between $f(d)$ and its smoothed first- and second-order derivatives $F'(d)$ and $F''(d)$.

3. Obtain the smoothed second-order derivative of $f(d)$ as a convolution of $f(d)$ and the second-order derivative of Gaussian function as follows.

$$F''(d) = \int_{-\infty}^{\infty} f(u) \frac{-(d-u)}{\sigma^3 \sqrt{2\pi}} e^{-(d-u)^2/2\sigma^2} du, \qquad (15.28)$$

where $f(d) = 1$ and $f(d) = N$ are used for $d < 0$ and $d > 1$ respectively. The smoothed first-order derivative $F'(d)$ of $f(d)$ represents spatial density of objects because it represents increase or decrease velocity of the objects induced by the change of dissimilarity. Therefore, by calculating its further derivative as $F''(d)$, we find a sparse region between two dense regions. Figure 15.11 illustrates relationship between $f(d)$ and its smoothed derivatives. The most sparse point d^* should take a local minimum of the density where the following conditions are satisfied.

$$F''(d^* - \Delta d) < 0 \text{ and } F''(d^* + \Delta d) > 0. \qquad (15.29)$$

Usually, there are some d^*s in $f(d)$ because $f(d)$ has multiple local minima. The value of σ in the above Gaussian function can be adjusted to eliminate meaningless small minima.

4. Choose the smallest d^* and object x_{j*} whose dissimilarity is the closest to but not larger than d^*. Finally, the dissimilarity threshold Th_{di} is obtained as $Th_{di} = d(x_i, x_{j*})$.

Evaluating Learning Algorithms to Support Human Rule Evaluation with Predicting Interestingness Based on Objective Rule Evaluation Indices

Hidenao Abe[1], Shusaku Tsumoto[1], Miho Ohsaki[2], and Takahira Yamaguchi[3]

[1] Department of Medical Informatics, Shimane University, School of Medicine
89-1 Enya-cho, Izumo, Shimane 693-8501, Japan
abe@med.shimane-u.ac.jp, tsumoto@computer.org
[2] Faculty of Engineering, Doshisha University
1-3 Tataramiyakodani, Kyo-Tanabe, Kyoto 610-0321, Japan
mohsaki@mail.doshisha.ac.jp
[3] Faculty of Science and Technology, Keio University
3-14-1 Hiyoshi, Kohoku Yokohama, Kanagawa 223-8522, Japan
yamaguti@ae.keio.ac.jp

Summary. In this paper, we present an evaluation of learning algorithms of a rule evaluation support method with rule evaluation models based on objective indices for data mining post-processing. Post-processing of mined results is one of the key processes in a data mining process. However, it is difficult for human experts to evaluate several thousands of rules from a large dataset with noises for finding out reraly included valuable rules. To reduce the costs in such rule evaluation task, we have developed the rule evaluation support method with rule evaluation models which learn from a dataset. This dataset comprises objective indices for mined classification rules and evaluations by a human expert for each rule. To evaluate performances of learning algorithms for constructing the rule evaluation models, we have done a case study on the meningitis data mining as an actual problem. Furthermore, we have also evaluated our method with twelve rule sets obtained from twelve UCI datasets. With regard to these results, we show the availability of our rule evaluation support method for human experts.

Keywords: Data Mining, Post-processing, Rule Evaluation Support, Objective Rule Evaluation Index

16.1 Introduction

In recent years, enormous amounts of data are stored on information systems in natural science, social science, and business domains. People have been able to obtain valuable knowledge due to the development of information technology. Besides, data mining techniques combine different kinds of technologies such as database technologies, statistical methods, and machine learning methods. Then, data mining has

H. Abe et al.: *Evaluating Learning Algorithms to Support Human Rule Evaluation with Predicting Interestingness Based on Objective Rule Evaluation Indices*, Studies in Computational Intelligence (SCI) **123**, 269–282 (2008)
www.springerlink.com

been well-known for utilizing data stored on database systems. In particular, if-then rules, which are produced by rule induction algorithms, are considered as one of the highly usable and readable outputs of data mining. However, to large datasets with hundreds of attributes including noise, the process often obtains many thousands of rules. From such a large rule set, it is difficult for human experts to find out valuable knowledge which are rarely included in the rule set.

To support such a rule selection, many efforts have done using objective rule evaluation indices such as recall, precision, and other interestingness measurements [11, 21, 23] (Hereafter, we refer to these indices as "objective indices"). Further, it is difficult to estimate the criterion of a human expert using a single objective rule evaluation index; this is because his/her subjective criterion such as interestingness and importance for his/her purpose is influenced by the amount of his/her knowledge and/or the passage of time.

With regard to the above mentioned issues, we have developed an adaptive rule evaluation support method for human experts with rule evaluation models. This method predicts the experts' criteria based on objective indices by re-using the results of the evaluations by human experts. In Section 16.2, we describe the rule evaluation model construction method based on objective indices. Then, we present a performance comparison of learning algorithms for constructing rule evaluation models in Section 16.3. With the results of the comparison, we discuss the availability of our rule evaluation model construction approach.

16.2 Rule Evaluation Support with Rule Evaluation Model Based on Objective Indices

At practical data mining situations, costly rule evaluation procedures are repeatedly done by a human expert. In these situations, useful experiences of each evaluation such as focused attributes, interesting their combinations, and valuable facts are not explicitly used by any rule selection system, but tacitly stored in the human expert. To these problems, we suggest a method to construct rule evaluation models based on objective rule evaluation indices as a way to describe criteria of a human expert explicitly, re-using the human evaluations. Combining this method with the rule visualization interface, we designed the rule evaluation support tool, which can carry out more certain rule evaluation with explicit rule evaluation models.

16.2.1 Constructing a Rule Evaluation Model

We considered the process of modeling rule evaluations of human experts as the process to clarify the relationships between the human evaluations and features of inputted if-then rules. Based on this consideration, we decided that the rule evaluation model construction process can be implemented as a learning task. Fig. 16.1 shows the rule evaluation model construction process based on the re-use of human evaluations and objective indices for each mined rule.

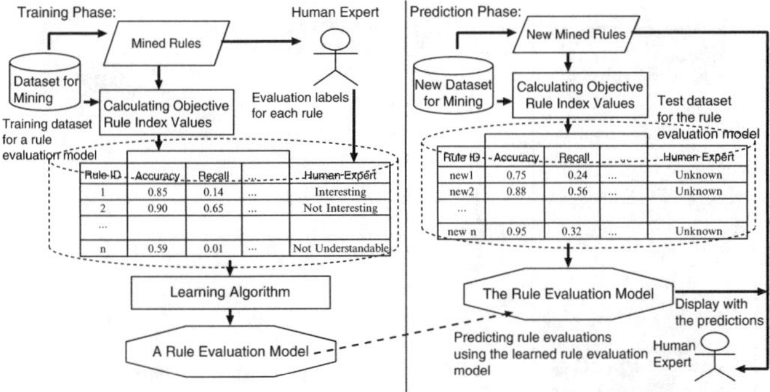

Fig 16.1. Overview of the construction method of rule evaluation models.

In the training phase, the attributes of a meta-level training data set are obtained by objective indices such as recall, precision and other rule evaluation values. The human evaluations for each rule are combined as classes of each instance. To obtain this data set, a human expert has to evaluate the whole or a part of the input rules at least once. After obtaining the training data set, its rule evaluation model is constructed by using a learning algorithm. At the prediction phase, a human expert receives predictions for new rules based on their objective index values. Since rule evaluation models are used for predictions, we need to choose a learning algorithm with high accuracy similar to the current classification problems.

16.2.2 A Tool to Support Rule Evaluation with Rule Evaluation Models

Our rule evaluation support tool implements an interactive support while a human expert evaluates rule sets from mining procedure.

At the first time to analyze a rule set with totally new task, a human expert can sort them based on some objective indices. Then he/she evaluates the whole or part of them. On the other hand, if there are some evaluation results of human experts for the same or similar problem of input rules, the predictions to the rules can be displayed to a human expert. To obtain the predictions of the rule set, this tool calls the procedure of the construction of rule evaluation models. Then a human expert corrects the displayed predictions as his/her evaluation. With the corrected evaluations by a human expert, the system rebuilds a rule evaluation model.

With above procedures, our rule evaluation support tool carries out rule evaluation support for a human expert as shown in Fig. 16.2.

A human expert can use this rule evaluation support tool as both of a passive support tool with sorting functions based on objective indices and an active support tool with predictions of rule evaluation models learned from a dataset based on objective indices.

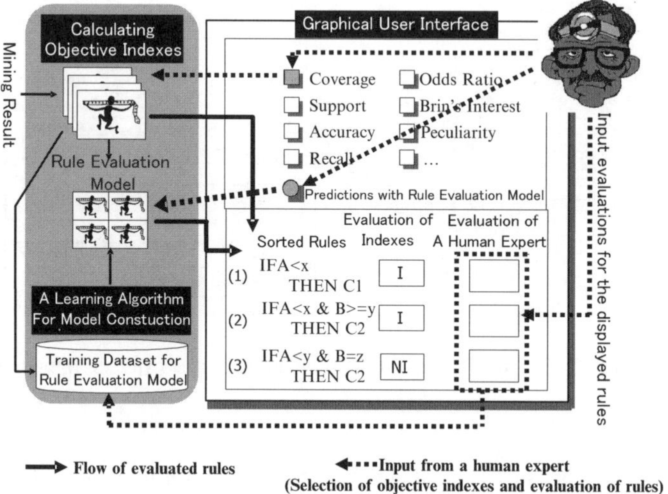

Fig 16.2. Overview of the visual rule evaluation support tool based on objective indices and rule evaluation models.

16.3 Performance Comparisons of Learning Algorithms for Rule Model Construction

To predict human evaluation labels of a new rule based on objective indices more accurately, we have to construct a rule evaluation model with a higher predictive accuracy.

In this section, we first present the result of an empirical evaluation with the dataset obtained from the result of a meningitis data mining [9]. Then, to confirm the performance of our approach on the other datasets, we evaluated five algorithms on twelve rule sets obtained from twelve UCI benchmark datasets [10]. Based on the experimental results, we discuss the followings: accuracy of rule evaluation models, analysis of learning curves of the learning algorithms, and contents of the learned rule evaluation models.

For evaluating the accuracy of the rule evaluation models, we have compared predictive accuracies on the entire dataset and Leave-One-Out validation. The accuracy of a validation dataset D is calculated with correctly predicted instances $Correct(D)$ as $Acc(D) = (Correct(D)/|D|) \times 100$, where $|D|$ is the size of the dataset. The recalls of class i on a validation dataset are calculated using correctly predicted instances about the class $Correct(D_i)$ as $Recall(D_i) = (Correct(D_i)/|D_i|) \times 100$, where $|D_i|$ is the size of instances of class i. Further, the precision of class i is calculated using the size of instances which are predicted i as $Precision(D_i) = (Correct(D_i)/Predicted(D_i)) \times 100$.

With regard to the learning curves, we obtained curves of accuracies of learning algorithms on the entire training dataset to evaluate whether each learning algorithm

Table 16.1. Objective rule evaluation indices for classification rules used in this research. **P:** Probability of the antecedent and/or consequent of a rule. **S:** Statistical variable based on P. **I:** Information of the antecedent and/or consequent of a rule. **N:** Number of instances included in the antecedent and/or consequent of a rule. **D:** Distance of a rule from the others based on rule attributes.

Theory	Index Name (**Abbreviation**) [Reference Number of Literature]
P	Coverage (**Coverage**), Prevalence (**Prevalence**) Precision (**Precision**), Recall (**Recall**) Support (**Support**), Specificity (**Specificity**) Accuracy (**Accuracy**), Lift (**Lift**) Leverage (**Leverage**), Added Value (**Added Value**) [21] Klösgen's Interestingness (**KI**) [14], Relative Risk (**RR**) [1] Brin's Interest (**BI**) [2], Brin's Conviction (**BC**) [2] Certainty Factor (**CF**) [21], Jaccard Coefficient (**Jaccard**) [21] F-Measure (**F-M**) [19], Odds Ratio (**OR**) [21] Yule's Q (**YuleQ**) [21], Yule's Y (**YuleY**) [21] Kappa (**Kappa**) [21], Collective Strength (**CST**) [21] Gray and Orlowska's Interestingness weighting Dependency (**GOI**) [7] Gini Gain (**Gini**) [21], Credibility (**Credibility**) [8]
S	χ^2 Measure for One Quadrant (χ^2-**M1**) [6] χ^2 Measure for Four Quadrant (χ^2-**M4**) [6]
I	J-Measure (**J-M**) [20], K-Measure (**K-M**) [15] Mutual Information (**MI**) [21] Yao and Liu's Interestingness 1 based on one-way support (**YLI1**) [23] Yao and Liu's Interestingness 2 based on two-way support (**YLI2**) [23] Yao and Zhong's Interestingness (**YZI**) [23]
N	Cosine Similarity (**CSI**) [21], Laplace Correction (**LC**) [21] ϕ Coefficient (ϕ) [21], Piatetsky-Shapiro's Interestingness (**PSI**) [16]
D	Gago and Bento's Interestingness (**GBI**) [5] Peculiarity (**Peculiarity**) [24]

can perform in the early stage of rule evaluation process. Accuracies of randomly sub-sampled training datasets are averaged with 10 trials on each percentage of the subset.

By observing the elements of the rule evaluation models on the meningitis data mining result, we consider the characteristics of the objective indices which are used in these rule evaluation models.

In order to construct a dataset to learn a rule evaluation model, the values of the objective indices have been calculated for each rule by considering 39 objective indices as shown in Table 16.1. Thus, each dataset for each rule set has the same number of instances as the rule set. Each instance has 40 attributes including those of the class.

We applied five learning algorithms to these datasets to compare their performances as a rule evaluation model construction method. We used the following learning algorithms from Weka [22]: C4.5 decision tree learner [Quinlan, 1993] called J4.8, neural network learner with back propagation (BPNN) [12], support vector machines (SVM)[1] [17], classification via linear regressions (CLR)[2] [3], and OneR [13].

---

[1] A polynomial kernel function was used.

[2] We set up the elimination of collinear attributes and the model selection with greedy search based on the Akaike information metric.

16.3.1 Constructing Rule Evaluation Models for an Actual Datamining Result

In this case study, we have considered 244 rules, which are mined from six datasets about six types of diagnostic problems as shown in Table 16.2. In these datasets, appearances of meningitis patients were considered as attributes and the diagnosis of each patient as a class. Each rule set was mined with its proper rule induction algorithm composed by a constructive meta-learning system called CAMLET [9]. For each rule, we labeled three evaluations (I: Interesting, NI: Not-Interesting, NU: Not-Understandable) according to evaluation comments provided by a medical expert.

Comparison of Classification Performances

In this section, we present the result of accuracy comparison over the entire dataset, recall of each class label, and precisions them. Since Leave-One-Out holds just one test instance and the remaining as the training dataset repeatedly for each instance of a given dataset, we can evaluate the performance of a learning algorithm to a new dataset without any ambiguity.

The results of the performances of the five learning algorithms to the entire training dataset and the results of Leave-One-Out are also shown in Table 16.3. All the Accuracies, Recalls of I and NI, and Precisions of I and NI are higher than those of the predicting default labels.

As compared to the accuracy of OneR, the other learning algorithms achieve equal or higher performances using combinations of multiple objective indices than by sorting with single objective index. With regard to the Recall values over class I,

Table 16.2. Description of the meningitis datasets and the results of datamining.

Dataset	#Attributes	#Class	#Mined rules	#'I' rules	#'NI' rules	#'NU' rules
Diag	29	6	53	15	38	0
C_Course	40	12	22	3	18	1
Culture+diag	31	12	57	7	48	2
Diag2	29	2	35	8	27	0
Course	40	2	53	12	38	3
Cult_find	29	2	24	3	18	3
TOTAL	—	—	244	48	187	9

Table 16.3. Accuracies (%), Recalls (%), and Precisions (%) of the five learning algorithms.

	Over the entire training dataset						Leave-One-Out						
		Recall			Precision				Recall			Precision	
	Acc.	I	NI	NU	I	NI	NU	Acc.	I	NI	NU	I	NI	NU
J4.8	85.7	41.7	97.9	66.7	80.0	86.3	85.7	79.1	29.2	95.7	0.0	63.6	82.5	0.0
BPNN	86.9	81.3	89.8	55.6	65.0	94.9	71.4	77.5	39.6	90.9	0.0	50.0	85.9	0.0
SVM	81.6	35.4	97.3	0.0	68.0	83.5	0.0	81.6	35.4	97.3	0.0	68.0	83.5	0.0
CLR	82.8	41.7	97.3	0.0	71.4	84.3	0.0	80.3	35.4	95.7	0.0	60.7	82.9	0.0
OneR	82.0	56.3	92.5	0.0	57.4	87.8	0.0	75.8	27.1	92.0	0.0	37.1	82.3	0.0

BPNN has achieved the highest performance. The other algorithms exhibit lower performance than that of OneR, because they tend to be learned classification patterns for the major class NI.

The accuracy of Leave-One-Out demonstrates the robustness of each learning algorithm. The Accuracy (%) of these learning algorithms ranges from 75.8% to 81.9%. However, these learning algorithms have not been able to classify the instances of class NU, because it is difficult to predict a minor class label in this dataset.

Learning Curves of the Learning Algorithms

Since the rule evaluation model construction method requires the mined rules to be evaluated by a human expert, we have investigated learning curves of each learning algorithm to estimate a minimum training subset to construct a valid rule evaluation model. The upper table in Fig. 16.3 shows the accuracies to the entire training dataset with each subset of training dataset. The percentage of achievements of each learning algorithm compared with their accuracy over the whole dataset are shown in the lower section of Fig. 16.3.

As observed in these results, SVM and CLR, which learn hype-planes, obtained an achievement ratio grater than 95% using less than 10% of training subset. Although a decision tree learner and BPNN could learn better classifiers to the entire dataset than the hyper-plane learners, they need more training instances to learn accurate classifiers.

In order to eliminate known ordinary knowledge from a large rule set, the noninteresting rules need to be classified correctly. The right upper table in Fig. 16.3 shows percentage of Recalls on NI. The right lower chart in Fig. 16.3 also shows

%training sample	10	20	30	40	50	60	70	80	90	100
J4. 8	73.4	74.7	79.8	78.6	72.8	83.2	83.7	84.5	85.7	85.7
BPNN	74.8	78.1	80.6	81.1	82.7	83.7	85.3	86.1	87.2	86.9
SMO	78.1	78.6	79.8	79.8	79.8	80.0	79.9	80.2	80.4	81.6
CLR	76.6	78.5	80.3	80.2	80.3	80.7	80.9	81.4	81.0	82.8
OneR	75.2	73.4	77.5	78.0	77.7	77.5	79.0	77.8	78.9	82.4

%training sample	10	20	30	40	50	60	70	80	90	100
J4. 8	82.7	85.3	92.8	88.7	93.2	92.7	93.2	92.9	94.0	97.9
BPNN	84.6	86.6	90.4	90.2	92.2	91.9	92.7	93.9	94.2	89.8
SMO	93.3	92.7	96.8	96.1	95.9	95.8	96.3	96.0	96.3	97.3
CLR	88.3	89.6	94.4	94.0	94.3	94.1	94.1	94.2	94.3	97.3
OneR	88.4	84.0	92.4	91.4	92.0	92.3	93.4	92.7	92.1	96.3

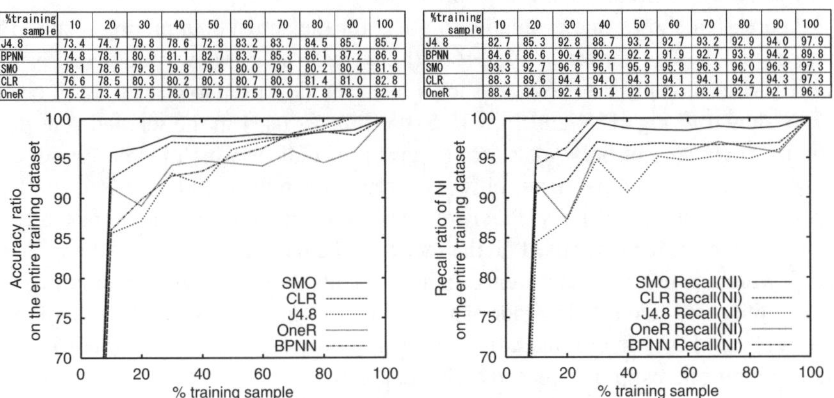

Fig 16.3. Learning curves of Accuracies (%) on the learning algorithms over subsampled training dataset: The left table shows accuracies (%) of each training dataset to the entire dataset. The left graph shows their achievement ratios (%). The right table shows recalls (%) and the graph shows their achievement ratios (%).

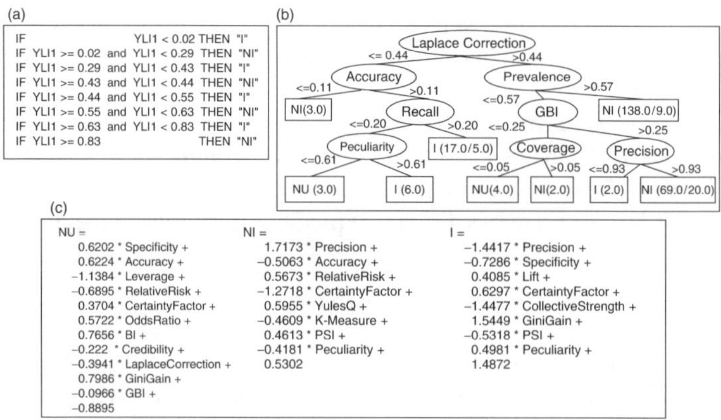

Fig 16.4. Learned models for the meningitis data mining result dataset.

the percentage of achievements of Recall of NI, and compares it with the Recall of NI of the entire training dataset. From this result, we can eliminate the NI rules with rule evaluation models from SVM and BPNN although there only 10% of rule evaluations are conducted by a human expert. This fact is guaranteed with no less than 80% precisions for all learning algorithms.

Rule Evaluation Models on the Actual Datamining Result Dataset

In this section, we present rule evaluation models for the entire dataset learned using OneR, J4.8 and CLR. This is because they are represented as explicit models such as a rule set, a decision tree, and linear model set.

Figure 16.4 shows rule evaluation models for the actual data mining result: The rule set of OneR is shown in Fig. 16.4 (a), Fig. 16.4 (b) shows the decision tree learned with J4.8, and Fig. 16.4 (c) shows linear models used to classify each class.

As shown in Fig. 16.4 and Fig. 16.5, the indices used in the learned rule evaluation models are not only taken from a group of indices which increase with correctness of a rule, but also from different groups of indices. Indices such as YLI1, Laplace Correction, Accuracy, Precision, Recall, Coverage, PSI and, Gini Gain are indices which are formerly used on the models. The latter indices are GBI and Peculiarity, which sums up the difference in antecedents between one rule and the other rules in the same rule set. This corresponds to the comment provided by the human expert. He said that he evaluated these rules not only according to their correctness but also their interestingness based on his expertise

16.3.2 Constructing Rule Evaluation Models on Artificial Evaluation Labels

We have also evaluated our rule evaluation model construction method using rule sets obtained from five datasets of the UCI machine learning repository to confirm the lower limit performances on probabilistic class distributions.

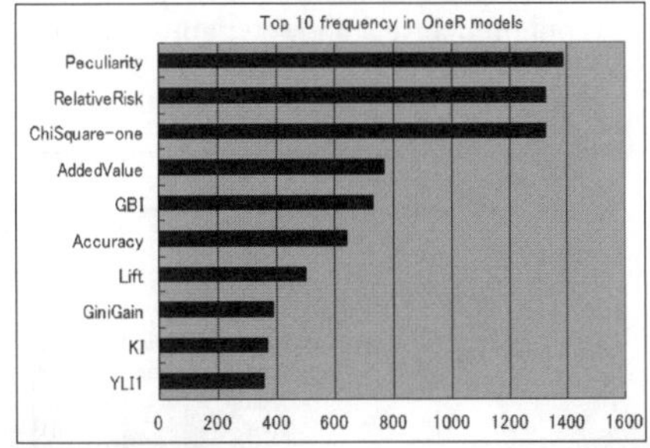

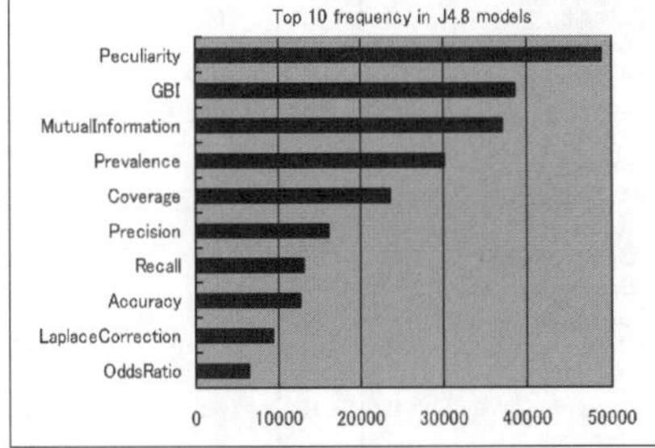

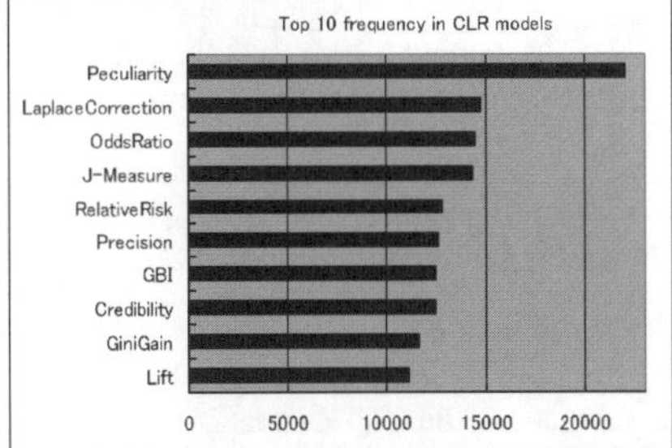

Fig 16.5. Top 10 frequencies of the indices used by the models of each learning algorithm with 10000 bootstrap samples of the meningitis datamining result dataset and executions.

We selected the following twelve datasets: Anneal, Audiology, Autos, Balance-scale, Breast-cancer, Breast-w, Colic, Credit-a, Credit-g, Diabetes, Waveform, and Letter. With these datasets, we obtained rule sets with bagged PART, which repeatedly executes PART [4] to the bootstrapped training subsample datasets.

For these rule sets, we calculated 39 objective indices as attributes of each rule. With regard to the classes of these datasets, we used three class distributions with multinomial distribution. Table 16.4 shows us the process flow diagram for obtaining these datasets and their description with three different class distributions. The class distribution for "Distribution I" is $P = (0.35, 0.3, 0.3)$ where p_i is the probability of class i. Thus, the number of class i instances in each dataset D_j become $p_i D_j$. Similarly, the probability vector of "Distribution II" is $P = (0.3, 0.5, 0.2)$ and that of "Distribution III" is $P = (0.3, 0.65, 0.05)$.

Accuracy Comparison on Classification Performances

To abovementioned datasets, we have used the five learning algorithms to estimate if their classification results reach or exceed the accuracies of that of just predicting each default class. The left table of Table 16.5 shows the accuracies of the five learning algorithms applied to each class distribution of the three datasets. As shown in Table 16.5, J4.8 and BPNN always perform better for just predicting a default class. However, their performances suffer from probabilistic class distributions for larger datasets such as Waveform and Letter.

Table 16.4. Flow diagram to obtain datasets and the datasets of the rule sets learned from the UCI benchmark datasets

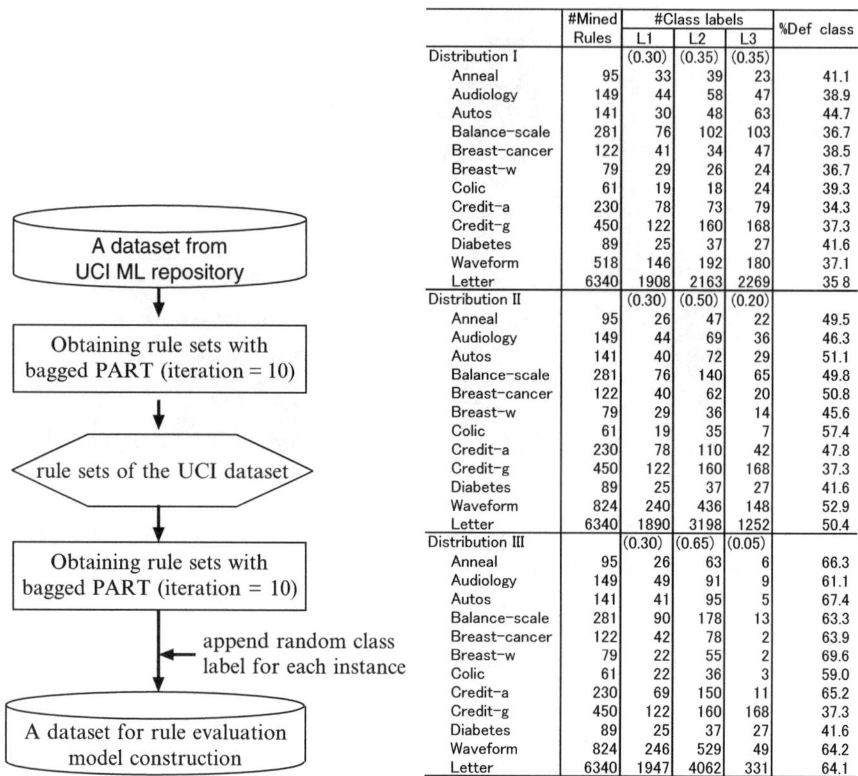

	#Mined Rules	#Class labels			%Def class
		L1	L2	L3	
Distribution I		(0.30)	(0.35)	(0.35)	
Anneal	95	33	39	23	41.1
Audiology	149	44	58	47	38.9
Autos	141	30	48	63	44.7
Balance-scale	281	76	102	103	36.7
Breast-cancer	122	41	34	47	38.5
Breast-w	79	29	26	24	36.7
Colic	61	19	18	24	39.3
Credit-a	230	78	73	79	34.3
Credit-g	450	122	160	168	37.3
Diabetes	89	25	37	27	41.6
Waveform	518	146	192	180	37.1
Letter	6340	1908	2163	2269	35 8
Distribution II		(0.30)	(0.50)	(0.20)	
Anneal	95	26	47	22	49.5
Audiology	149	44	69	36	46.3
Autos	141	40	72	29	51.1
Balance-scale	281	76	140	65	49.8
Breast-cancer	122	40	62	20	50.8
Breast-w	79	29	36	14	45.6
Colic	61	19	35	7	57.4
Credit-a	230	78	110	42	47.8
Credit-g	450	122	160	168	37.3
Diabetes	89	25	37	27	41.6
Waveform	824	240	436	148	52.9
Letter	6340	1890	3198	1252	50.4
Distribution III		(0.30)	(0.65)	(0.05)	
Anneal	95	26	63	6	66.3
Audiology	149	49	91	9	61.1
Autos	141	41	95	5	67.4
Balance-scale	281	90	178	13	63.3
Breast-cancer	122	42	78	2	63.9
Breast-w	79	22	55	2	69.6
Colic	61	22	36	3	59.0
Credit-a	230	69	150	11	65.2
Credit-g	450	122	160	168	37.3
Diabetes	89	25	37	27	41.6
Waveform	824	246	529	49	64.2
Letter	6340	1947	4062	331	64.1

Evaluation of Learning Curves

Similar to the evaluations of the learning curves on the meningitis rule set, we have estimated minimum training subsets for a valid model, which works better for just predicting a default class.

The right table in Table 16.6 shows the sizes of the minimum training subsets, which can help construct more accurate rule evaluation models than percentages of a default class formed by each learning algorithm. With smaller dataset, these learning algorithms have been able to construct valid models with less than 20% of the given training datasets. However, to larger dataset such as Waveform and Letter, they need more training subsets to construct valid models, because their performances with whole training dataset fall to the percentages of default class of each dataset as shown in the left table in Table 16.6.

Table 16.5. Accuracies (%) on entire training datasets labeled with three different distributions.

	Distribution I				
	J4.8	BPNN	SVM	CLR	OneR
Anneal	74.7	71.6	47.4	56.8	55.8
Audiology	47.0	51.7	40.3	45.6	52.3
Autos	66.7	63.8	46.8	46.1	56.0
Balance–scale	58.0	59.4	39.5	43.4	53.0
Breast–cancer	55.7	61.5	40.2	50.8	59.0
Breast–w	86.1	91.1	38.0	46.8	54.4
Colic	91.8	82.0	42.6	60.7	55.7
Credit–a	57.4	48.7	35.7	39.1	54.8
Credit–g	49.6	48.2	27.6	39.3	54.9
Diabetes	64.0	78.7	41.6	42.7	53.9
Waveform	46.5	46.4	37.6	39.8	54.9
Letter	36.8	36.4	30.1	36.6	52.1
	Distribution II				
	J4.8	BPNN	SVM	CLR	OneR
Anneal	68.4	66.3	56.8	60.0	56.8
Audiology	60.4	61.1	43.6	55.0	56.4
Autos	63.1	64.5	52.5	53.2	57.4
Balance–scale	61.6	57.7	49.8	55.2	58.0
Breast–cancer	68.0	70.5	47.5	58.2	59.8
Breast–w	89.9	93.7	49.4	58.2	62.0
Colic	77.0	78.7	57.4	62.3	67.2
Credit–a	61.3	59.1	41.3	52.6	56.1
Credit–g	61.3	59.1	41.3	52.6	56.1
Diabetes	79.8	84.3	52.8	53.9	60.7
Waveform	61.2	57.8	52.9	53.0	59.7
Letter	51.0	51.0	50.4	50.4	57.0
	Distribution III				
	J4.8	BPNN	SVM	CLR	OneR
Anneal	74.7	70.5	67.4	70.5	73.7
Audiology	65.8	67.8	63.8	64.4	67.1
Autos	85.1	73.8	68.1	70.2	73.8
Balance–scale	70.5	69.8	64.8	65.8	69.8
Breast–cancer	71.3	77.0	66.4	65.6	77.9
Breast–w	74.7	86.1	73.4	68.4	74.7
Colic	70.5	77.0	65.6	60.7	73.8
Credit–a	70.9	70.0	65.2	65.2	71.3
Credit–g	69.6	68.9	64.7	64.9	68.0
Diabetes	82.0	88.8	67.4	67.4	73.0
Waveform	74.4	69.3	64.2	64.2	69.3
Letter	64.1	64.3	64.1	64.1	68.3

16.4 Conclusion

In this paper, we have described the evaluation of five learning algorithms for a rule evaluation support method with rule evaluation models to predict evaluations for an if-then rule based on objective indices by re-using evaluations by a human expert.

Based on the performance comparison of the five learning algorithms, rule evaluation models have achieved higher accuracies for just predicting each default class. Considering the difference between the actual evaluation labeling and the artificial evaluation labeling, it is shown that the evaluation of the medical expert considered particular relations between an antecedent and a class/another antecedent in each rule. By using these learning algorithms for estimating the robustness of a new rule

Table 16.6. Number of minimum training subsamples for outperforming the Accuracy (%) of default class.

	Distribution I				
	J4.8	BPNN	SVM	CLR	OneR
Anneal	20	14	17	29	29
Audiology	21	18	65	64	41
Autos	38	28	76	77	70
Balance–scale	12	14	15	15	32
Breast–cancer	16	17	22	41	22
Breast–w	7	10	10	18	14
Colic	8	8	9	22	14
Credit–a	9	12	16	30	28
Credit–g	40	49	0	87	84
Diabetes	14	10	24	33	20
Waveform	60	52	46	355	152
Letter	189	217	–	955	305
	Distribution II				
	J4.8	BPNN	SVM	CLR	OneR
Anneal	29	20	16	42	46
Audiology	36	45	–	61	67
Autos	49	39	49	123	88
Balance–scale	81	84	69	221	168
Breast–cancer	31	28	102	40	46
Breast–w	14	11	23	30	26
Colic	24	20	36	42	36
Credit–a	51	74	–	134	109
Credit–g	112	245	–	273	275
Diabetes	33	25	47	55	54
Waveform	251	355	763	–	533
Letter	897	>1000	451	–	>1000
	Distribution III				
	J4.8	BPNN	SVM	CLR	OneR
Anneal	54	58	64	76	–
Audiology	64	73	45	76	107
Autos	66	102	84	121	98
Balance–scale	118	103	133	162	156
Breast–cancer	50	31	80	92	80
Breast–w	44	36	31	48	71
Colic	28	24	46	30	42
Credit–a	118	159	–	–	173
Credit–g	251	283	353	18	383
Diabetes	50	42	60	–	72
Waveform	329	425	191	–	601
Letter	>1000	>1000	998	>1000	>1000

with Leave-One-Out, we have achieved an accuracy greater than 75.8%. By evaluating learning curves, SVM and CLR were observed to have achieved an achievement ratio greater than 95% using less than 10% of the subset of the training dataset, which includes certain human evaluations. These results indicate the availability of this rule evaluation support method for a human expert.

In the future, we will introduce a selection method of learning algorithms to construct a proper rule evaluation model according to each situation. We also apply this rule evaluation support method to estimate other data mining result such as decision tree, rule set, and combination of them with objective indices, which evaluate all the mining results.

References

1. Ali, K., Manganaris, S., Srikant, R.: Partial Classification Using Association Rules. Proc. of Int. Conf. on Knowledge Discovery and Data Mining KDD-1997 (1997) 115–118
2. Brin, S., Motwani, R., Ullman, J., Tsur, S.: Dynamic itemset counting and implication rules for market basket data. Proc. of ACM SIGMOD Int. Conf. on Management of Data (1997) 255–264
3. Frank, E., Wang, Y., Inglis, S., Holmes, G., and Witten, I. H.: Using model trees for classification. Machine Learning, Vol.32, No.1 (1998) 63–76
4. Frank, E, Witten, I. H., Generating accurate rule sets without global optimization. Proc. of the Fifteenth International Conference on Machine Learning, (1998) 144–151
5. Gago, P., Bento, C.: A Metric for Selection of the Most Promising Rules. Proc. of Euro. Conf. on the Principles of Data Mining and Knowledge Discovery PKDD-1998 (1998) 19–27
6. Goodman, L. A., Kruskal, W. H.: Measures of association for cross classifications. Springer Series in Statistics, 1, Springer-Verlag (1979)
7. Gray, B., Orlowska, M. E.: CCAIIA: Clustering Categorical Attributes into Interesting Association Rules. Proc. of Pacific-Asia Conf. on Knowledge Discovery and Data Mining PAKDD-1998 (1998) 132–143
8. Hamilton, H. J., Shan, N., Ziarko, W.: Machine Learning of Credible Classifications. Proc. of Australian Conf. on Artificial Intelligence AI-1997 (1997) 330–339
9. Hatazawa, H., Negishi, N., Suyama, A., Tsumoto, S., and Yamaguchi, T.: Knowledge Discovery Support from a Meningoencephalitis Database Using an Automatic Composition Tool for Inductive Applications. Proc. of KDD Challenge 2000 in conjunction with PAKDD2000 (2000) 28–33
10. Hettich, S., Blake, C. L., and Merz, C. J.: UCI Repository of machine learning databases [http://www.ics.uci.edu/~mlearn/MLRepository.html], Irvine, CA: University of California, Department of Information and Computer Science, (1998)
11. Hilderman, R. J. and Hamilton, H. J.: Knowledge Discovery and Measure of Interest. Kluwe Academic Publishers (2001)
12. Hinton, G. E.: "Learning distributed representations of concepts", *Proceedings of 8th Annual Conference of the Cognitive Science Society*, Amherest, MA. REprinted in R.G.M. Morris (ed.) (1986)
13. Holte, R. C.: Very simple classification rules perform well on most commonly used datasets, Machine Learning, Vol. 11 (1993) 63–91
14. Klösgen, W.: Explora: A Multipattern and Multistrategy Discovery Assistant. in Fayyad, U. M., Piatetsky-Shapiro, G., Smyth, P., Uthurusamy R. (Eds.): Advances in Knowledge Discovery and Data Mining. AAAI/MIT Press, California (1996) 249–271
15. Ohsaki, M., Kitaguchi, S., Kume, S., Yokoi, H., and Yamaguchi, T.: Evaluation of Rule Interestingness Measures with a Clinical Dataset on Hepatitis. Proc. of ECML/PKDD 2004, LNAI3202 (2004) 362–373
16. Piatetsky-Shapiro, G.: Discovery, Analysis and Presentation of Strong Rules. In Piatetsky-Shapiro, G., Frawley, W. J. (eds.): Knowledge Discovery in Databases. AAAI/MIT Press (1991) 229–248
17. Platt, J.: Fast Training of Support Vector Machines using Sequential Minimal Optimization. In: B. Schölkopf, C. Burges, and A. Smola (eds.): Advances in Kernel Methods - Support Vector Learning, MIT Press (1999) 185–208
18. Quinlan, R.: C4.5: Programs for Machine Learning. Morgan Kaufmann Publishers, (1993)

19. Rijsbergen, C.: Information Retrieval, Chapter 7, Butterworths, London, (1979)
    http://www.dcs.gla.ac.uk/Keith/Chapter.7/Ch.7.html
20. Smyth, P., Goodman, R. M.: Rule Induction using Information Theory. In Piatetsky-
    Shapiro, G., Frawley, W. J. (eds.): Knowledge Discovery in Databases. AAAI/MIT Press
    (1991) 159–176
21. Tan, P. N., Kumar V., Srivastava, J.: Selecting the Right Interestingness Measure for Asso-
    ciation Patterns. Proc. of Int. Conf. on Knowledge Discovery and Data Mining KDD-2002
    (2002) 32–41
22. Witten, I. H and Frank, E.: DataMining: Practical Machine Learning Tools and Tech-
    niques with Java Implementations. Morgan Kaufmann, (2000)
23. Yao, Y. Y. Zhong, N.: An Analysis of Quantitative Measures Associated with Rules. Proc.
    of Pacific-Asia Conf. on Knowledge Discovery and Data Mining PAKDD-1999 (1999)
    479–488
24. Zhong, N., Yao, Y. Y., Ohshima, M.: Peculiarity Oriented Multi-Database Mining. IEEE
    Trans. on Knowledge and Data Engineering, 15, 4, (2003) 952–960

Risk Mining for Infection Control

Shusaku Tsumoto[1], Kimiko Matsuoka[2], and Shigeki Yokoyama[3]

[1] Department of Medical Informatics,
Shimane University, School of Medicine
89-1 Enya-cho, Izumo 693-8501 Japan
[2] Osaka Prefectural General Hospital, Osaka, Japan
[3] Department of Medical Information, Koden Industry, Tokyo, Japan

Summary. This paper proposes risk mining, where data mining techniques were applied to detection and analysis of risks potentially existing in the organizations and to usage of risk information for better organizational management. We applied this technique to the following two medical domains: risk aversion of nurse incidents and infection control. The results show that data mining methods were effective to detection of risk factors.

17.1 Introduction

It has passed about twenty years since clinical information are stored electronically as a hospital information system since 1980's. Stored data include from accounting information to laboratory data and even patient records are now started to be accumulated: in other words, a hospital cannot function without the information system, where almost all the pieces of medical information are stored as multimedia databases. Especially, if the implementation of electronic patient records is progressed into the improvement on the efficiency of information retrieval, it may not be a dream for each patient to benefit from the personal database with all the healthcare information, "from cradle to tomb". However, although the studies on electronic patient record has been progressed rapidly, reuse of the stored data has not yet been discussed in details, except for laboratory data and accounting information to which OLAP methodologies are applied. Even in these databases, more intelligent techniques for reuse of the data, such as data mining and classical statistical methods has just started to be applied from 1990's [1, 2].

Human data analysis is characterized by a deep and short-range investigation based on their experienced "cases", whereas one of the most distinguished features of com-puter-based data analysis is to enable us to understand from the different viewpoints by using "cross-sectional" search. It is expected that the intelligent reuse of data in the hospital information system provides us to grasp the all the characteristics of univer-sity hospital and to acquire objective knowledge about how the hospital management should be and what kind of medical care should be served in the university hospital.

S. Tsumoto et al.: *Risk Mining for Infection Control*, Studies in Computational Intelligence (SCI) **123**, 283–297 (2008)
www.springerlink.com

This paper focuses on application of data mining to medical risk management. To err is human. However, medical practice should avoid as many errors as possible to achieve safe medicine. Thus, it is a very critical issue in clinical environment how we can avoid the near misses and achieve the medical safety. Errors can be classified into the following three type of erros. First one is systematic errors, which occur due to problems of system and workflow. Second one is personal errors, which occur due to lack of expertise of medical staff. Finally, the third one is random error. The important point is to detect systematic errors and personal errors, which may be prevented by suitable actions, and data mining is expected as a tool for analysis of those errors.

For this purpose, this paper proposes *risk mining* where data including risk information is analyzed by using data mining methods and mining results are used for risk prevention. We assume that risk mining consists of three major processes: risk detection, risk clarification and risk utilization, as shown in Section 2.

We applied this technique to the following two medical domains: risk aversion of nurse incidents, infection control. The results show that data mining methods were effective to detection of risk factors.

This paper is organized as follows. Section 2 shows background of our studies. Section 3 proposes three major processes of risk mining. Section 4 gives an illustrative application of risk mining. Finally, Section 5 concludes this paper.

17.2 Background

A hospital is a very complicated organization where medical staff, including doctors and nurses give a very efficient and specialized service for patients. However, such a complicated organization is not robust to rapid changes. Due to rapid advances in medical technology, such as introduction of complicated chemotherapy, medical workflow has to be changed in a rapid and systematic way. Such rapid changes lead to malpratice of medical staff, sometimes a large-scale accident may occur by chain reaction of small-scale accidents.

Medical accidents include not only careless mistakes of doctors or nurses, but also prescription errors, intrahospital infections or drug side-effects. The cause for such accidents may not be well investigated and it is unknown whether such accidents can be classified into systematic errors or random errors. Since the occurrence of severe accidents is very low, case studies are used for their analysis. However, in such investigations, personal errors tend to be the cause of the accidents. Thus, it is very important to discover knowledge about how such accidents occur in a complicated organization and knowledge about the nature of systematic erors or random errors.

On the other hand, clinical information have been stored electronically as a hospital information system(HIS). The database stores all the data related with medical actions, including accounting information, laboratory examination, treatement and patient records described by medical staffs. Incident or accident reports are not exception: they are also stored in HIS as clinical data. Thus, it is now expected that mining such combined data will give a new insight to medical accidents.

17.3 Risk Mining

In order to utilize information about risk extracted from information systems, we propose risk mining which integrates the following three important process: risk detection, risk clarification and risk utilization.

17.3.1 Risk Detection

Patterns or information unexpected to domain experts may be important to detect the possiblity of large scale accidents. So, first, mining patterns or other types of information which are unexpected to domain experts is one of the important processes in risk mining. We call this process *risk detection*, where acquired knowdedge is refered to as *detected risk information*.

17.3.2 Risk Clarification

Focusing on detected risk information, domain experts and data miners can focus on clarification of modelling the hidden mechanism of risk. If domain experts need more information with finer granularity, we should collect more data with detailed information, and apply data mining to newly collected data. We call this process *risk clarification*, where acquired knowdedge is refered to as *clarified risk information*.

17.3.3 Risk Utilization

We have to evaluate clarified risk information in a real world environment to prevent risk events. If risk information is not enough to prevention, then more analysis is required. Thus, additional data collection is evoked for a new cycle of risk mining process. We call this process *risk utilization*. where acquired knowdedge is refered to as *clarified risk information*.

Figure 17.1 shows the overview of risk mining process.

17.3.4 Elemental Techiques for Risk Mining

Mining Unbalanced Data

A large scale accident rarely occur: usually such it can viewed as a large deviation of small scale accidents, called incidents. Since even the occurence of incidents is very low, the probability of large accidents is nearly equal to 0. On the other hand, most of the data mining methods depend on "frequency" and mining such unbalanced data with small probabilities is one of the difficult problems in data mining research. Thus, for risk mining, techiques for mining unbanced data are very important to detect risk information.

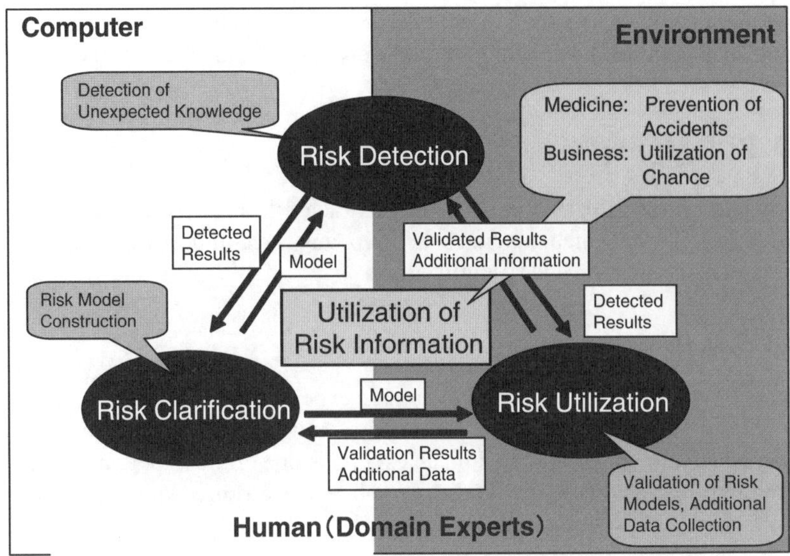

Fig 17.1. Risk Mining Proces: Overview

17.3.5 Interestingness

In convetional data mining, indices for mining patterns are based on frequency. However, to extract unexpected or interesting knowledge, we can introduce measures for unexpectedness or interestingness to extract patterns from data, and such studies have been reported in data mining literature.

17.3.6 Uncertainty and Granularity: Granular Computing

Since incident reports include information about human actions, these data are described by subjective information with uncertainty, where we need to deal with coarseness and fineness of information (information granularity). Granular computing, including fuzzy sets and rough sets, are closely related with this point.

17.3.7 Visualization

Visualizing cooccurence events or items may enable domain experts to detect risk information, to clarify the mechanism of risk, or to utilize risk information.

17.3.8 Structure Discovery: Graph Mining

Risk may be detected or clarified only by relations between several items in a large network structure. Thus, exracting partial structure from network hidden in data is a very important techique, focusing on risk information based on relations between items.

17.3.9 Clustering

Similarity may find relations between similar objects which seems not to be similar. Or events which seems to occur independently can be grouped into several "similar" events, which enables us to find dependencies between events. For this purpose, clustering is a very important techique.

17.3.10 Evaluation of Risk Probablity

Since probability is formally defined as a Lebegue measure on a fixed sample space, its performance is very unstable when the definition of sample space is unstable. Especially, when we collect data dynamically, such unstablility frequently occurs. Thus, deep reflection on evaluation of risk probability is very important.

17.3.11 Human Computer Interaction

This process is very important for risk mining process because of the following reasons. First, risk information may be obtained by deep discussions on mining results among domain experts because mining results may show only small part of the total risk information. Since domain experts have knowledge, which is not described in a datasets, they can compensate for insufficient knowledge to obtain a hypothesis or explanation of mining results. Second, mining results may lead to domain experts' deep understanding of workflow, as shown in Section 17.4. Interpretation of mining results in risk detection may lead to new data collection for risk clarification. Finally, human computer interaction gives a new aspect for risk utilization. Domain experts can not only performance of risk clarification results, but also look for other possiblities from the rules which seems to be not so important, compared with rules for risk clarification and also evalute the possibility to design a new data collection.

17.4 Application I: Prevention of Medication Errors

As an illustrative example, we applied risk mining process to analysis of nurses' incident data. First, data collected in 6 months were analyzed by rule induction methods, which detects several important factors for incidents (risk detection). Since data do not include precise information about these factors, we recollect incident data for 6 months to collect precise information about incidents. Then, rule induction is applied to new data. Domain experts discussed all the results obtained and found several important systematic errors in workflow (risk clarification). Finally, nurses changed workflow to prevent incidents and data were recollected for 6 months. Surprisingly, the frequency of medication errors has been reduced to one-tenth (risk utilization).

17.4.1 Background

A hospital is a very complicated organization where medical staff, including doctors and nurses give a very efficient and specialized service for patients. However, such a complicated organization is not robust to rapid changes. Due to rapid advances in medical technology, such as introduction of complicated chemotherapy, medical workflow has to be changed in a rapid and systematic way. Such rapid changes lead to malpratice of medical staff, sometimes a large-scale accident may occur by chain reaction of small-scale accidents.

Medical accidents include not only careless mistakes of doctors or nurses, but also prescription errors, intrahospital infections or drug side-effects. The cause for such accidents may not be well investigated and it is unknown whether such accidents can be classified into systematic errors or random errors. Since the occurrence of severe accidents is very low, case studies are used for their analysis. However, in such investigations, personal errors tend to be the cause of the accidents. Thus, it is very important to discover knowledge about how such accidents occur in a complicated organization and knowledge about the nature of systematic erors or random errors.

On the other hand, clinical information have been stored electronically as a hospital information system(HIS). The database stores all the data related with medical actions, including accounting information, laboratory examination, treatement and patient records described by medical staffs. Incident or accident reports are not exception: they are also stored in HIS as clinical data. Thus, it is now expected that mining such combined data will give a new insight to medical accidents.

17.4.2 Risk Detection

Dataset

Nurses' incident data were collected by using the conventional sheet of incident reports during 6 months from April, 2001 to September, 2001 at the emergency room in Osaka Prefectural General Hospital.

The dataset includes the types of the near misses, the patients' factors, the medical staff's factors and the shift (early-night, late-night, and daytime) and the number of items of incidents collected was 245.

We applied C4.5 [3], decision tree induction and rule induction to this dataset.

Rule Induction

We obtained a decision tree shown in Figure 17.2 and the following interesting rules.

```
(medication error):
If late-night and lack of checking,
then medication errors occur: probability
(53.3%, 8/15).
```

***** Decision tree : First 6 Months *****

```
┌Injection error—Injection route trouble(an obstruction due to the bending・reflow,
│                the disconnection) = Yes: early-night work (2—>2)
└Injection error—Injection route trouble(an obstruction due to the bending・reflow,
    │           the disconnection) = No
  ├Injection error—Pulled out. (accident and self) = Yes: early-night (2—>2)
  ├Injection error—Pulled out. (accident and self) = No
     ├Injection error— Interrupts for the work = Yes: late-night (5—>3)
     ├Interrupts for the work = No
        ├Injection error—Lack of knowledge for drugs and injection
        │          = Yes: late-night (5—>3)
        ├Injection error—Lack of knowledge for drugs and injection = No
           ├Injection error—Lack of command on the serious patients
           │          = Yes: late-night (3—>2)
           ├Injection error— Lack of command on the serious patients = No
              ├Injection error—Lack of attention and confirmation
              │ ( drug to, dosage by, patient at, time in, route ) = No: day-time(6—>4)
              ├Injection error—Lack of attention and confirmation = Yes
                 ├Injection error—Wrong IV rate of flow = Yes: early-night work (4—>2)
                 ├Injection error—Wrong IV rate of flow = No: day-time (28—>15)
```

Fig 17.2. Decision Tree in Risk Detection

```
(injection error):
If daytime and lack of checking,
then injection incidents occur: probability
(53.6%, 15/28).
```

```
(injection error):
If early-night, lack of checking,
and error of injection rate,
then injection incidents occur: probability
(50%, 2/4)
```

Those rules show that the time shift of nurse and lack of checking were the principal factors for medication and injection errors. Interestingly, lack of expertise (personal errors) was not selected. Thus, time shift and lack of checking could be viewed as risk factor for these errors. Since the conventional format of incident reports did not include furture information about workflow, we had decided to ask nurses' to fill out new report form for each incident. This is the next step in risk clarification.

17.4.3 Risk Clarification

Dataset

Just after the first 6 months, we had found that the mental concentration of nurses may be important factors for medical errors. During the next 6 months from October

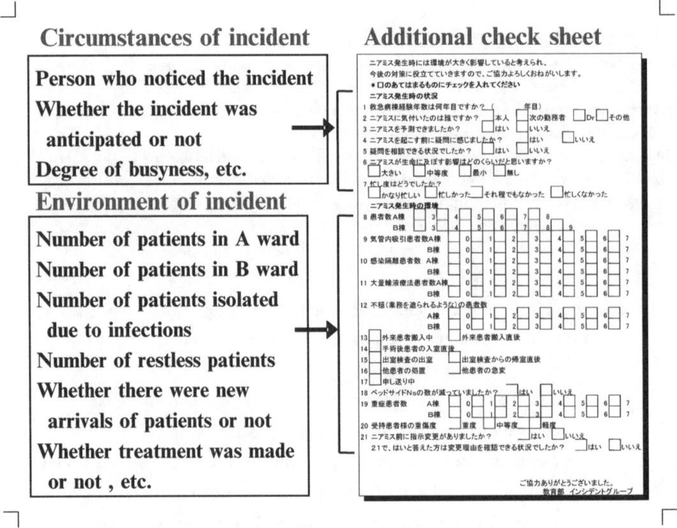

Fig 17.3. Sheet for Additional Information

2001 to March 2002, the detailed interference factors were included in the additional incident report form as the items of "environmental factors".

Figure 17.3 shows a sheet for additional information. The additional items included the duration of experience at the present ward, the number of nurse, the degree of business, the number of serious patients whether the nursing service was interrupted or not and so on.

We applied C4.5 [3], decision tree induction and rule induction to this dataset.

Rule Induction

The following rules were obtained:

```
(medication error):
If the number of disturbing patients
    is one or more,
    then medi-cation errors occur:
    probability (90%, 18/20).


(medication error):
If nurses' work interrupted,
then medication errors occur:
probability (80%, 4/5).
```

By addition of "the environmental factors", these high probability rules of medication errors were extracted.

Rule Interpretation

With these results, the nurses discussed their medication check system.

At the emergency room, the nurses in charge of the shift prepared the medication (identification, quantity of medicines, etc.). The time of preparation before the beginning of the shift was occasionally less than 30 minutes when the liaison conference between shifts took time. In such cases, the sorting of medicines could not be made in a advance and must be done during the shift.

If nurses' concentration was disturbed by the restless patients in such situations, double check of the preparation for medicine could not be made, which leads to medication errors.

17.4.4 Risk Utilization

Therefore, it was decided that two nurses who had finished their shifts would prepare medicines for the next shift, and one nurse in charge of the medication would check the dose and identification of medicines alone (triple check by a total of 3 nurses). (However, heated discussions among domain experts (nurses) needed for this decision, as shown in Section 17.4.5.) Improvement was applied to the check system as a result of their discussion. During the last 6 months (April 2002 to October 2002), incident reports were collected.

After introducing the triple check system, the total number of the medication errors during the last 6 months decreased to 24 cases. It was considered that the nurses' medication work was improved by the triple check system during the last 6 months.

17.4.5 Discussion for Case Study

Risk Utilization as Information Sharing

For discussion among domain experts, mining results were presented to medical staffs as objective evidence. Discussion on mining results give a very interactive discussion among the staff of the department of emergency and finally achieve common understanding of the problem on its workflow. Then, it is found that changes in workflow is required for solving the problem: If the staff assigned to the shift cannot prepare medicines, other members who are free should cooperate. However, this idea met a fierce objection in the department at first because of disagreement among nurses about the responsibility of those who prepare medicines. After repeated discussions, it was decided that nurses in charge of medication were responsible for mistakes rather than those who made preparations and nurses in the preceding shift should prepare medicines for the next shift.

During the last 6 months, medication errors were reduced markedly by creating the common perception that liaison (overlapping of shift margins, or paste margins) is important among nurses, and the initial opposition completely subsided. Following this nursing example, we could extend this policy of "paste margins", i.e. mutual support by free staff members, to the entire department.

This process also shows that information granularity is a very important issue for risk clarification.

Items in a conventional report form, such as "lack of checking, lack of attention, etc." are too coarse for risk clarification. Rather, detailed description of environmental factors are much more important to evoke domain experts' discussion and their risk utilizaiton.

17.5 Application II: Infection Control (Risk Detection)

17.5.1 Objective

For the prevention of blood stream infection, we analyzed the effects of lactobacillus therapy and the background risk factors of bacteria detection on blood cultures. For the purpose of our study, we used the clinical data collected from the patients, such as laboratory results, isolated bacterium, anti-biotic agents, lactobacillus therapy, various catheters, departments, and underlying diseases.

Material

The population for this study consisted of 1291 patients with blood stream infection who were admitted to our center between January and December, 2002. The subjects were divided into two groups by the absence or presence of lactobacillus therapy. Lactobacillus group patients were administrated lactobacillus preparation or yoghurt within 5 days from microbial detection in blood cultures, and control group patients never took those preparations. Table 17.1 shows all the components of this dataset.

Table 17.1. Attributes in a Dataset on Infection Control

Item	Attributes (63)
Patient's Profile	ID, Gender, Age
Department	Department, Ward, Diagnosis(3)
Order	Background Diseases, Sampling Date, Sample, No.
Symptom	Fever, Cathether(5), Traheotomy, Endotracheal intubation, Drainage(5)
Examination Data	CRP, WBC, Urin data, Liver/Kidney Function, Immunology
Therapy	Antibiotic agents(3), Steroid, Anti-cancer drug, Radiation Therapy, Lactobacillus Therapy
Culture	Colony count, Bacteria, Vitek biocode, $\beta-$lactamase
Susceptibility	Cephems, Penicillins, Aminoglycoside, Macrolides, Carbapenums, Chloramphenicol,Rifanpic, VCM, etc.

Table 17.2. Contingency table of bacteria detection and lactobacillus therapy

	Lactobacillus therapy		
	N (Absence)	Y (Presence)	Total
Bacteria Y (Detection)	247	55	302
Bacteria N (No detection)	667	322	989
Total	914	377	1291

Analytic Method

As the analytical methods, we used decision tree, chi-square test and logistic regression. "If-then rules" were extracted from the decision trees. The chi-square test and logistic regression were applied in order to analyze the effect of lactobacillus therapy.

17.5.2 Results

Chi-square Test and Odds Ratio

Chi-square test was applied to evaluate the association between lactobacillus therapy and blood stream infection (bacteria detection on blood cultures). Table 17.2 shows the cross table of the bacteria detection on blood samples and the lactobacillus therapy. In this cross table, its p-value was $0.000000159 < 0.01$. Therefore, the effect of lactobacillus presence to lactobacillus absence was considered statistically significant. Odds ratio was calculated as the relative risk of lactobacillus absence to lactobacillus presence. Probability of bacteria detection on lactobacillus absence is $p = 247/914$. Probability of bacteria detection on lactobacillus presence is $q = 55/377$. Odds ratio (OR) of lactobacillus absence to lactobacillus presence is 2.17 and 95% CI (confidence interval) was between 1.57 and 2.99.

Since the bacteria detection risk of lactobacillus absence was about 2 (95%CI: 1.57-2.99) to lactobacillus presence, lactobacillus therapy might be significantly effective to prevent the bacteria detection on blood sample.

Thus, these results showed that lactobacillus therapy might have the effect to the prevention of blood stream infection.

Decision Tree

The following decision tree shown in Figure 17.4 and 17.5 was obtained as the relationship between the bacteria detection and the various factors, such as diarrhea, lactobacillus therapy, antibiotics, surgery, tracheotomy, CVP/IVH catheter, urethral catheter, drainage, other catheter.

Figure 17.4 shows the sub-tree of the decision tree on lactobacillus therapy = Y (Y means its presence.) and Figure 17.5 shows the sub-tree of the decision tree on lactobacillus therapy = N (N means its absence). The target variable of the

```
Lactobacillus therapy (Y/N) = Y:
  Diarrhea (Y/N) = N:
     Catheter (Y/N) = N: Bacteria : N (74->71)
     Catheter (Y/N) = Y:
        Surgery (Y/N) = N: Bacteria : N (173->149)
        Surgery (Y/N) = Y:
           CVP/IVH (Y/N) = Y: Bacteria : N (12->12)
           CVP/IVH (Y/N) = N:
              Drainage (Y/N) = N: Bacteria : N (17->13)
              Drainage (Y/N) = Y:
                 Urethral catheter(Y/N) = N: Bacteria : N (3->2)
                 Urethral catheter(Y/N) = Y: Bacteria : Y (2->2)
  Diarrhea (Y/N) = Y:
     Catheter (Y/N) = Y: Bacteria : N (75->63)
     Catheter (Y/N) = N:
        Anti-biotics (Y/N) = N: Bacteria : N (7->7)
        Anti-biotics (Y/N) = Y: Bacteria : Y (14->9)
```

Fig 17.4. Sub-tree on lactobacillus therapy(Y/N) = Y

```
Lactobacillus therapy (Y/N) = N:
  Tracheotomy (Y/N) = N:
     Diarrhea (Y/N) = N:
        Anti-biotics (Y/N) = Y: Bacteria : N (428->356)
        Anti-biotics (Y/N) = N:
           Catheter (Y/N) = N: Bacteria : N (170->129)
           Catheter (Y/N) = Y:
              Surgery (Y/N) = N:
                 Drainage (Y/N) = Y: Bacteria : N (9->7)
                 Drainage (Y/N) = N:
                    Urethral catheter(Y/N) = N: Bacteria : N (124->75)
                    Urethral catheter(Y/N) = Y:
                       CVP/IVH (Y/N) = N: Bacteria : Y (27->16)
                       CVP/IVH (Y/N) = Y: Bacteria : N (18->11)
              Surgery (Y/N) = Y:
                 Drainage (Y/N) = N: Bacteria : N (5->5)
                 Drainage (Y/N) = Y: Bacteria : Y (6->4)
     Diarrhea (Y/N) = Y:
        Drainage (Y/N) = Y: Bacteria : N (14->13)
        Drainage (Y/N) = N:
           Urethral catheter(Y/N) = N:
              CVP/IVH (Y/N) = Y: Bacteria : Y (17->12)
              CVP/IVH (Y/N) = N:
                 Anti-biotics (Y/N) = N: Bacteria : Y (15->8)
                 Anti-biotics (Y/N) = Y: Bacteria : N (15->9)
           Urethral catheter(Y/N) = Y:
              Anti-biotics (Y/N) = N:
                 Surgery (Y/N) = N: Bacteria : N (5->5)
                 Surgery (Y/N) = Y: Bacteria : Y (3->2)
              Anti-biotics (Y/N) = Y:
                 CVP/IVH (Y/N) = N: Bacteria : N (2->2)
                 CVP/IVH (Y/N) = Y: Bacteria : Y (12->7)
  Tracheotomy (Y/N) = Y:
     Surgery (Y/N) = N: Bacteria : N (19->16)
     Surgery (Y/N) = Y:
        Diarrhea (Y/N) = Y: Bacteria : N (2->2)
        Diarrhea (Y/N) = N:
           Drainage (Y/N) = Y: Bacteria : Y (13->12)
           Drainage (Y/N) = N:
              Anti-biotics (Y/N) = Y: Bacteria : N (3->3)
              Anti-biotics (Y/N) = N:
                 CVP/IVH (Y/N) = N: Bacteria : N (2->2)
                 CVP/IVH (Y/N) = Y: Bacteria : Y (5->5)
```

Fig 17.5. Sub-tree on lactobacillus therapy(Y/N) = Y

decision tree is bacteria(Y/N). The first node of the decision tree is lactobacillus therapy(Y/N). Therefore, lactobacillus therapy might be the most significant factor for prevention of blood stream infection. In the sub-tree on lactobacillus therapy(Y/N) = Y (Figure 17.4), the second branch is diarrhea(Y/N), and the third branch is catheter(Y/N).

On the other hand, in the sub-tree on lactobacillus therapy(Y/N) = N (Figure 17.5), the second branch is tracheotomy(Y/N), and the third branch is diarrhea(Y/N) or surgery(Y/N). The decision tree showed that bacteria(Y/N) have the strong relationship with lactobacillus therapy(Y/N), diarrhea(Y/N), catheter(Y/N) and tracheotomy(Y/N), etc.

If-then Rules from the Decision Tree

The following significant "If-then rules" were extracted from the above decision tree between the bacteria detection(Y/N) and the various factors.

The following significant "If-then rules" were extracted from the sub-tree.

```
If-then rule -1:
If Lactobacillus therapy (Y/N) = Y and
      Diarrhea (Y/N) = N and Catheter (Y/N) = Y and
      Surgery (Y/N) = Y and CVP/IVH (Y/N) = Y,
then Bacteria = N. (1.00 = 12/12)
```

If-then rule-1 showed that lactobacillus therapy presence might prevent bacteria detection from blood sample when patient has not diarrhea and has central venous pressure (CVPjcatheter and intravenous hyper-alimentation (IVHjcatheter after the surgery.

```
If-then rule -2:
If Lactobacillus therapy(Y/N) = Y and
Diarrhea (Y/N) = Y and Catheter (Y/N) = Y,
then Bacteria = N. (0.84 = 63/75)
```

It was considered that lactobacillus therapy presence might prevent bacteria detection from blood sample when patient has diarrhea and catheter inserted into the blood vessel. That is, even though patient has diarrhea, lactobacillus therapy might protect patient's normal bacterial flora.

```
If-then rule -3:
If Lactobacillus therapy (Y/N) = Y and
      Diarrhea (Y/N) = Y and Catheter (Y/N) = N and
      Antibiotics (Y/N) = Y,
then Bacteria = Y. (0.64 = 9/14)
```

If-then rule-3 showed that lactobacillus therapy presence might not prevent bacteria detection from blood sample when patient has diarrhea and has no catheter and anti-biotics. When patient might have diarrhea by anti-biotics, lactobacillus therapy could not protect patient's normal bacterial flora.

```
If-then rule -4:
If Lactobacillus therapy (Y/N) = N and
      Tracheotomy (Y/N) = Y and Surgery (Y/N) = Y and
      Diarrhea (Y/N) = N and Drainage (Y/N) = Y,
then Bacteria = Y.
(Confidence:0.92 = 12/13)
```

If-then rule-4 shows that lactobacillus therapy absence might not prevent bacteria detection from blood sample when patient has tracheotomy, no diarrhea, central venous pressure (CVPjcatheter, intravenous hyper-alimentation (IVHjcatheter and drainage after the surgery.

```
If-then rule -5:
If Lactobacillus therapy (Y/N) = N and
      Tracheotomy (Y/N) = Y and Surgery (Y/N) = Y and
      Diarrhea (Y/N) = N and CVP/IVH (Y/N) = Y and
      Anti-biotics (Y/N) = N,
then Bacteria = Y.
  (Confidence1.00 = 5/5)
```

It was considered that lactobacillus therapy absence might not prevent bacteria detection from blood sample when patient has tracheotomy, no diarrhea, central venous pressure (CVPjcatheter, intravenous hyper-alimentation (IVHjcatheter and no anti

```
If-then rule -6:
If Lactobacillus therapy (Y/N) = N
      and Tracheotomy (Y/N) = N
      and Diarrhea (Y/N) = N and
Antibiotics (Y/N) = Y,
then Bacteria = N.
(Confidence: 0.83 = 428/356)
```

If-then rule-6 shows that bacteria detection from blood sample might be prevented by antibiotics when patient has lactobacillus therapy absence, no tracheotomy and no diarrhea.

From these rules, there might be the strong relationship between treatment and bacteria detection from blood samples in case of lactobacillus therapy absence.

17.5.3 Discussion

We had an empirical rule that lactobacillus therapy (probiotic product) is effective in patient prognosis. Currently, lactobacillus preparation is used in the most departments of our center. This analysis was conducted to extract background risk factors of blood stream infection in a year data of 2002, by chi-square test, decision tree, If-then rules and logistic regression. Anti-biotics preparation has antibiotic properties, but it tends to get off balance of the normal bacteria flora and to cause diarrhea. On the other hand, lactobacillus therapy regulates the functions of the intestines and has no side-effects. From the results of chi-square test (Table 17.2), its p-value was 0.000000159 ¡ 0.01. The odds ratio of lactobacillus absence to lactobacillus presence showed that bacteria detection risk of lactobacillus absence was about 2 (95On the other hand, the first node of the decision tree was lactobacillus therapy(Y/N). Therefore, lactobacillus therapy might be the most significant factor for prevention of blood stream infection. Various significant If-then rules were extracted from the decision tree. From (If-then rule-1), lactobacillus therapy presence might prevent bacterial translocation when patient has not diarrhea and has central venous pressure (CVPjcatheter and intravenous hyper-alimentation (IVHjcatheter after the surgery. From (If-then rule-2), it was considered that lactobacillus therapy presence might protect patient's normal bacteria flora and might prevent bacterial translocation from the intestinal tract even though patient has diarrhea. Furthermore, (If-then rule-4) and (If-then rule-5) showed that tracheotomy might caused bacteria detection from blood sample when patient has intravenous hyper-alimentation (IVHjcatheter on lactobacillus therapy absence. As the above mentioned, it was reported that bacterial translocation might be caused by antibiotics administration and intravenous hyper-alimentationiIVH). Patient, who has tracheotomy, could not almost swallow down. Furthermore, when the patient has also intravenous hyperalimentation (IVHjcatheter, bacteria in patient's oral cavity might increased abnormally and the patient's intestinal tract might lost its functions. Therefore, bacterial translocation from oral cavity or intestinal tract might be yielded.

17.6 Conclusion

Since all the clinical information have been stored electronically as a hospital information system (HIS), it is now expected that mining such combined data will give a new insight to medical accidents.

In order to utilize information about risk extracted from information systems, we propose risk mining which integrates the following three important process: risk detection, risk clarification and risk utilization. Risk Detection discovers patterns or information unexpected to domain experts, which can be viewed as a sign of large scale accidents. In risk clarification, domain experts and data miners construct the model of the hidden mechanism of risk, focusing on detected risk information. If domain experts need more information with finer granularity, we should collect more data with detailed information, and apply data mining to newly collected data. Risk utilization evaluated clarified risk information in a real world environment to prevent risk events. If risk information is not enough to prevention, then more analysis is required. Thus, additional data collection is evoked for a new cycle of risk mining process. We applied this process to the following two medical domains: risk aversion of nurse incidents, infection control. The results show that data mining methods were effective to detection of risk factors.

References

1. Tsumoto, S.: Knowledge discovery in clinical databases and evaluation of discovered knowledge in outpatient clinic. Information Sciences (124) (2000) 125–137
2. Tsumoto, S.: G5: Data mining in medicine. In Kloesgen, W., Zytkow, J., eds.: Handbook of Data Mining and Knowledge Discovery. Oxford University Press, Oxford (2001) 798–807
3. Quinlan, J.: C4.5 - Programs for Machine Learning. Morgan Kaufmann, Palo Alto (1993)

Evaluating the Error Risk of Email Filters Based on ROC Curve Analysis

Wenbin Li[1,3], Ning Zhong[1,2] and Chunnian Liu[1]

[1] The International WIC Institute, Beijing University of Technology Beijing 100022, P. R. China
   ai@bjut.edu.cn
[2] Department of Information Engineering, Maebashi Institute of Technology
   460-1 Kamisadori-Cho, Maebashi-City 371-0816, Japan
   zhong@maebashi-it.ac.jp
[3] Department of Information Engineering, Shijiazhuang University of Economics
   Shijiazhuang, Hebei Province 050031, P. R. China
   oh_my_sun2003@hotmail.com

Summary. Email filtering is a cost-sensitive task, because missing a legitimate message is more harmful than the opposite error. Therefore, how to evaluate the error risk of a filter which is trained from a given labeled dataset is significant for this task. This paper surveys the researches on the Receiver Operation Characteristic (ROC) curve analysis. And, with the experimental results of four compared filters on four public available corpus, we discuss how to use the techniques of ROC curve analysis to evaluate the risk of email filters. In our view, this work is useful for designing a bread-and-butter filter.

18.1 Introduction

Email is one of the most successful computer applications yet devised. As email becomes a more prevalent salesmanship of e-business, dealing with **spam** is becoming more and more costly and time consuming. There is now a great deal of interest in email clients that allow a technically naive user to easily construct a personalized client for filtering spam.

Although the filtering task seems to be a simple instance of the more general text categorization task, it shows an obvious characteristic, e.g., the spam filtering errors are not of equal importance. Individuals usually prefer a conservative filter that tends to classify spam as legitimate, because missing a legitimate message is more harmful than the opposite [8,10]. A cost model is imperative to avoid the risk of missing legitimate emails. In our view, an ideal filter is the one which never *classifies legitimate as spam* (i.e., **false negative error**) and hardly ever makes the *inverse errors* (i.e., **false positive error**). However, how to judge whether a filter has such a characteristic is a difficult task.

W. Li et al.: *Evaluating the Error Risk of Email Filters Based on ROC Curve Analysis*, Studies in Computational Intelligence (SCI) **123**, 299–314 (2008)
www.springerlink.com

In tradition, *precision* and *recall* are used to evaluate the performance of a filter [11]. While, the two criteria cannot reflect the different cost of the false positive and false negative errors. That is to say, a filter with a higher macro precision (or recall) may not be a better one. Therefore, other criteria or methods needs to be developed to bridge this gap [2,9]. In this chapter, we introduce the technique of Receiver Operation Characteristic (ROC) curve analysis to meet such needs [7]. ROC curve analysis is a statistical method which combines *sensitivity* and *specificity* to appraise the practicability of a filter. We survey such technique, and discuss how to use it to evaluate the risk of email filters. In our view, this work is useful for designing a bread-and-butter filter.

18.2 Email Filtering: A Challenging Task

18.2.1 Formal Description

Email messages can be modeled as semi-structured documents that consist of a set of classes and a number of variable length free-text. Thus, many text mining techniques, especially Automated Text Categorization (ATC), can be used to develop an email filtering and management system. Under the framework of ATC, email filtering is viewed as a 2-class categorization task. Two kinds of errors will occur when a filter labels new emails, i.e., the false positive error and the false negative error. The former is mislabeling a legitimate email as a spam and the latter is mislabeling a spam as a legitimate email. The costs of the two types of errors are different. Following the definition of ATC in [16], we give a formal definition of the Automated Cost-Sensitive Email Filtering problem.

Suppose $D = \{d_1, d_2, ..., d_{|D|}\}$ is the training set and $C = \{c_0 = \text{``legitimate''},$ $c_1 = \text{``spam''}\}$ is the set of their classes or categories, where $|D|$ denotes the cardinality of D. Each email $d_i \in D$ belongs only to one of the email categories. Formally, this can be expressed as a function $\phi : D \times C \longrightarrow \{true, false\}$, where $D \times C$ is the Cartesian product of D and C. The function ϕ assigns $true$ to (d_i, c_j) if c_j is the real category of d_i and $false$ otherwise. The key task of email filtering is to learn the classification function ϕ. In general, without any constraints on the form and properties of the classification function ϕ, this is almost an impossible task. In what follows, we use a learning algorithm ℓ to obtain an approximation $\hbar$: $D \times C \longrightarrow \{true, false\}$ of the unknown target function ϕ. The function $\hbar$ is called a $filter$ and should be as close to ϕ as possible.

Unlike the cost-insensitive classifiers which minimize zero-one loss or an error rate, cost-sensitive email filters choose the class that minimizes the expected cost of a prediction as given by [5]:

$$c(c_i|\overrightarrow{x}) = \sum_{j=0}^{1} P(c_j|\overrightarrow{x})c(i,j) \tag{18.1}$$

where $\overrightarrow{x}$ is the vector representation of an email, $c(i,j)$ ($i,j \in \{0,1\}$) denotes the cost of classifying an email of c_i into c_j, $P(c_j|\overrightarrow{x})$ is the conditional probability that

$\overrightarrow{x}$ belongs to c_j, and $c(c_i|\overrightarrow{x})$ is the expected cost predicting $\overrightarrow{x}$ to c_i. A filter $\hbar$ is obtained by using Bayes optimal prediction, which guarantees that it achieves the lowest possible overall cost.

Cost-sensitive email filtering can be defined as a task that learns an approximate classification function $\hbar : D \times C \longrightarrow \{true, false\}$ from full or partial data in D with a cost-sensitive algorithm ℓ_c, and it minimizes the expected cost of prediction. The main issues for building an email filter include training set preparation, email representation, feature selection, the filtering model learning with ℓ_c and filter evaluation.

Email Representation and Feature Selection

As a prerequisite for building a filter, one must represent each message so that it can be accepted by a learning algorithm ℓ_c. The commonly used representation method is the term/feature weight vector in the Vector Space Model (VSM) in information retrieval [15]. Suppose $V = \{f_1, f_2, ..., f_{|V|}\}$ is the set of vocabulary which consists of features (i.e., words or phrases) appeared in D, where $|V|$ is the size of V. A vector representation of an email is defined as a real-value vector $\overrightarrow{x} \in \Re^{|V|}$, where each component x_j (also called weight) is statistically related to the occurrence of the jth vocabulary entry in the email. The value of x_j can be computed based on two types of frequencies: the absolute feature frequency and the relative feature frequency. The absolute frequency is simply the count of f_j appearing in the email. Based on absolute frequency, one can easily obtain a binary weight scheme. That is, $x_j \in \{0, 1\}$ simply indicates the absence or the presence of feature f_j in the email. A very popular weighting scheme is the so-called $tf \times idf$ weighting defined by:

$$\chi_j = tf_j \cdot log_2(\frac{N}{df}) \tag{18.2}$$

where tf_j is the number that f_j occurs in the email, N is the number of training emails, and df is the number of training emails in which f_j occurs.

For a moderate-sized email test collection, the size of V will reach tens or hundreds of thousands. The size of V is prohibitively high for many learning algorithms, such as Support Vector Machine (SVM), Artificial Neural Network (ANN), Decision Tree (DT) and so on. A method that automatically reduces the dimensions without sacrificing filtering accuracy is highly desirable. This process is called feature selection (FS for short). In general, FS mainly includes two steps. The first step is to calculate feature weight for each entry in V with an FS function φ, and the second step is to rank all features and extract the top M (in general $M \ll |V|$) features from V. Formally, FS can be defined as a mapping process from V to V' by using φ, i.e., $V \xrightarrow{\varphi} V'$, where $V' \subset V$, V' is a substitute of V.

Many functions are available for completing the above mapping process, such as Information Gain (IG), χ^2-test (CHI), Document Frequency (DF), Mutual Information (MI), Term Strength (TS), Odds Ratio (OdR) and so on [17]. By comparing the effectiveness of the five functions in the context of general ATC tasks, the reported experimental results show that IG, CHI and DF are more effective than MI and TS.

18.3 The Groundwork of ROC Analysis

It is well-known that the filtering results depend on two factors: the unlabeled email e (whose represented vector is $\overrightarrow{x}$) and the filter $\hbar$. The two factors will result in two types errors, i.e., the false positive error and the false negative error. Suppose that $p = \hbar(c_1|e)$ is the posterior probability that e belongs to the spam class. In general, if $p >= \alpha$ ($\alpha = 0.5$), then $\hbar$ labels e as a spam. In this occasion, the filter equally views legitimate messages and spam. Thus, $\hbar$ will make above two kinds of errors with an equal probability. It is obvious that users would not accept such a filter. In order to reduce the chance making the false negative error, some researchers suggest that we should set the α to a higher value [3]. However, on the one hand, it is hard to determine the value of α. On the other hand, $\hbar$ inevitably makes those two kinds of errors, even that we set the α to 1.

Table 18.1 is the contingency table which provides a convenient display of the prediction behavior of $\hbar$. For example, TP and TN are the count denoting how often the prediction is correct. FN is the times of the false negative error of the filter. TN is the times of the false positive error of the filter.

When the α is set to different values, TP, FP, FN, and TN will change at the same time, as shown in Figure 18.1(a). Thus, the *sensitivity* (see Definition 1) and the *specificity* (see Definition 2) of $\hbar$ will also change [7]. Suppose the α is changed from α_1 to α_2, we will get many pairs $< se, sp >$. Furthermore, we can use the points (1-sp, se) of these pairs to draw a curve of ROC. Figure 18.1(b) provides an example of such a curve.

Table 18.1. Contingency Table

	Actual Legitimate	Actual Spam	Total
Predict Legitimate	TP	FP	TP + FP
Predict Spam	FN	TN	FN + TN
Total	TP + FN	FP + TN	TP + FP + FN +TN

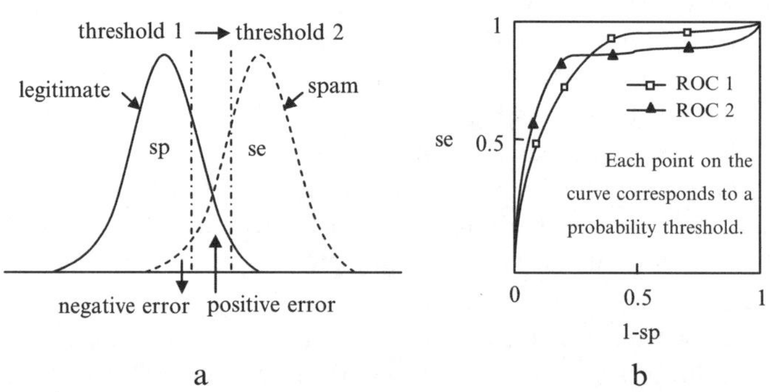

Fig 18.1. The relationship between sensitivity and specificity, and ROC curves.

Definition 1. *(Sensitivity) Sensitivity is defined as follows.*

$$se = TP/(TP + FN). \tag{18.3}$$

Definition 2. *(Specificity) Specificity is defined as follows.*

$$sp = TN/(FP + TN). \tag{18.4}$$

According to the above two definitions, it seems that more higher se a filter has, more better the filter is. However, this modus operandi is dogmatical. First, when α is set to the different value, se will be changed. That is, comparing a filter with another one at a fixed α is unilateral. Second, if the first filter has a higher se, and another filter has a higher sp, it is difficult to compare them with others. Third, the above two definitions do not consider the cost of different errors. Although, there are some cost-sensitive evaluation criteria, such as *slope* [13], *TEC* [9], *WErr* [2], it is difficult to estimate the cost matrix, in the filtering task, until the application time. Therefore, in this work, we describe how to use the ROC analysis technique to evaluate the error risk of email filters.

ROC analysis is a dynamic method which contrasts the ROC curve of a filter with another one's at multiple and different α.

18.4 Traditional Criteria for Evaluating Email Filters

Based on the data given in Table 18.1, except *sensitivity* (see Definition 1) and *specificity* (see Definition 2), many cost-insensitive metrics can be defined: (1) the false negative rate, $FNR = FN/(TP + FN)$; (2) the false positive rate, $FPR = FP/(TN + FP)$; (3) precision, $Precision = TP/(TP + FP)$; (4) recall, $Recall = sensitivity$; (5) break-even, $Bre = (Precision + Recall)/2$; (6) F_1 measure, $F_1 = (Precision * Recall)/Bre$.

In some degree, the criteria reflect the frequency of that a filter makes errors. However, the important thing for email filtering tasks is not to obtain the "filter" with fewer errors but with the lowest cost. And, TNR, FPR, $Precision$ etc. cannot be used for the aim of a cost-sensitive evaluation. Therefore, the metric *slope* (see Definition 3) is used. For the task of email filtering, the value "slope" is sufficient to determine which filter is the best. For example, as shown in Figure 18.2, we can compute FNR, FPR and sl of $filter1$ and $filter2$, respectively. For $filter1$, $FNR = 2/(300 + 2) = 0.66\%$, $FPR = 20/(20 + 50) = 28.57\%$, $sl = (2000 * 70)/(4000 * 302) = 0.116$. For $filter2$, $FNR = 1/(200 + 1) = 0.5\%$, $FPR = 30/(30 + 40) = 42.86\%$, $sl = (3000 * 70)/(2000 * 201) = 0.522$. Furthermore, we can work out the cost per unit for each filter. The value of $filter1$ is 1*0.66%+0.116*28.57%=0.0397, and the value of $Filter2$ is 1*0.5%+0.522*42.86%=0.2287. Therefore, $filter1$ is better than $filter2$.

Although the metric *slope* is sufficient for the cost-sensitive evaluation. Unfortunately, as mentioned above, it is difficult to estimate the cost matrix. Below, we describe how to use the ROC analysis method to solve this problem.

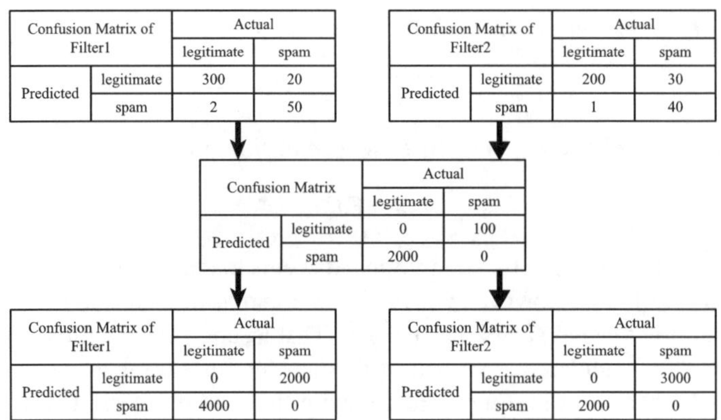

Fig 18.2. An example of evaluating email filters.

Definition 3. *(Slope) Slope is defined as follows.*

$$sl = FP * c_{NP} * (FP + TN)/\{FN * c_{PN} * (TP + FN)\} \tag{18.5}$$

where c_{NP} is the cost of the false positive error, and c_{PN} is the cost of the false negative error. TP, FP, FN and TN are shown in Table 18.1, respectively.

18.5 Applying ROC Analysis to Evaluate Email Filters

In practice, it is often required to select a best filter (or some better filters) from multiple ones, or to set optimal running parameters for a filter. In this section, we depict how to reach these two goals, respectively.

18.5.1 The Metrics for ROC Curves

There are three evaluation criteria of ROC curve: (1) sensitivity and specificity, (2) the area under the ROC curve (AUC), and (3) the part area under the ROC curve. Among those three metrics, AUC is the most common one. As mentioned in Section 2, the performance of each filter can be denoted by a ROC curve. The AUC reflects the overlapped degree of the distribution of positive and negative classification results. A filter with a greater AUC is better. In general, AUC should be in the range (0.5, 1].

AUC is the average value of se when sp changes from 0 to 1. When a ROC curve intersects with another one at a point β (see Figure 18.1(b), $\beta \in (0, 1)$). AUC cannot be used to judge which filter is better. To solve this problem, the part area under the ROC curve is introduced. For example, as shown in Figure 18.1(b), in the range (β, 1], the filter of ROC1 is better than the filter of ROC2. While, in the range [0, β), the filter of ROC2 is better.

18.5.2 Selecting a Better one from Multiple Filters

Given several cost-insensitive filters which do not require the cost matrix in the training and evaluation phase, and a fixed threshold. And these filters predict a class among {"legitimate", "spam"} without a confidence value, we call these filters as *crisp filters*. We train each filter by using a training dataset at first. Then, we use these filters to classify the messages of a testing dataset. After that, we can get each pair (1-*sp*, *se*) of each filter. In a same ROC space, we construct the convex hull of those points (1-*sp*, *se*) as well as the two trivial filters (0, 0) and (1, 1). We will find that some filters are below the ROC curve. Because there is no combination of a class distribution and a cost matrix for which these filters could be optimal, we discard them. In the application time, after the user sets the cost matrix, we can choose the best filter according to the cost per unit of each filter from those remaining to use.

Based on the above idea, we propose a method of dynamic combining multiple filters. The goal of combining multiple filters is to obtain a more accurate prediction than obtained from any single source alone. Richard Segal, et al. pointed out that "We believe that *no one anti-spam solution is the "right" answer*, and that *the best approach is a multifaceted one*, combining various forms of filtering ..." [14].

Suppose that there are m filters, they are $\hbar_1, ..., \hbar_m$. Thus, there are 2^m combination solutions. Furthermore, suppose each combiner is denoted by $\Psi_1, ..., \Psi_{(2^m)}$, respectively. If these combiners are viewed as a single filter, the method described in the first paragraph of this section can be used to select the best solution when the cost matrix is given in the application time. When the cost matrix is changed, the choice will be changed at the same time. However, when m is large, the training process of this method is very time-consuming.

18.5.3 Constructing a ROC Curve for Soft Filters

A "soft" filter predicts a class with an estimation of the reliability. In fact, most of learning methods can be adapted to generate this confidence value. When such a filter classifies an unlabeled email, we often use a threshold to convert this filter to a crisp one. As mentioned in Section 2.1, for example, if $p(spam|e) > 0.99$, then e is labeled as a spam. Here, $p(spam|e)$ is the posterior probability of that e belongs to a spam, which is predicted by the filter.

Thus, with different thresholds, we will get "different" filters, which give more or less relevance to the classes "legitimate" and "spam", for a learning method. Each filter has a performance pair, i.e., (1-*sp*, *se*). If all these points are drawn in a same ROC space, a ROC curve will be got for this learning method. On the one hand, this curve can be used to select an optimal threshold for this method. On the other hand, by comparing this curve with other curves got for different learning methods, we can know which method is the best at a specified threshold according to the part area under ROC curves.

For some cost-sensitive methods, we often are confused by how to set a cost matrix. That is, what cost of false positive errors and false negative errors should be adopted to obtain the best performance of a cost-sensitive method. To implement this goal, we can get its performance pairs (1-sp, se) at different cost combinations for a method. After that, its ROC curve can be drawn. According to this curve, as mentioned in Section 4.2, a best filter can be chosen. Furthermore, we can know which cost solution results in this filter. When a best one needs to be selected from multiple methods, we can compare their curves.

18.6 Experimental Results

18.6.1 Data Corpus

In general, a user unwillingly releases his/her own legitimate emails because of his/her own and/or senders' privacy issue. Thus, collecting benchmark datasets for the tasks of spam filtering is more difficult than collecting datasets for ATC tasks. Fortunately, researchers provide some public available datasets such as PU1, Ling-Spam, SpamAssassin, Spambase, TREC05 and so on. Among them, PU1[1], Ling-Spam[2], SpamAssassin[3], Spambase[4] are used in this work.

PU1 consists of 481 "real" legitimate emails, and 618 spams. Header fields except subject and html tags of messages in PU1 are removed. In addition, in order to bypass privacy issues, each token was mapped into a unique integer. PUA, PU2, PU3, together with PU1 are called as PU collection. Although the size of these four datasets is different, the former three ones still are viewed as other versions of PU1 because they just adopted a different pre-processing method with PU1.

Ling-Spam consists of 481 spam, and 2412 legitimate messages retrieved from the archives of a mail list. According to its providers, legitimate messages in Ling-Spam are more topic-specific than the legitimate messages most users receive. Therefore, the performance of a learning-based anti-spam filter on Ling-Spam may be an over-optimistic estimate of the performance that can be achieved on the incoming messages of a real user, where the topic-specific terminology may be less dominant among legitimate messages [1]. However, Ling-Spam is the currently best (although not perfect) candidate for the evaluation of spam filtering [1]. Like the PU collection, Ling-Spam also has other three versions.

SpamAssassin contains 6047 messages, 4150 of which are marked as legitimate and 1897 of which are spams. Legitimate emails in SpamAssassin are messages collected from BBS or real emails donated by personal users. Androutsopoulos and colleagues [1] claimed that the performance of a learning-based filter on the SpamAssassin may be an under-estimate of the performance that a personal filter can achieve.

Spambase only distributes information about each message rather than the messages themselves for avoiding privacy issues. With 57 pre-selected features, each real email was represented by a vector. This corpus contains 4601 vectors about emails.

---

[1] http://www.iit.demokritos.gr/~ionandr/publications/

[2] http://www.iit.demokritos.gr/~ionandr/publications/

[3] http://spamassassin.org/publiccorpus/

[4] http://www.ics.uci.edu/~mlearn/databases/spambase/

Among these vectors, 2788 ones are about legitimate messages and 1813 are about spam. It is obvious that Spambase is much more restrictive than the above three corpus. On the one hand, its messages are not available in the raw form; on the other hand, it is impossible to experiment with features other than those chosen by its creators.

18.6.2 The Compared Algorithms

Naive Bayes (NB). A NB filter assumes that an email is generated by a mixture model with parameters θ, consisting of components $C = \{c_0, c_1\}$ that corresponds to the classes. An email is generated by first selecting a component $c_j \in C$ ($j = 0, 1$) according to the prior distribution $P(c_j|\theta)$, and then choosing an email e_i (whose represented vector is $\vec{x_i}$) according to the parameters of c_j with a distribution $P(\vec{x_i}|c_j; \theta)$. The likelihood of an email is given by the total probability:

$$p(\vec{x_i}|\theta) = \sum_{j=0}^{1} p(c_j|\theta)p(\vec{x_i}|c_j; \theta). \tag{18.6}$$

However, the true parameters θ of the mixture model are not known. Therefore, the parameters need to be estimated from the training emails. In this work, we adopt *Multi-variate Bernoulli Model* [12] to estimate θ and compute the probability of an email given a class from the probabilities of the words given the class.

Given a new email e and its vector $\vec{x}$, classification of e is performed by computing the posterior probability of each class by applying Bayes' rule:

$$p(c_j|\vec{x}; \theta) = \frac{p(c_j|\theta)p(\vec{x}|c_j; \theta)}{p(\vec{x}|\theta)}. \tag{18.7}$$

The classifier simply selects the class with the highest posterior probability. Note the $p(\vec{x}|\theta)$ is the same for all classes, thus e can be classified by computing:

$$c_{\vec{x}} = \arg \max_{c_j \in C}[p(c_j|\theta)p(\vec{x}|c_j; \theta)]. \tag{18.8}$$

k-NN. The idea of k-NN is very easy. Suppose D is a dataset, and x is a new sample. Furthermore, each example including the new one is represented as a vector. In order to label x, the k-NN algorithm first computes the distance (in general, euclidian distance) between x and each training example. Then, it sorts all the training examples in descend order according to the computed distances. Among top k examples, if the examples with c_i(i=1, 2, ..., $|C|$) class are maximum, then k-NN outputs c_i as x's label.

Bagging. Bagging is introduced by Breiman (1996). "Bagging" stands for "Boostrap **aggregating**". It is an ensemble method, i.e., a method of combining multiple predictors. Let D be a training set, in the training phase, the Bagging algorithm repeats to do the following two things for T times: (1) Getting a boostrap sample D_k from D; (2) Training a predictor using D_k.

The training phase gets T predictors. Then, in the usage phase, the Bagging algorithm combines T predictors to label a new example. The combining methods include Voting (for classification problem) and Averaging (for estimation problem).

AdaBoost. Algorithm 1 shows the generalized AdaBoost algorithm. From the description about Bagging and Algorithm 1, we can see that the most difference between the two methods is the way of selecting samples in each round. Another difference is the means for combining multiple predictors.

Algorithm 1: Generalized AdaBoost Algorithm

Data: $\{(x_1, y_1), ..., (x_m, y_m)\}$ //m is the size of training dataset; $x_i \in X$,
    $y_i \in \{-1, +1\}$
Process:
Initialize $D_1(i)=1/m$;
for *t=1, ..., T* **do**
    Train weak learner using distribution D_t;
    Get weak hypothesis $h_t : X \to \Re$
    Choose $\alpha_t \in \Re$.
    Update
$$D_(t+1)(i) = \frac{D_t(i)exp(-\alpha_t y_i h_t(x_i))}{Z_t}$$
    where Z_t is a normalization factor (chosen so that $D_(t+1)$ will be a distribution)
end
Output the final hypothesis:
$$H(x) = sign(\textstyle\sum_{t=1}^{T} \alpha_t h_t(x)).$$

18.6.3 Traditional Methods for Evaluating the Four Filters

This section uses some traditional methods to evaluate the compared algorithms. They are *precision* (see Eq. 18.9), *recall* (see Eq. 18.10) and *TCR* (see Eq. 18.11). Note that although we just show *precision* and *recall* with respect to c_1, the definitions with respect to c_0 are similar as the ones of c_1.

$$precision(c_1) = \frac{TN}{TN + FP} \tag{18.9}$$

where TN is the number of spam that is labeled correctly, and FP is the number of false positive errors.

$$recall(c_1) = \frac{TN}{TN + FN} \tag{18.10}$$

where FN is the number of false negative errors.

$$TCR = \frac{N_s}{\lambda \cdot FP + FN} \tag{18.11}$$

where, FP and FN are the numbers of the false positive and false negative errors; N_s denotes the total number of training spam, λ denotes that a false positive error is λ times more costly than a false negative error.

18.6.4 Experiment Setup

We conduct the compared experiments on the four corpus as mentioned above. For simplification, D_1 is used to denote the PU1 dataset, D_2 is used to denote the Lingspam dataset, D_3 s used to denote SpamAssasin, and D_4 is used to denote Spambase. The aims of the experiments are that firstly we want to compare some filtering algorithms using tradition evaluation methods. Furthermore, we want to know the disadvantages of those methods. Then, we can compare those methods with ROC one. In addition, we want to compare the datasets.

The parameters of our experiments are as follows:

- The compared algorithms are NB, k-NN, Bagging and AdaBoostM1;
- Our experiments are implemented based on Weka package;
- The feature selection method is Information Gain (IG);
- The size of vector space is 150;
- The test method is 10-cross validation;
- k in k-NN is set to 15;
- The times of round in Bagging is 10, weak learner in Bagging is C4.5;
- The times of round in AdaBoostM1 is 10, weak learner in it is C4.5;
- λ in TCR is set to be 9.

18.6.5 The Results by Using the Traditional Evaluating Methods

Figure 18.3 shows precisions about c_0 and c_1 of the compared algorithms on D_1, D_2, D_3 and D_4. The higher $precision(c_0)$ denotes that a filter classifies legitimate emails more accuracy. Figure 18.3(a) shows that AdaBoostM1 gets the highest $precision(c_0)$ on all the corpus. On D_1, D_2 and D_3, the values of Bagging are larger than other algorithms'. Except D_3, NB gets better performance than k-NN. Roughly, based on Figure 18.3(a), we can get the following partial order: AdaBoostM1 $\succeq$ Bagging $\succeq$ NB $\succeq$ k-NN just according to $precison(c_0)$ criterion. (Note: Suppose x and y are filters, $x \succeq y$ denotes that x has better performance than y). Figure 18.3(b) gives $precision(c_1)$ of the four methods. The more higher $precision(c_1)$ denotes that a filter classifies spam more actuary. On all the datasets, based on Figure 18.3(b), we can see that firstly AdaBoostM1 gets better performance than Bagging. Secondly, their $precision(c_1)$s on each dataset is higher than 92%. NB and k-NN shows unstable performance on the datasets. Concretely, NB gets $precision(c_1)$s more than 95% on D_1 and D_2. However, on D_3 and D_4, its effect is not good enough. On the first two datasets, k-NN has the highest values than any other algorithms'. And also, it gets 97.2% on D_3 that is very close to AdaBoostM1's. While, it just shows 88.9% on D_4. Thus, just according to $precision(c_1)$, we can roughly sort the four algorithms: AdaBoostM1, k-NN $\succeq$ Bagging $\succeq$ NB.

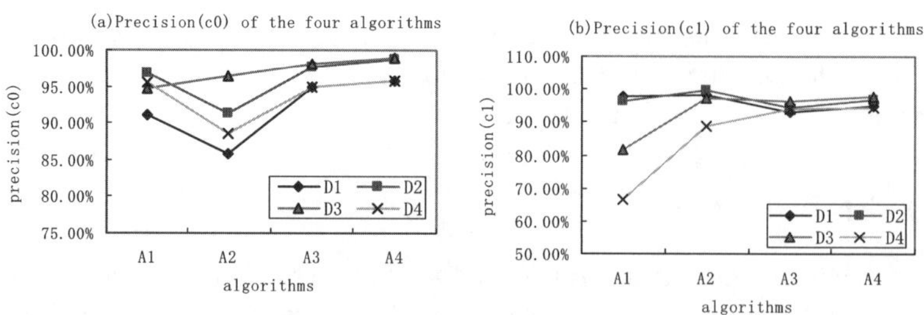

Fig 18.3. The precisions of the compared algorithms. In the figure, A1 denotes NB; A2 denotes k-NN; A3 denotes Bagging; A4 denotes AdaBoostM1.

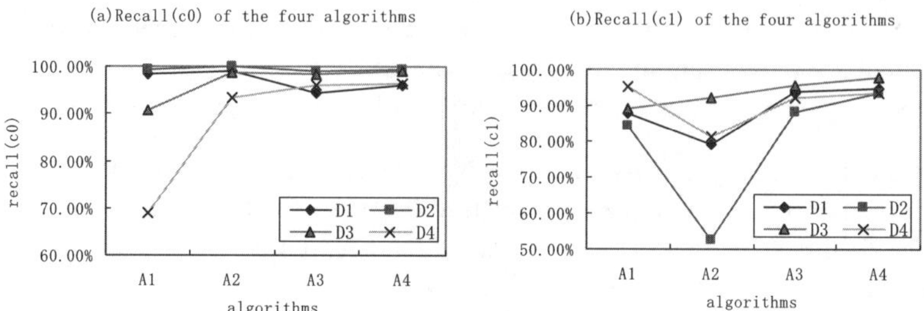

Fig 18.4. The recalls of the compared algorithms. In the figure, A1 denotes NB; A2 denotes k-NN; A3 denotes Bagging; A4 denotes AdaBoostM1.

Figure 18.4 shows the $recall(c_0)$ and $recall(c_1)$ of the compared algorithms on four corpus individually. More higher $recall(c_0)$ denotes, fewer false positive errors. On all the datasets, according to Figure 18.4(a), all the $recall(c_0)$s of AdaBoostM1, Bagging, k-NN are higher than 93%. Except D_4, NB also gets $recall(c_0)$ that is greater than 90%. However, NB just gets 69% $recall(c_0)$ on D_4. It shows that the performance of NB greatly depends on the training and testing dataset. Thus, just according to $recall(c_0)$, we can roughly sort the four algorithms: AdaBoostM1, k-NN $\succeq$ Bagging $\succeq$ NB. More higher $recall(c_1)$ denotes fewer false negative errors. Figure 18.4(b) shows that firstly AdaBoostM1 gets more than 93% $recall(c_1)$ on each dataset. Secondly, Bagging also has perfect performance on all the datasets. Thirdly, NB and k-NN shows unstable $recall(c_1)$ on the used corpus. For example, on D_4, the $recall(c_1)$ of NB is 95.1%, while the value is 84.2% on D_2. The same thing occurs with k-NN. On D_3, the $recall(c_1)$ of k-NN is above 92%, while the value on D_2 is only 52.8%. Therefore, just according to the values of $recall(c_1)$, we can sort as follows: AdaBoostM1 $\succeq$ Bagging $\succeq$ NB $\succeq$ k-NN.

In the mass, the four criteria of AdaBoostM1 on all the datasets are above 93%. All Bagging's values are above 88%. Thus, we can draw the first conclusion from the results: the ensemble learning methods have more accuracy than a single filter. The second observation is that NB and k-NN show unstable performance. That is, the quality of training and testing dataset is a fatal factor that affects NB and k-NN's capabilities.

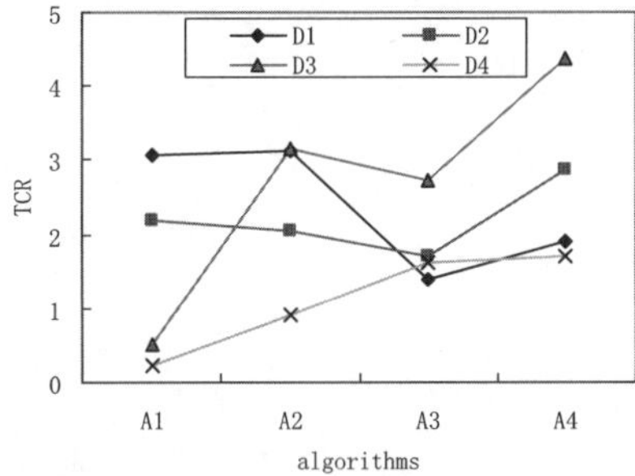

Fig 18.5. The TCRs of the compared algorithms. In the figure, A1 denotes NB; A2 denotes k-NN; A3 denotes Bagging; A4 denotes AdaBoostM1.

Table 18.2. The areas under ROC of the four algorithms on different dataset

	NB	k-NN	Bagging	AdaBoostM1
D_1	0.9919	0.9844	0.9844	0.9896
D_2	0.9971	0.9703	0.9815	0.9946
D_3	0.9491	0.951	0.9927	0.985
D_4	0.9393	0.9537	0.9776	0.9843

Figure 18.5 shows the compared TCR results of the four algorithms. A filter with higher TCR will be more useful in practice. And, an available filter should has TCR value that is greater than 1. On the four corpus, NB's TCR ranges from 0.23 to 3.057. k-NN's TCR is in the range $[0.91, 3.147]$. Bagging's TCR is in $[1.391, 2.72]$. AdaBoostM1's TCR ranges from 1.706 to 4.374. Thus, on the one hand, the same conclusions drawn before can be got too when we compare those algorithms just according to TCR. On the other hand, the third conclusion can be got: the algorithm will show optimistic performance when we adopt D_1 or D_2 as the dataset. While, D_3 and D_4 will result in pessimistic value of criterion.

18.6.6 ROC-based Analysis on the Four Filters

Table 18.2 shows the compared areas under ROC (AUR) of the four algorithms on different dataset. The ROC curves of each algorithm are given in Figures 18.6 and 18.7, respectively. NB shows its highest AUR on D_1 and D_2. However, it shows its worse AUR on other datasets. That is, from the AUR criterion, we can see that NB's performance greatly depends on the used dataset. As mentioned above, D_1 and D_2 could result in optimistic performance. So, we cannot say that NB is a best filter. Except NB, AdaBoostM1 gets highest AUR values on each dataset than other algorithms. Note that, although that D_3 and D_4 are pessimistic datasets, AdaBoostM1 still get above 0.984 AUR. Bagging also shows the closed performance to AdaBoostM1. k-NN shows stable performance on the four datasets. This conclusion is different from the before observation. Roughly, based on Table 18.2, we can sort: AdaBoostM1 $\succeq$ Bagging $\succeq$ k-NN $\succeq$ NB.

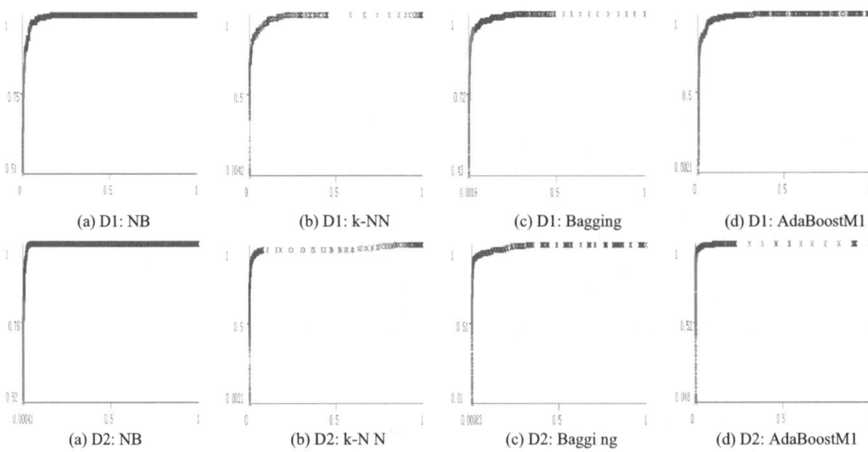

(a) D1: NB   (b) D1: k-NN   (c) D1: Bagging   (d) D1: AdaBoostM1

(a) D2: NB   (b) D2: k-N N   (c) D2: Baggi ng   (d) D2: AdaBoostM1

Fig 18.6. The ROC curves of the compared algorithms on D_1 and D_2.

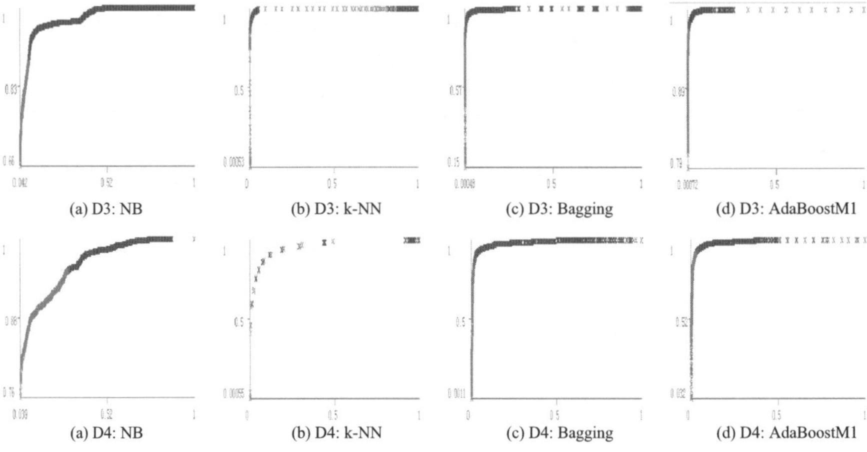

(a) D3: NB   (b) D3: k-NN   (c) D3: Bagging   (d) D3: AdaBoostM1

(a) D4: NB   (b) D4: k-NN   (c) D4: Bagging   (d) D4: AdaBoostM1

Fig 18.7. The ROC curves of the compared algorithms on D_3 and D_4.

Comparing with the traditional evaluation methods, ROC is an easy way to estimate the performance of a filter. Concretely, if you use precision and recall to do that, you need to analyze $precision(c_0)$, $precision(c_1)$, $recall(c_0)$ and $recall(c_1)$. And, these values often show a different (even conflict) partial sort of the compared algorithms. Although TCR is a good replacement of those criteria, λ that affects TCR greatly is very difficult to be set. It is fortunate that all these problems existed in the traditional methods are overcame in the ROC approach.

18.7 Conclusions

Spam filtering is a cost-sensitive task. A filter trained from a given training dataset will make two types errors in the application time. In order to evaluate the error risk of email filters, this chapter introduced a ROC analysis-based method. We provided a systematic survey on ROC researches and discussed how to use the techniques of ROC analysis in email filtering task. However, we did not describe LROC, GROC, SROC, FROC here. With the experimental results of four compared filters on four public available corpus, we discussed how to use the techniques of ROC curve analysis to evaluate the risk of email filters. From the experiments, we found that ROC is useful for designing a bread-and-butter filter.

Acknowledgments

This work is partially supported by the NSFC major research program: "Basic Theory and Core Techniques of Non-Canonical Knowledge" (NO. 60496322), NSFC research program (NO. 60673015), the Open Foundation of Beijing Municipal Key Laboratory of Multimedia and Intelligent Software Technology, the Project (NO. 07213507D and 06213558) of Dept. of Science and Technology of Hebei Province, and the Project (NO. Y200606) of Shijiazhuang University of Economics.

References

1. Androutsopoulos, I., Georgios, P. and Michelakis, E. "Learning to filter unsolicited commercial e-mail". *Technical Report 2004/2*, NCSR Demokritos00, (2004).
2. Androutsopoulos, I., Koutsias, J., Chandrinos, K.V., Paliouras, G. and Spyropoulos, C.D. "An evaluation of naive Bayesian anti-spam filtering". In: *Proc. of the Workshop on Machine Learning in the New Information Age, 11th European Conference on Machine Learning*, (2000) 9-17.
3. Androutsopoulos, I., Koutsias, J., Chandrinos, K.V. and Spyropoulos, C.D. "An experimental comparison of naive Bayesian and keyword-based anti-spam filtering with personal e-mail messages". In: *Proc. of the 23rd ACM SIGIR Conference on Research and Development in Information Retrieval*, (2000) 160-167.
4. Breiman, L. "Bagging predictors". *Machine Learning*, 24(2) (1996) 123-140.
5. Duda, R.O. and Hart, P.E. *Pattern Classification and Scene Analysis.* (1973).
6. Freund, Y. "Boosting a weak algorithm by majority". *Information and Computation*, 121(2) (1995) 256-285.
7. Hanley, J.A and Mcneil, B.J. "The meaning and use of the area under a ROC curve". *Radiology*, (143) (1982) 29-36.
8. Jos, M.G.H., Manuel, M.L. and Enrique, P.S. "Combining text and heuristics for cost-sensitive spam filtering". In: *Proc. of the 2nd Workshop on Learning Language in Logic and the 4th Conference on Computational Natural Language Learning*, (7) (2000) 99-102.
9. Li, W.B., Liu, C.N. and Chen, Y.Y. "Combining multiple email filters of naive Bayes based on GMM". *ACTA ELECTRONICA SINICA*, 34(2) (2006) 247-251.

10. Li, W.B., Zhong, N. and Liu, C.N. "Combining multiple email filters based on multivariate statistical analysis". In: *Proc. of the 15th International Symposium on Methodologies for Intelligent Systems*, (2006) 729-738.
11. Li, W.B., Zhong, N. and Liu, C.N. "Design and implementation of an email classifier". In: *Proc. of International Conference on Active Media Technology*, (2003) 423-430.
12. McCallum, A. and Nigam, K. "A comparison of event models for naive Bayes text classification". In: *Proc. of AAAI-98 Workshop on Learning for Text Categorization*, (1998) 41-48.
13. Peter, A.F. "The many faces of ROC analysis in machine learning". In: *Proc. of The Twenty-First International Conference on Machine Learning*, (2004).
14. Segal, R., Crawford, J., Kephart, J. and Leiba, B. "SpamGuru: an enterprise anti-spam filtering system". In: *Proc. of the First Conference on Email and Anti-Spam*, (2004).
15. Salton, G. *Automatic text processing: the transformation, analysis, and retrieval of information by computer.* (1989).
16. Sebastiani, F. "Machine learning in automated text categorization". *ACM Computing Surveys*, 34(1) (2002) 1-47.
17. Yang, Y. and Pedersen, J.O. "A comparative study on feature selection in text categorization". In: *Proc. of 14th International Conference on Machine Learning*, (1997) 412-420.

19

Categorized and Integrated Data Mining of Medical Data

Akinori Abe[1,2], Norihiro Hagita[1,3], Michiko Furutani[1], Yoshiyuki Furutani[1], and Rumiko Matsuoka[1]

[1] International Research and Educational Institute for Integrated Medical Science (IREIIMS), Tokyo Women's Medical University
8-1 Kawada-cho, Shinjuku-ku, Tokyo 162-8666 Japan
[2] ATR Knowledge Science Laboratories
2-2-2, Hikaridai, Seika-cho, Soraku-gun, Kyoto 619-0288 Japan
[3] ATR Intelligent Robotics and Communication Laboratories
2-2-2, Hikaridai, Seika-cho, Soraku-gun, Kyoto 619-0288 Japan
`{ave, kogure}@atr.jp, ave@ultimaVI.arc.net.my`

Summary. In this chapter, we introduce categorized and integrated data mining. Because of recent rapid progress in medical science as well as medical diagnosis and treatment, integrated and cooperative research among medical researches, biology, engineering, cultural science, and sociology is required. Therefore, we propose a framework called Cyber Integrated Medical Infrastructure (CIMI) which is a framework of integrated management of medical data on computer networks consisting of a database, a knowledge base, and an inference and learning component, which are connected to each other in the network. Within this framework, we can deal with various types of data. Consequently, we need to integratedly analyze various types of data. In this study, for medical science, we analyze the features and relationships among various types of data and show the possibility of categorized and integrated data mining.

19.1 Introduction

Medical science as well as clinical diagnosis and treatment has progressed rapidly in recent years with each field becoming more specialized and independent. As a result, cooperation and communication among researchers in the two fields has decreased which has led to problems between both communities, not only in terms of medical research but also with clinical treatment. Therefore, an integrated and cooperative approach to research between medical researchers and biologists is needed. Furthermore, we are living in a changing and quite complex society, so important knowledge is always being updated and is becoming more complex. Therefore, integrated and cooperative research needs to be extended to include engineering, cultural science, and sociology. As for medical research, the integration of conventional (Western) and unconventional (Eastern) medical research, which should be fundamentally the same but in fact are quite different, has been suggested.

A. Abe et al.: *Categorized and Integrated Data Mining of Medical Data*, Studies in Computational Intelligence (SCI)
123, 315–330 (2008)
`www.springerlink.com`

With this situation in mind, we propose a framework called Cyber Integrated Medical Infrastructure (CIMI) which is a framework of integrated management of clinical data on computer networks consisting of a database, a knowledge base, and an inference and learning component, which are connected to each other in the network. In this framework, medical information (e.g. clinical data) is collected, and analyzed or data mined to build a knowledge base for predicting all possible diseases and to support medical diagnosis.

For medical data mining, several techniques such as Inductive Logic Programming (ILP), statistical methods, decision tree learning, Rough Sets and KeyGraph have been applied (e.g. [Ichise and Numao, 2005, Tsumoto, 2004] and [Ohsawa, 2003]) and acceptable results have been obtained. For data mining, generating plausible or correct results is of course important, but if the results are trivial or well-known, they are not so important for physicians. Thus, the research focus has recently shifted from how to obtain proper and plausible results to how to obtain interesting results. In this context, "interesting" means "interesting for physicians." That is, it is better to discover knowledge which doctors were unaware of or previously ignored and which represents the mechanisms which lead to serious disease. Of course, the generated knowledge should be correct. In fact, in the knowledge discovery field, we focus on discovering not only frequently occurring trends but also rare or novel events. The aims of research in "active mining" [Tsumoto et al., 2005] and "chance discovery" [Ohsawa and McBurney, 2003] are to establish techniques or procedures for discovering interesting, rare or novel knowledge or events. Previously, we applied C4.5 [Quinlan, 1993] to medical data to discover hidden relations [Abe, Kogure and Hagita, 2003] and pointed out the importance of discovering knowledge covering gray areas. For medical data analysis, we adopted C4.5 because it does not require any background knowledge, which is very difficult to create for us, and can generate a decision tree that can be regarded as logical clause sets or logical relationships. This type of knowledge is suitable to logical diagnosis. In fact, general induction or statistical analysis such as C4.5 might discovers not only frequently occuring knowledge but also rare or novel knowledge, but such rare or novel knowledge tends to be hidden among general knowledge when we conduct general data mining.

In the above analyses, we dealt with clinical data that consists of mostly the same data types, so we could analyze the data regardless of the relationships between data from different categories. However, if we deal with multiple categorized data, it is rather difficult to discover such hidden or potential knowledge. Furthermore, even the discovery of normal knowledge can be disturbed by influential data in other categories. It might be necessary to analyze the features of data sets to determine their relationships and influence patterns. In this study, we analyze actual clinical data, consisting of multiple categorized items collected by the International Research and Educational Institute for Integrated Medical Science (IREIIMS) project, to show features of the data and suggest methods to discover their hidden or potential features. In Section 2, we introduce Cyber Integrated Medical Infrastructure (CIMI). In Section 3, we describe the features of the clinical data. In section 4, we analyze the clinical data to find relationships among the differently categorized data. In section 5, we overview the possibility of integrated data mining.

19.2 Cyber Integrated Medical Infrastructure

Recently, for medical research, integration and cooperation among the various fields has been advocated. Therefore, we propose a framework called Cyber Integrated Medical Infrastructure (CIMI) which is a framework of integrated management of clinical data on computer networks. Figure 19.1 is an image of CIMI. As shown, CIMI consists of a database, a knowledge base, and an inference and learning component which are connected to each other in the network. Various types of data are collected and stored in the database to be analyzed by machine learning techniques. Usually, medical data to be analyzed are collected during general clinical examination. To save costs, physicians do not collect unnecessary data. For instance, if certain data are not related to the patient's situation, physicians will not collect them. Accordingly, such data sets collected during general clinical examination are incomplete which lack many parts of the data. If parts of the data are missing, even if we can collect many data sets, some of them are ignored during simple data mining procedures. To prevent this situation, we need to supplement missing data sets. However, it is difficult to automatically supplement the missing data sets. In fact, to supplement missing data sets, Ichise proposed a non-linear supplemental method [Ichise and Numao, 2003], but when we collected data from various patients it was difficult to guess relationship among the data, so we gave up to the idea of introducing such a supplemental method. Instead, we introduced a boosting method which estimates the distribution of original data sets from incomplete data sets and increases data by adding Gaussian noise [Abe et al., 2004]. We obtained results with robustness but we could not guarantee the results. In addition, when we used data sets collected in clinical inspections, we could only collect a small number of incomplete data sets. Therefore, in the International Research and Educational Institute for Integrated Medical Science (IREIIMS) project, we decided to collect complete medical data sets. In addition, we collect as many items as possible, because according to

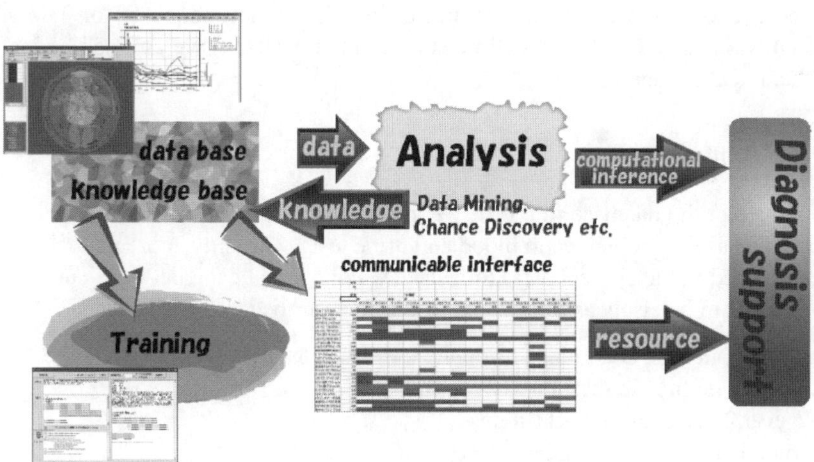

Fig 19.1. Cyber Integrated Medical Infrastructure

physician's experience, if they have insufficient items, they cannot discover serious disease. For instance, if physician can review only 20 items, he/she will diagnoses the patient has stomach ulcer, but if he/she can review more than 100 items, he/she can diagnoses the patient has stomach cancer. Therefore, we decided to collect more than 100 item sets of clinical data. In fact, for 24 items which is adopted by a certain medical university for clinical examinations, cumulative contribution ratio with regard to 119 items is 29.2%.

Then, the analyzed results are regarded as medical knowledge and stored in the knowledge base to be used in computational inferences to support medical diagnosis. In addition, the knowledge base is used with the training of medical students and doctors (To be described elsewhere).

Thus, in the CIMI framework, medical, clinical and other information (e.g. personal information, interview) is analyzed or data mined to discover relationships between the medical, clinical and other data and all possible diseases. Although CIMI includes various types of data, as a first step we mainly deal with clinical data. The clinical data includes liver, pancreas, and kidney test results, tumor markers, and blood test results. In addition, clinical interviews, Ryodouraku results and plethysnographic analysis (Eastern medicine) data will be added. Thus, the database will contain data from both Western and Eastern medicine. As shown in the previous section, usually Western and Eastern medical research is conducted separately. However, for instance, a certain medicine has the same element but different names. Thus they should have certain relationships. One of our aims is to discover relationships between Western and Eastern medical treatment. We will not discuss relationships between Western and Eastern medical in this chapter. We focus on relationships between clinical data such as tumor marker and health levels.

19.3 Features of the Clinical Data

In this section, we describe the features of the clinical data collected for CIMI. Although various types of data will be stored in the database, we are mainly dealing with clinical data here.

19.3.1 Clinical Data

To construct the database in CIMI, we are now collecting various types of clinical data, such as those obtained in blood and urine tests. Currently, more than 100 items are included in the clinical data. In fact, they are clinical data, but they can be categorized more precisely as follows (for the data described in this chapter):

1) liver, pancreas, and kidney test data: 24 items
2) metabolic function test data: 29 items
3) general urine test data: 11 items
4) blood and immunity test data: 31 items
5) tumor markers: 36 items

These categories include the following items.

1) Total protein, albumin, serum protein fraction-α1-globulin
2) Na, K, Ferritin, total acid phosphatase
3) Urobilinogen, urine acetone
4) Mycoplasma pneumoniae antibody, cellular immunity
5) Immunosuppressive acidic protein, Sialyl Le X-i antigen, urine β2-microglobulin

In addition, data from clinical interviews, family tree, and lifestyle are collected. Although these data are relevant to the health status, many factors which were neither formalized nor coherent were included in the interview data, so we did not fully analyze this data, but will do so in a future study.

Currently, we have collected data from about 1800 persons (In some, the data were collected more than once.) and are still collecting the data. It is quite hazardous to directly analyze such a large amount of data. Therefore, first, we analyzed data from only 77 persons. In addition, health levels are assigned by doctors resulting from the clinical data and by an interview. Health levels that express the health status of patients are defined based on *Tumor stage* [Kobayashi and Kawakubo, 1994] and modified by Matsuoka[1]. Categorization of the health levels is shown in Fig. 19.2 ("%" represents a standard distribution ratio of persons in the level.). Persons at level I and II can be regarded as being healthy, but those at levels III, IV, and V can possibly develop cancer. In [Kobayashi and Kawakubo, 1994], level III is defined as the stage

Health Level		Health Condition	(%)
I		Excellent	0
II		Good	10
III		Fair	60
IV		Needs an improvement in lifestyle	25
V		Needs a precise examination and therapy	5

Fig 19.2. Health levels (standard)

---

[1] Matsuoka categorized health levels into 8 categories which are 1, 2, 3, 4a, 4b, 4c, 5a, and 5b, since levels 4 and 5 include many clients' data. But, in this chapter, we used the original categorization

before the shift to preclinical cancer, level IV is defined as conventional stage 0 cancer (G0), and level V is defined as conventional stages 1–4 cancer (G1–G4).

Thus, the cutoff values for each markers were manually decided by doctors, but some of them are doubtful. Our aim also includes to check the cutoff values for each markers as well as discover complex relationships between the markers.

19.3.2 Features of the Clinical Data

The ratio of persons in each health level is shown in Table 19.1. The ratio pattern of Table 19.1 is quite different from that of Fig. 19.2 and the data shown by [Kobayashi and Kawakubo, 1994]. Especially, the ratio of persons in level II is quite low and that of persons in level IV is quite high. Thus in Table 19.1, the health level distribution pattern seems to shift to higher levels in parallel. This tendency is similar to that seen in the data of 1750 persons', In fact, persons in health level 4 increased in ratio, but a general tendency of the distribution is similar. This because we still collect data from the similar group which includes few younger persons. Besides the problems shown below, this imbalance should influence the results, but in this chapter, we do not deal with this negative influence. We will deal with the problem in the other paper.

The distribution of age is shown in Fig. 19.3. As shown, the ratio of persons in their 50's is very high which might be the reason for the different distribution

Table 19.1. Health levels.

health level	1	2	3	4	5
ratio (77) (%)	0.0	0.0	12.0	58.7	29.3
ratio (1750) (%)	0.0	0.0	3.1	83.4	13.5

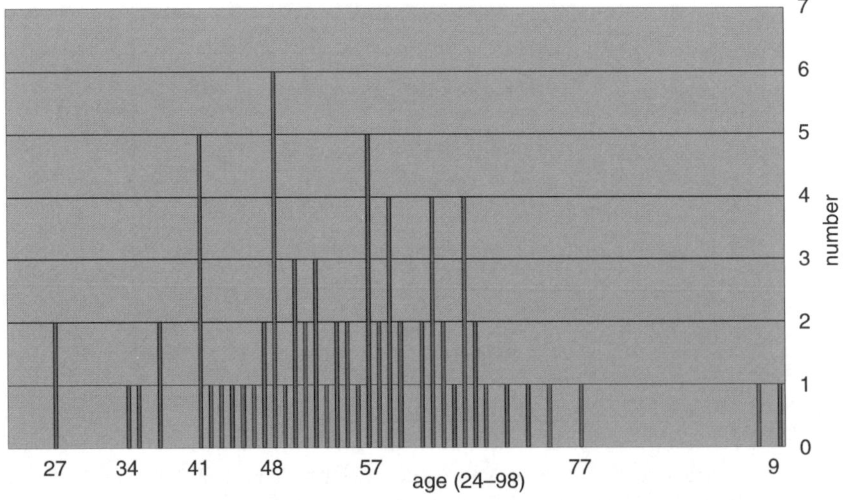

Fig 19.3. Distribution of age (Clinical data)

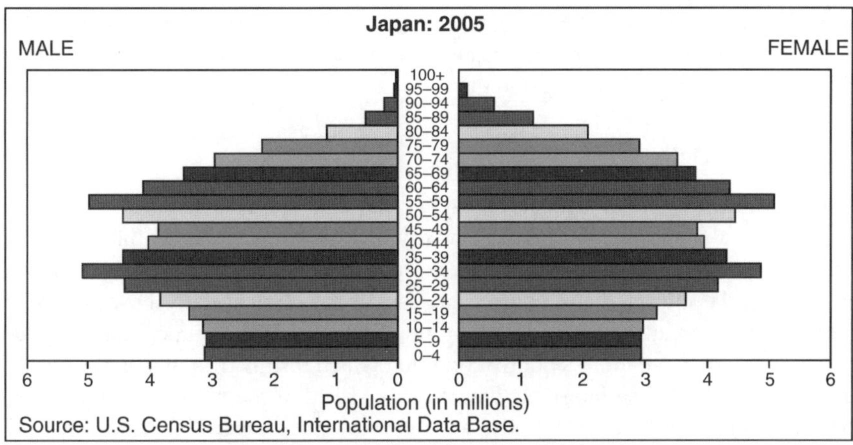

Fig 19.4. Population Pyramids for Japan (2005)

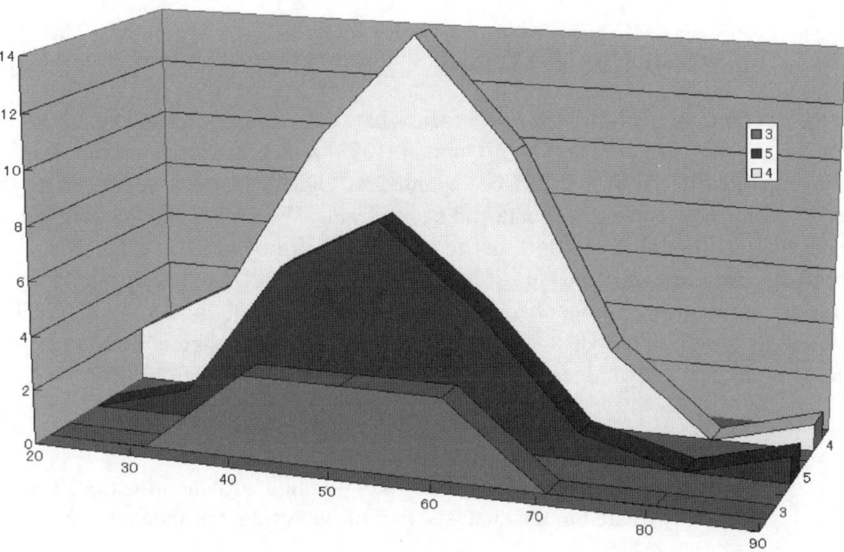

Fig 19.5. Distribution of age according to health levels.

of health levels. Currently, we collect data from office workers (aged 40 to 50 years old) but not from students or younger persons. Therefore, our results might not reflect social tendencies.

The age distribution pattern is not the same as the Japanese standard age distribution pattern (Fig. 19.4). This is because we did not collect samples from persons from all over Japan but only from those who applied to take part in our study. Most of our subjects are office workers and live in or close to Tokyo. Figure 19.5 shows the distribution pattern of age according to health. For health levels IV and V, the peak is in the 50's. Although the data is not sufficiently large, the distribution pattern seems to be a normal distribution. Therefore, taking account of the pattern shown in Fig. 19.3, if we would collect more clinical data from persons in their teen or twenties, the distribution pattern might be different.

19.3.3 Other Data

Although not fully dealt with in this study, the data set also includes information from a clinical interview. For instance, worrying symptoms, nonessential foods (coffee, alcohol etc.), medicine, length of exercise, meal style, and family history were included in the interview. Of course, analysis of the clinical data shown above together with these interview data is very important, since the clinical data includes information about a patient's bodily condition which results from his or her lifestyle and family history. The interview data set is a record of the patients' daily life and their family history. Thus the clinical and interview data sets explain each other. It is quite significant to analyze both data sets as results and reasons, but in this study, we focused on the relationships between a person's clinical data and health status.

19.4 Analysis of Clinical Data

In this section, we analyzed the relationships between clinical data and health levels. For the analysis, we applied C4.5 [Quinlan, 1993] which is a decision tree learner. The generated decision tree can be regarded as logical formulae representing the relationships between clinical data and health levels. We apply C4.5 because it generates logical formulae and does not require any background knowledge. Since we try to discover new relationships, it is rather difficult to prepare background knowledge and sometimes improper background knowledge leads to incorrect results. Thus a learning system that does not need any background knowledge is suitable for our aim to build new knowledge base and discover rare or novel relationships.

19.4.1 Relationship between Health Levels and Clinical Data

First, we analyzed all the data described in 19.3.1 without any modification. That is, all data are equally treated in the analysis. Part of the results are shown below.

```
β2-microglobulin (mg/l) > 1.8 : 5
β2-microglobulin (mg/l) <= 1.8 :
|   γ-GTP (U/l) > 119 : 5
|   γ-GTP (U/l) <= 119 :
|   |   Creatinine (mg/dl) > 1 : 3
|   |   Creatinine (mg/dl) <= 1 :
|   |   |   γ-seminoprotein (ng/ml) <= 0.8 :...
|   |   |   |   CA72-4 (U/ml) > 3 : 5
|   |   |   |   CA72-4 (U/ml) <= 3 :....
|   |   |   |   |   PIVKA-II (mAU/ml) > 17 : 4
|   |   |   |   |   PIVKA-II (mAU/ml) <= 17 :....
```

The results are almost acceptable, but have been generated by analyzing large numbers of items from various categories. Therefore, certain items might be too influential and hide the effects of less influential ones. As shown below, we analyzed relationships between health levels and data in each category.

1) *liver, pancreas, and kidney test data*

```
    Cholinesterase (U/l) <= 4811 :
    |   Creatinine (mg/dl) > 0.9 : 3
    |   Creatinine (mg/dl) <= 0.9 :
    |   |    TP (g/dl) > 6.9 : 4
    |   |    TP (g/dl) <= 6.9 :
    |   |   |    Serum protein fraction-γ-globulin (%) <= 15 : 4
    |   |   |    Serum protein fraction-γ-globulin (%) > 15 : 3
    Cholinesterase (U/l) > 4811 :
    |   Serum protein fraction-α1-globulin (%) > 2.5 : 5
    |   Serum protein fraction-α1-globulin (%) <= 2.5 :
    |   |   Creatinine (mg/dl) > 1 : 5
    |   |   Creatinine (mg/dl) <= 1 :......
```

2) *metabolic function test data*

```
    Total acid phosphatase <= 9.5 :
    |   Non-esterified fatty acid (mEq/l) <= 0.3 :
    |   |    Fe (μg/dl) <= 69 : 4
    |   |    Fe (μg/dl) > 69 : 3
    |   Non-esterified fatty acid (mEq/l) > 0.3 :
    |   |    Mg (mEq/l) <= 2.1 : 4
    |   |    Mg (mEq/l) > 2.1 :
    Total acid phosphatase > 9.5 :
    |   Cl (mEq/l) > 104 : 3
    |   Cl (mEq/l) <= 104 :
    |   |   Apolipoprotein B (mg/dl) <= 72 : 3
    |   |   Apolipoprotein B (mg/dl) > 72 :..
```

3) *general urine test data*

```
    Urine acetone > 0 : 4
    Urine acetone <= 0 :
    |   Urine sedimentary test, squamous epithelium > 1 : 4
    |   Urine sedimentary test, squamous epithelium <= 1 :
    |   |   Urine sediment-bacteria <= 2 :
    |   |   |   Urine sedimentary test, squamous epithelium <= 0 :
    |   |   |   |   pH <= 5.5 :
    |   |   |   |   |    Urine occult blood > 0 : 4
    |   |   |   |   |    Urine occult blood <= 0 :
```

```
|   |   |   |   |   |   |   pH <= 5 : 5
|   |   |   |   |   |   |   pH > 5 : 4
|   |   |   |   pH > 5.5 :
|   |   |   |   |   Urine sediment-protein quantitative > 0 : 5
|   |   |   |   |   Urine sediment-protein quantitative <= 0 :....
```

4) *blood and immunity test data*

```
    C3 (mg/dl) <= 105 :
|    Cellular immunity (T CELL CD2) (%) <= 84 :
|    |    Cellular immunity (T CELL CD2) (%) > 76 : 4
|    |    Cellular immunity (T CELL CD2) (%) <= 76 :
|    |    |    Leukocyte classification Mono (%) <= 5.8 : 4
|    |    |    Leukocyte classification Mono (%) > 5.8 : 3
|    Cellular immunity (T CELL CD2) (%) > 84 :
|    |    Leukocyte counts (×10³/µl) <= 4000 : 4
|    |    Leukocyte counts (×10³/µl) > 4000 : 3
C3 (mg/dl) > 105 :
|    Leukocyte counts (×10³/µl) > 7000 : 5
|    Leukocyte counts (×10³/µl) <= 7000 :
|    |    EB-VCA (IgG) <= 40 :
|    |    |    Leukocyte counts (×10³/µl) <= 5200 : 3
|    |    |    Leukocyte counts (×10³/µl) > 5200 : 4
|    |    EB-VCA (IgG) > 40 :.....
```

5) *tumor markers*

```
    β2-microglobulin (mg/l) > 1.8 : 5
β2-microglobulin (mg/l) <= 1.8 :
|    Carcinoembryonic antigen (ng/ml) <= 4.1 :
|    |    CA72-4 (U/ml) > 3 : 5
|    |    CA72-4 (U/ml) <= 3 :
|    |    |    Squamous cell carcinoma antigen (ng/ml) <= 1.1 :
|    |    |    |    CA15-3 (U/ml) <= 3.7 : 3
|    |    |    |    CA15-3 (U/ml) > 3.7 :
|    |    |    |    |    PIVKA-II (mAU/ml) > 17 : 4
|    |    |    |    |    PIVKA-II (mAU/ml) <= 17 :.....
```

The above results represent relationships between the data in each category and health levels. The first classification of the analysis of whole data and that of tumor markers is the same. This means that tumor markers obviously influence the classification results. To obtain influential power relationships of each category, we analyzed relationships between health levels and mixed category data. Although we analyzed all the possible combinations (2 and 3 categorizations), only typical relationships are shown below.

♣ *liver, pancreas, and kidney test data+metabolic function test data*

```
Cholinesterase (U/l) <= 4811 :
|   Creatinine (mg/dl) > 0.9 : 3
|   Creatinine (mg/dl) <= 0.9 :
|   |    TP (g/dl) > 6.9 : 4
|   |    TP (g/dl) <= 6.9 :
|   |   |    Serum protein fraction-γ-globulin (%) <= 15 : 4
|   |   |    Serum protein fraction-γ-globu! lin (%) > 15 : 3....
```

♣ *liver, pancreas, and kidney test data+general urine test data*

```
Urine acetone > 0 : 4
Urine acetone <= 0 :
|   Urine sedimentary test, squamous epithelium > 1 : 4
|   Urine sedimentary test, squamous epithelium <= 1 :
|   |   Urine sediment-bacteria <= 2 :
|   |   |   Urine sedimentary test, squamous epithelium <= 0 :....
```

♣ *liver, pancreas, and kidney test data+blood and immunity test data*

```
C3 (mg/dl) <= 105 :
|   Cellular immunity(T CELL CD2) (%) <= 84 :
|   |    Cellular immunity(T CELL CD2) (%) > 76 : 4
|   |    Cellular immunity(T CELL CD2) (%) <= 76 :
|   |   |    Leukocyte classification Mono (%) <= 5.8 : 4
|   |   |    Leukocyte classification Mono (%) > 5.8 : 3 ...
```

♣ *liver, pancreas, and kidney test data+tumor markers*

```
β2-microglobulin (mg/l) > 1.8 : 5
β2-microglobulin (mg/l) <= 1.8 :
|    γ-GTP (U/l) > 119 : 5
|    γ-GTP (U/l) <= 119 :
|   |    Creatinine (mg/dl) > 1 : 3
|   |    Creatinine (mg/dl) <= 1 :....
```

♣ *metabolic function test data+general urine test data*

```
Total acid phosphatase <= 9.5 :
|   Urine sediment-bacteria <= 2 :
|   |   Urine sediment-protein quantitative > 0 : 4
|   mid   Urine sediment-protein quantitative <= 0 :
|   |   |   Non-esterified fatty acid (mEq/l) > 0.5 : 4
|   |   |   Non-esterified fatty acid (mEq/l) <= 0.5 :....
```

♣ *blood and immunity test data+tumor markers*

```
β2-microglobulin (mg/l) > 1.8 : 5
β2-microglobulin (mg/l) <= 1.8 :
|   Carcinoembryonic antigen (ng/ml) <= 4.1 :
|   |   CA72-4 (U/ml) > 3 : 5
|   |   CA72-4 (U/ml) <= 3 :
|   |   |   Erythrocyte counts (×10⁶/μl) > 526 : 5
|   |   |   Erythrocyte counts (×10⁶/μl) <= 526 :....
```

The number of items in each category is quite different (from 11 to 36). This imbalance might influence the results. In addition, we only dealt with a small number of data as a pre-examination. It would be hazardous to determine relationships among the categories by analyzing the results of pre-examinations. However, we can determine a simple relationship, such as the influence of the category on health levels, by comparing the root of decision trees, since results from the most influential factors usually come at or near the root of decision trees. Thus we found the following influential order of the health levels:

metabolic function test data $\prec$
liver, pancreas, and kidney test data $\prec$ *general urine test data* $\prec$
blood and immunity test data $\prec$ *tumor markers*

Health levels are assigned according to the possibility of the presence of disease, for instance, cancer. Therefore, it would be reasonable that *tumor markers* is the most influential factor. In addition, the diagnosis (assignment of health level) is performed for those who are not believed to be suffering from cancer. As a result, factors such as internal organs play a less influential role in health levels. That is, the internal organs would not be badly damaged. As for the decision trees of factors other than tumor markers, health level 5 cannot be observed and the classification points occur within normal values. In addition, the concept of health levels has been introduced to pinpoint the period before a patient's condition becomes a disease during which the patient is moving toward disease (presymptomatic stage).

19.4.2 Relationship Between Health Levels and Interview Data

For the preview experiment, we add the interview data to the clinical data to analyze relationships between health levels, and clinical data and interview data. In fact, the interview data contains various types of information such as family history, food, and lifestyle data. Currently, it is neither well formalized nor coherent, so it is difficult to analyze without any modification. In fact, if we use the data without proper modification, we can only obtain meaningless results. Some of the results obtained after applying data cleaning technique (data that causes meaningless results is removed.) are shown below:

```
type(alcohol) = sour: 5
type(alcohol) = beer: 4
type(alcohol) = wine: 4
type(alcohol) = sake: 4
type(alcohol) = sake, shochu: 4
type(alcohol) = 0:
|   Diabetes(mother) = 1: 5
|   Diabetes(mother) = 0:
|   |   dinner = 0: 4
|   |   dinner = 1:
|   |   |   start_age(health food(else(1))) <= 30 : 4
|   |   |   start_age(health food(else(1))) > 30 : 5 ....
```

The mother's diabetes might effect her children's health and alcohol intake might effect health. However, it is difficult to evaluate the results. Also the values in interview data are mostly discrete. In contrast, values from the clinical data are mostly continuous. When analyzing by C4.5, discrete values can easily be classified. Thus, most of the results from clinical data disappear when we add the interview data. It is rather difficult, therefore, to determine whether the clinical data is less influential than interview data on health levels, and further studies are needed.

19.5 Toward Integrated Data Mining

In the previous section, we analyzed clinical data according to categories such as tumor markers. In addition, we analyzed the influence of these categories on health levels. The results suggest that it would be better to analyze the clinical data according to categories or by considering the influence of the powerful categories. To classify the clinical data, we applied Principal Component Analysis, but we could not find any significant classifications. Thus, it is rather difficult to automatically determine meaningful classification of the clinical data. However, as mentioned before, the influence on the health levels differs according to the categories. In [Abe et al., 2008], we refer to the categorization to perform a categorized and integrated data mining. In the analysis, we pointed out that it is necessary to introduce categorized data mining when certain factors are hidden by more influential factors. For instance, we showed that LDH which is categorized into the liver, pancreas, and kidney test data is easy to hidden by the influential factors such those in tumor markers, and that when we apply categorized data mining, an exact health level can be determined. Thus at least, categorized data mining is necessary to determine hidden factors.

In addition, we discussed an influence power order in the previous section. Agrawal proposed an association rule that represents relationships between items in databases [Agrawal et al., 1993]. The association rule is frequently used when analyzing POS data to discover tendencies of users' shopping patterns (basket analysis). However, from the analysis, we can only discover frequently co-occurring patterns.

Also, relational data mining has recently been proposed [Džroski and Lavrač, 2001]. This paradigm also discovers relationships between items in a (relational) database by using ILP techniques. Their approaches are important for complex data mining. However our major aim is not to discover relationships between each category but to determine an effective classification for data mining while considering influential power. By conducting pre-experiments, we found the following needs for (integrated) data mining of clinical data sets.

- We need to classify the data not according to statistical or associated patterns but rather according to their influence on health levels. Currently we do not have a clear answer to the problem. Actually, we applied Principal Component Analysis to the medical data, but we could not obtained sufficient results. By applying partial data mining more than once and comparing the results, we might be able to solve the problem.
- Zheng proposed committee learning [Zheng and Webb, 1998] which divide data set into several parts and perform data mining for each divided data set and generates a result after comparison of each result. The classification strategy is different from categorized data mining and needs many data sets, but if we can collect sufficient number of medical data, it will be better to introduce committee learning. In addition, the concept of committee learning can be applied to a multiple partial data mining. Actually, in [Abe et al., 2008], we conducted a majority rule criterion application for health level estimation and obtained the result that even a simple strategy such as majority rule criterion can generate better result than simple application of data mined rules.
- After proper classification, we could determine the influence of category on health levels. By removing more powerful influential categories, we could then find hidden, potential, rare or novel relationships between the clinical data and health levels. Therefore, we need to conduct multiple partial data mining.
- In addition, we need to integrate the results from partial data mining. The integration would enable us to discover complex relationships between the clinical data and health levels. Similarly, we can combine the analysis of other types data such as interview data and that from Eastern medicine (e.g. Ryodouraku). Then relationship Western and Eastern medicine can be also discovered by observed the results by persons from various field.

19.6 Conclusions

In this chapter, we showed features of clinical data and analyzed relationships between clinical data and health levels. Although we have collected 1800 persons' data, for the pre-experiment, we used a small subset of the clinical data to discover their features. In fact, for the data, we used given (authorized) categorizations such as blood test data and tumor markers. We found that the most influential category for the health level is tumor markers. As a result, when the other categories are mixed with tumor markers, relationships between items in the other categories might be hidden by tumor markers. We, therefore, suggest data mining that categorizes the clinical data into multiple categories to discover relationships between the items in each category and the health levels and integrates the results. As for the authorized categorization of clinical data, in [Abe et al., 2008], we showed that due to categorization we can use necessary rule sets, which cannot be generated without categorization, to determine health levels. For certain situation, categorized data mining is necessary to conduct sufficient medical diagnosis.

We can, of course, add data from Eastern medicine. In addition, the integrated data mining can be applied to chance discovery to discover rare or novel events that might cause serious disease, as powerfully influential factors that might disturb the influence of powerless factors can be removed by the categorization. We currently have data sets from about 1800 persons and data are still being collected from some of these persons and from new subjects. Therefore, we will be able to conduct temporal data mining. We think the discovery of temporal patterns in the health levels is very important to protect the subjects from developing diseases. In addition to integrated data mining, we should take account of temporal data mining to discover a temporal pattern to disease.

Acknowledgments

This research was supported in part by the Program for Promoting the Establishment of Strategic Research Centers, Special Coordination Funds for Promoting Science and Technology, Ministry of Education, Culture, Sports, Science and Technology (Japan).

References

[Abe, Kogure and Hagita, 2003] Abe A., Kogure K. and Hagita N.: Discovery of Hidden Relations from Medical Data, *Proc. of HCI2003 3rd. Int'l Workshop on Chance Discovery, pp. 37–43,* 2003.

[Abe et al., 2004] Abe A., Naya F., Kogure K., and Hagita N.: Rule Acquisition from small and heterogeneous data set, *Technical Report of JSAI, SIG-KBS-A304-32, pp. 189–194* (2004) in Japanese

[Abe et al., 2008] Abe A., Hagita N., Furutani M., Furutani Y., and Matsuoka R.: Data mining of Multi-categorized Data, *Mining Complex Data, Post-Proceedings of the ECML/PKDD'07 Third International Workshop, MCD2007 (Ras Z.W., Tsumoto S., Zighed D. eds.) (LNAI), Springer Verlag* 2008. to appear

[Agrawal et al., 1993] Agrawal R., Imielinski T., and Swami A.: Mining association rules between sets of items in large databases, *Proc. of ACM SIGMOD Int'l Conf. on Management of Data, pp. 207–216,* 1993.

[Džroski and Lavrač, 2001] Džroski S and Lavrač N eds.: *Relational Data Mining,* Springer Verlag, 2001.

[Ichise and Numao, 2003] Ichise R., Numao M.: A Graph-based Approach for Temporal Relationship Mining, *Technical Report of JSAI, SIG-FAI-A301, pp. 121–126* (2003)

[Ichise and Numao, 2005] Ichise R. and Numao M.: First-Order Rule Mining by Using Graphs Created from Temporal Medical Data, *LNAI, Vol. 3430, pp. 112–125*, 2005.

[Ohsawa, 2003] Ohsawa Y., Okazaki N., and Matsumura N.: A Scenario Development on Hepatics B and C, *Technical Report of JSAI, SIG-KBS-A301, pp. 177–182*, 2003.

[Ohsawa and McBurney, 2003] Osawa Y. and McBurney P. eds.: *Chance Discovery*, Springer Verlag, 2003.

[Tsumoto, 2004] Tsumoto S.: Mining Diagnostic Rules from Clinical Databases Using Rough Sets and Medical Diagnostic Model, *Information Sciences, Vol. 162, No. 2, pp. 65–80*, 2004.

[Tsumoto et al., 2005] Tsumoto S., Yamaguchi T., Numao M., and Motoda H.: Active Mining Project: Overview, *LNAI, Vol. 3430, pp. 1–10*, 2005.

[Kobayashi and Kawakubo, 1994] Kobayashi T. and Kawakubo T.: Prospective Investigation of Tumor Markers and Risk Assessment in Early Cancer Screening, *Cancer, Vol. 73, No. 7, pp. 1946–1953*, 1994.

[Quinlan, 1993] Quinlan J.R.: C4.5: Programs for Machine Learning, *Morgan Kaufman*, 1993.

[Zheng and Webb, 1998] Zheng Z. and Webb G.I.: Stochastic Attribute Selection Committees, *Proc. of AI98, pp. 321–332* (1998)

Privacy-Preserving Data Mining for Medical Data: Application of Data Partition Methods

Yi Peng[1], Gang Kou[2,5], Yong Shi[3,4], and Zhengxin Chen[4]

[1] School of Management, University of Electronic Science and Technology of China, No. 4, Section 2, North Jianshe Road, Chengdu, China
pengyicd@gmail.com
[2] Thomson Corporation, R&D, 610 Opperman Drive, Eagan, MN 55123, USA
kougang@yahoo.com
[3] Chinese Academy of Sciences Research Center on Data Technology & Knowledge Economy, Graduate University of the Chinese Academy of Sciences, 100080, China
yshi@gucas.ac.cn
[4] College of Information Science & Technology, University of Nebraska at Omaha
[5] The corresponding author. Tel: ++1-402-4030269.

Summary. Medical data mining has been a popular data mining topic of late. Compared with other data mining applications, medical data mining has some unique characteristics. Since medical records are related to human subjects, privacy protection is taken more seriously than other data mining tasks. This paper applied two data separation techniques – vertical and horizontal partition - to preserve privacy in medical data classification. In the vertical partition approach, each site uses a portion of the attributes to compute its results and the distributed results are assembled at a central trusted party using majority-vote ensemble method. In the horizontal partition approach, data are distributed among several sites. Each site computes its own data and a central trusted party integrate these results using ensemble. We implement these two approaches using medical datasets from UCI Machine Learning archive and report the experimental results.

Keywords: classification, privacy-preserving data mining, horizontal partition, vertical partition

20.1 Introduction

The advances in digital data collection devices and data storage technology allow companies and organizations to store up huge amount of data. Data mining field, which focuses on the extraction of useful knowledge from large databases, has made great progress during the last two decades. Since these electronic data cover almost all aspects of our life (e.g. credit card purchases, news, medical files, maps or terrain visualizations, human genome), how to ensure that sensitive data are kept from misuse is always an important issue in data mining. As early as 1989, when the first KDD

Y. Peng et al.: *Privacy-Preserving Data Mining for Medical Data: Application of Data Partition Methods*, Studies in Computational Intelligence (SCI) **123**, 331–340 (2008)
www.springerlink.com

workshop was held in Detroit, Michigan, privacy protection has been brought up. In recent years, research interest in privacy preserving data mining (PPDM) is increasing (See, for example: Agrawal and Srikant, 2000; Agrawal and Aggarwal, 2001; Lindell and Pinkas, 2002; Vaidya and Clifton, 2002; Kantarcioglu and Clifton, 2004). Many countries have enacted laws to protect data privacy. For instance, U.S. federal rules set guidelines to conceal individual patient identifiers (Cios and Moore, 2002).

Privacy preserving is an especially important issue in medical data mining. Medical data include diagnosis, treatments, images, laboratory data, and observations. These data are highly sensitive. Compared with other data mining applications, medical data mining has some unique characteristics. Cios and Moore (2002) organized these characteristics into four groups: heterogeneity; statistical philosophy; special status of medicine; ethical, legal, and social issues. Medical data are heterogeneous because they exist in various forms, such as high-dimensional images, text, numerical, and even signals. Statistical philosophy refers to the fact that the basic assumptions of statistics may be fundamentally different for medical data. Medicine has a special status because the outcomes of medical care concern human life. Ethical, legal, and social issues include data ownership, lawsuits, privacy and security of human data, expected benefits, and administrative issues. Privacy and security of human data belongs to the ethical, legal, and social category.

Currently, there are three major classes of privacy-preserving techniques: data modification, summarization, and data separation (Clifton, 2002; Verykios et al., 2004). Data modification alters raw data before applying mining methods, including methods and techniques like perturbation, blocking, swapping and sampling. Perturbation methods change attributes' values; blocking hides sensitive values; swapping methods interchange record values; sampling techniques take a portion of the entire dataset (Verykios et al., 2004). Summarization protects privacy by providing only overall statistics of the whole data and limiting query functions. Data separation, also known as data distribution, divides data to trusted parties.

The objective of this paper is to apply data separation-based techniques to preserve privacy in classifying medical data. We take two approaches to protect privacy: vertical partition and horizontal partition. In the vertical partition approach, each site uses a portion of the attributes to compute its results and the distributed results are assembled at a central trusted party using majority-vote ensemble method. In the horizontal partition approach, data are distributed among several sites. Each site computes its own data and a central trusted party integrate these results. These two approaches are implemented using two medical datasets from UCI Machine Learning repository: Wisconsin prognostic breast cancer dataset (Wolberg and Mangasarian, 1990) and heart-disease dataset (UCI Machine Learning repository, 2006). The experimental results demonstrate that through the majority-vote ensemble method, it is possible to protect data privacy and maintain satisfactory classification accuracies simultaneously.

This paper is structured as follows. The next section explains why and how we use vertical and horizontal separation techniques to protect privacy of medical data. The third section describes the classification experiments. The last section concludes the paper.

20.2 Privacy-preserving Medical Data Mining: Data Separation Techniques

In this section we use two scenarios to illustrate how vertical and horizontal techniques can be applied to protect medical data privacy.

Vertical separation techniques can be used in situations when a data owner wants to extract useful information from the data but lacks the capability to analyze them. Thus data owner needs a third party to analyze the data. A data owner may be a hospital or a medical center. Though the third party involved is trusted, the data privacy will be more reliably guarded if the raw data are processed before releasing the data. Various methods can be applied to process the raw data, such as sampling, summarization, obfuscation, and separation. Vertical separation is one of the simple and straightforward techniques. The basic idea is to vertically divide the dataset into subsets (i.e., remove one or more attributes/columns from each subset) and each third party has only a portion of the whole data. As a result of vertical partition, only the data owner has the entire data and data privacy can be protected. Take the Wisconsin prognostic breast cancer dataset as an example. In order to improve the prediction accuracy for breast tumor diagnosis, the University of Wisconsin Hospitals send their data to several third parties for analysis. To protect patients' privacy, identifiers or personal information are removed from the data. In addition, one or more attributes can be removed to create subsets. Each subset will be analyzed at separate sites and the results will be returned back to the data owner. Figure 20.1 illustrates this vertical separation scenario.

Horizontal separation techniques are used when the datasets are distributed among multiple data owners and the data structure at each site is the same. Each data owner has the ability to analyze his/her data. The problem is that each dataset has limited data objects and classifiers generated by small datasets are often less

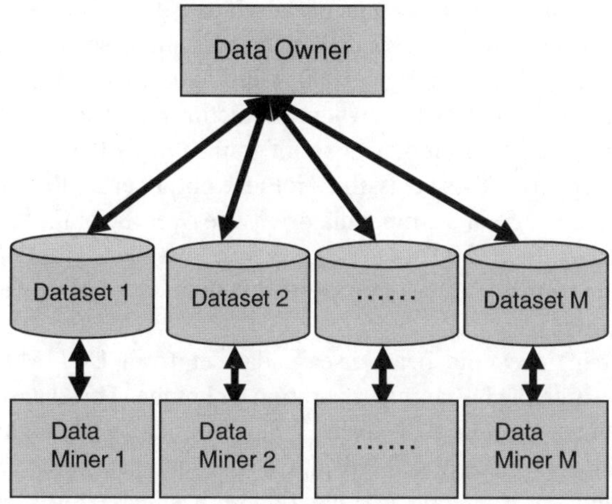

Fig 20.1. Vertical Separation Technique (Adopted from Clifton 2002)

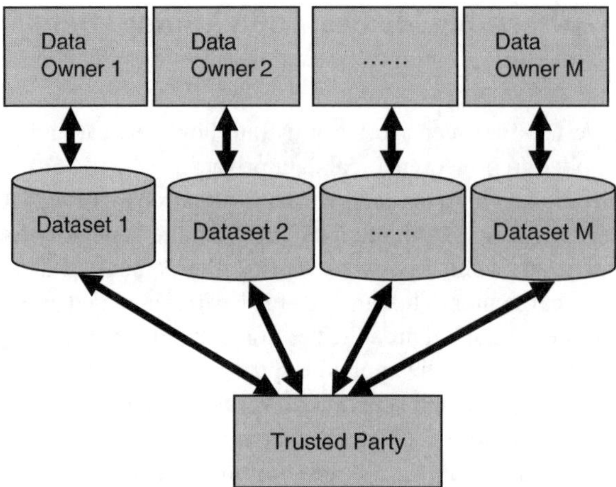

Fig 20.2. Horizontal Separation Technique

generalizable and classification accuracy than those obtained from large datasets. To increase the prediction accuracy, one possible choice is to combine small datasets into a large data and build classifiers based on the large data. However, sometimes data owners may not willing to share their sensitive data. Horizontal partition provides a solution to this problem. It does not require data owners to share their data; rather, it increases the classification accuracy by sharing mined knowledge among data owners. Figure 20.2 illustrates this process.

Depending on the types of data mining tasks, the mined knowledge exists in a variety of forms. In classification, the mined knowledge may take the form of classification rule, decision tree, and so on. The mined knowledge from each site can be integrated through different methods. In this paper, majority-vote ensemble procedure is used to combine classification results of individual site and obtain the final results. Previous studies have shown that ensemble can help increase accuracy and stability (Opitz and Maclin, 1999; Dietterich, 2000; Kim, 2002). An ensemble consists of a set of individually trained classifiers whose predictions are combined. There are two fundamental elements of ensembles: a set of trained classifiers and an aggregation mechanism that organizes these classifiers into the output ensemble. The aggregation process can be an average or a simple majority vote over the output of the ensembles (Zenobi and Cunningham, 2002). In majority-vote, the class label of each record is determined by the majority votes of classifiers. A majority-vote method requires odd number of classifiers.

The experiment uses the heart-disease dataset from UCI Machine Learning repository to illustrate the horizontal separation technique. This dataset was collected from three locations, including Cleveland Clinic Foundation, Hungarian Institute of Cardiology, and Long Beach V.A. Medical Center. Each data source has limited data records: Cleveland has 303 records; Hungarian set has 294 records; and Long Beach set has 200 records. The data structure of each dataset is the same. Each data owner builds a classifier using their data and sends only the resulting classifiers to a trusted party. The trusted party then combines these classifiers and produces final prediction.

The next section describes the datasets, the experimental procedures, and classification results.

20.3 Experiments

To implement the vertical and horizontal data separation techniques, we select two datasets from UCI repository: Wisconsin prognostic breast cancer dataset and heart-disease dataset (UCI Machine Learning repository, 2006; Bennett and Mangasarian, 1992). Wisconsin prognostic breast cancer dataset is used for vertically partitioned analysis and the heart-disease dataset is used for horizontally partitioned analysis.

20.3.1 Vertical Data Separation Experiment

Wisconsin prognostic breast cancer dataset has 699 records and 9 variables. These records belong to either benign or malignant class. As mentioned in section 2, we create nine sub-datasets by removing one attribute at a time. Each sub-dataset has only eight variables. These sub-datasets are classified separately using See5 (Rulequest Research, 2003) software with ensemble and 10-fold cross validation. The results are integrated following the Privacy-preserving classification (vertical) process:

Privacy-preserving Classification (Vertical) Process   @

Input: The Medical dataset $M = \{M_1, M_2, M_3, ..., M_n\}$ with n records, each record has m attributes

Output: Average classification accuracies for benign and malignant in 10-fold cross-validation; scores for all records; decision trees ensemble.

Step 1 Generate m subsets with one different attribute removed from M at each time.

Step 2 Training each subset with See5 with adaptive boosting and 10-fold cross validation to get m decision trees $D_1, D_2, D_3, ..., D_m$.

Step 3 Ensemble the final decision function $D = \{D_1, D_2, D_3, ..., D_m\}$, via majority-vote of the m decision trees from step 2.

Step 4 Classify M by the final decision function.

END

For comparison purpose, we also classify the whole dataset (i.e., with 9 variables) using See5 with adaptive boosting, 10-fold cross validation and majority-vote ensemble method. The results are summarized in Figure 20.3 and Table 20.1 (Appendix).

Malignant error indicates the percentage of malignant records that have been misclassified as benign; benign error indicates the percentage of benign records that have been misclassified as malignant. On the X axis, number 1 through 9 refers to the

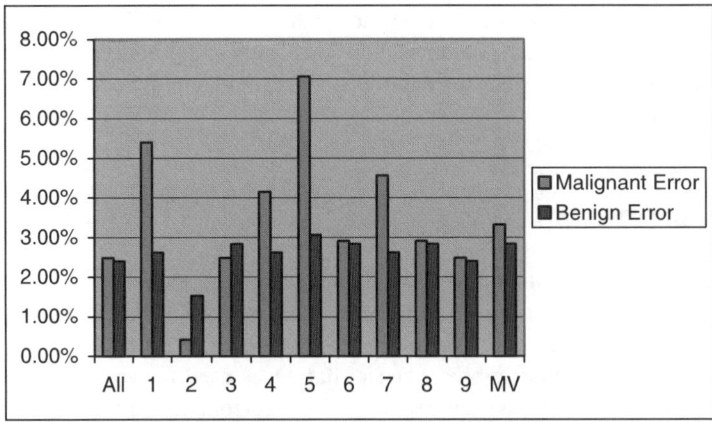

Fig 20.3. Wisconsin Prognostic Breast Cancer Dataset Classification Results

malignant and benign errors of each sub-dataset; "All" refers to the error rates using the entire dataset; "MV" refers to the error rates of majority-vote classification result. Figure 20.3 shows that the classification results using the whole dataset is better than using sub-datasets and the majority-vote result, except for number 2 dataset. The majority-vote result is slightly higher than the average of 9 sub-datasets for malignant class (3.32% vs. 3.59%) and slight lower than the average of 9 sub-datasets for benign class (2.84% vs. 2.6%).

This experiment shows that the performance of vertical partitioned data varies: malignant class error rates range from 0.41% to 7.05% and benign class error rates range from 1.53% to 3.56%. Majority-vote ensemble can reduce the variation and produce above average classification accuracy for vertically partitioned data.

20.3.2 Horizontal Data Separation Experiment

The heart-disease dataset has 797 records and 13 variables. These records belong to either heart-disease or normal class. The data were collected from Cleveland Clinic Foundation, Hungarian Institute of Cardiology, University Hospital of Zurich, and Long Beach V.A. Medical Center. The subset from Zurich was dropped because it is highly imbalanced. These datasets have the same set of attributes but different records: Cleveland has 303 records; Hungarian set has 294 records; and Long Beach set has 200 records. Each dataset was classified separately using See5 with adaptive boosting, 10-fold cross validation and majority-vote ensemble. The results of three datasets are integrated following the Privacy-preserving classification (horizontal) process:

Privacy-preserving Classification (Horizontal) Process   @

Input: The Medical datasets from r different sources, $M^1 = \{M_1^1, M_2^1, M_3^1,...., M_{n1}^1\}$, $M^2 = \{M_1^2, M_2^2, M_3^2,..., M_{n2}^2\}$, ..., $M^r = \{M_1^r, M_2^r, M_3^r,..., M_{nr}^r\}$. Each dataset has n records with m attributes.

Output: Average classification accuracies for Normal and Heart-disease of the dataset in 10-fold cross-validation; scores for all records; decision trees ensemble.

Step 1 Training r datasets using See5 with adaptive boosting and 10-fold cross validation to get r decision trees $D_1, D_2, D_3, ..., D_r$.

Step 2 Ensemble the final decision function $D = \{D_1, D_2, D_3, ..., D_r\}$, via majority vote of the r decision trees from step 1.

Step 3 Classify all r datasets by the final decision function.

END

For comparison purpose, we also classify the combined dataset (include all three datasets) using See5 with adaptive boosting, 10-fold cross validation and majority-vote ensemble. The results are summarized in Figure 20.4 and Table 20.2 (Appendix).

Heart-disease error indicates the percentage of heart-disease records that have been misclassified as normal. Normal error indicates the percentage of normal records that have been misclassified as heart-disease. On the X axis, C, H, and V refer to Cleveland dataset, Hungarian dataset, and Long Beach dataset, respectively; "All" refers to the classification result using the combination of three datasets; "MV" refers to the majority-vote classification result. Figure 20.4 indicates that the classification result using the combined dataset has lower error rates than using individual dataset and the majority-vote result. The majority-vote result is better than the average of three individual dataset for both classes (heart-disease error rate: 15.99% vs. 23.35%; normal error rate: 16.63% vs. 18.28%). To summarize, using horizontal data separation techniques, we can both protect data privacy and achieve higher classification accuracy.

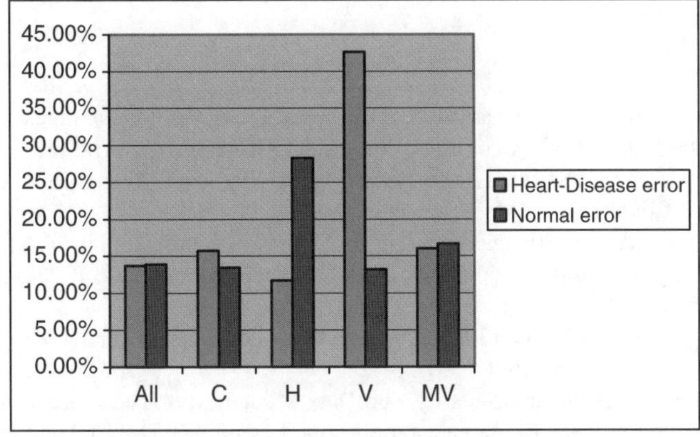

Fig 20.4. Heart-disease Dataset Classification Results

20.4 Conclusion

Privacy-preserving is an important issue in medical data mining. This paper applied data separation techniques in medical data classification task. The experiments demonstrate that data separation techniques can not only protect data privacy, but also increase classification accuracy sometimes (e.g., horizontally partitioned data).

The techniques used in our experiments are straightforward and there are a lot of rooms for improvement. For instance, in vertically partitioned data situation, simply remove one or several variables from datasets can not ensure that data can not be traced to an individual record. In such a case, more sophisticated methods and techniques can be used.

20.5 Acknowledgment

A portion of this paper has been published as an invited chapter in Advanced Topics in Data Warehousing and Mining 2007 by the IGI Publisher. This research has been partially supported by grants #70621001, #70531040,#70472074, National Natural Science Foundation of China; 973 Project #2004CB720103, Ministry of Science and Technology, China; and BHP Billiton Co., Australia. The authors would like to thank the principal investigators responsible for the data collection for heart-disease and breast cancer datasets: Hungarian Institute of Cardiology, Budapest: Andras Janosi, M. D.; University Hospital, Zurich, Switzerland: William Steinbrunn, M. D.; University Hospital, Basel, Switzerland: Matthias Pfisterer, M. D.; V.A. Medical Center, Long Beach and Cleveland Clinic Foundation: Robert Detrano, M. D., Ph. D.; David W. Aha (data donor of heart-disease dataset); the University of Wisconsin Hospitals, Madison: Dr. William H. Wolberg; Dr. Olvi Mangasarian (data donor of breast-cancer dataset).

References

1. D. Agrawal and C. C. Aggarwal (2001) On the Design and Quantification of Privacy Preserving Data Mining Algorithms, Proc. ACM SIGMOD, pp. 247-255.
2. R. Agrawal and R. Srikant (2000) Privacy-Preserving Data Mining, Proc. of the ACM SIGMOD Conference on Management of Data.
3. K. P. Bennett & O. L. Mangasarian (1992) Robust linear programming discrimination of two linearly inseparable sets, *Optimization Methods and Software* 1, 23-34, Gordon & Breach Science Publishers.
4. K. J. Cios & G. W. Moore (2002) Uniqueness of medical data mining, *Artificial Intelligence in Medicine*, Vol. 26, Issue 1-2, 1-24.
5. C. Clifton (2002) Privacy, Security, and Data Mining, presented at the combined conference 13th European Conference on Machine Learning (ECML'02) and 6th European Conference on Principles and Practice of Knowledge Discovery in Databases (PKDD'02), Helsinki, Finland, pp. 19-23.

6. T.G. Dietterich (2000) Ensemble methods in machine learning. First international workshop on multiple classifier systems, Springer Verlag, New York, pp. 1-15.

7. M. Kantarcioglu and C. Clifton (2004) Privacy-Preserving Distributed Mining of Association Rules on Horizontally Partitioned Data, IEEE Transactions on Knowledge and Data Engineering, vol. 16, no. 9, pp. 1026-1037.

8. J. Kim (2002) Ensemble methods for data mining, Probability and data mining lab, Feb 4, 2002, Available online: http://srccs.snu.ac.kr/VerII/Activity/Tutorial/ensemble.pdf

9. Y. Lindell and B. Pinkas (2002) Privacy Preserving Data Mining, Journal of Cryptology, Volume 15, Number 3, pp. 177-206, Springer New York.

10. D. Opitz and R. Maclin (1999) Popular ensemble methods: an empirical a) study, *Journal of Artificial Intelligence Research 11*, pp. 169-198.

11. Rulequest Research (2003) Retrieved April 29, 2006, from http://www.rulequest.com/see5-info.html.

12. UCI Machine Learning repository (2006) Retrieved April 29, 2006, from a) http://www.ics.uci.edu/~mlearn/databases/

13. J. Vaidya and C. Clifton (2002) Privacy preserving association rule mining in vertically partitioned data, Proceedings of the eighth ACM SIGKDD, pp. 639 – 644.

14. V.S. Verykios, E. Bertino, I.N. Fovino, L.P. Provenza, Y. Saygin, and Y. Theodoridis (2004) State-of-the-Art in Privacy Preserving Data Mining, ACM SIGMOD Record, vol. 3, no. 1, pp. 50-57.

15. W. H. Wolberg and O.L. Mangasarian (1990) Multisurface method of pattern separation for medical diagnosis applied to breast cytology, Proceedings of the National Academy of Sciences, U.S.A., Volume 87, December 1990, pp. 9193-9196.

16. G. Zenobi and P. Cunningham (2002) An Approach to Aggregating Ensembles of Lazy Learners That Supports Explanation, Lecture Notes in Computer Science, Vol. 2416, pp. 436-447.

Appendix

Table 20.1. Wisconsin Prognostic Breast Cancer Dataset Classification Error Rates

	All	1	2	3	4	5	6	7	8	9	MV
Malignant Error rate	2.49%	5.39%	0.41%	2.49%	4.15%	7.05%	2.90%	4.56%	2.90%	2.49%	3.32%
Benign Error rate	2.40%	2.62%	1.53%	2.84%	2.62%	3.06%	2.84%	2.62%	2.84%	2.40%	2.84%

Best: Best of all variables used; 1: Remove Clump Thickness; 2: Remove Uniformity of Cell Size; 3: Remove Uniformity of Cell Shape; 4: Remove Marginal Adhesion; 5: Remove Single Epithelial Cell Size; 6: Remove Bare Nuclei; 7: Remove Bland Chromatin; 8: Remove Normal Nucleoli; 9: Remove Mitoses

Table 20.2. Heart-disease Dataset Classification Error Rates

	All	C	H	V	MV
Heart-Disease error rate	13.71%	15.74%	11.68%	42.64%	15.99%
Normal error rate	13.90%	13.40%	28.29%	13.15%	16.63%

C: Cleveland Clinic Foundation; H: Hungarian Institute of Cardiology, Budapest; V: V.A. Medical Center, Long Beach, CA (long-beach-va.data)